EMPOWERMENT
EVALUATION

To Our Mothers

who have nurtured and cared for us
and brought us into this world
to make a difference

Elsie Fetterman
Hasmik Khachatourian
Hadassah Wandersman

EMPOWERMENT EVALUATION
Knowledge and Tools for Self-Assessment & Accountability

David M. Fetterman
Shakeh J. Kaftarian
Abraham Wandersman
EDITORS

SAGE Publications
International Educational and Professional Publisher
Thousand Oaks   London   New Delhi

For information address:

 SAGE Publications, Inc.
2455 Teller Road
Thousand Oaks, California 91320
E-mail: order@sagepub.com

SAGE Publications Ltd.
6 Bonhill Street
London EC2A 4PU
United Kingdom

SAGE Publications India Pvt. Ltd.
M-32 Market
Greater Kailash I
New Delhi 110 048 India

Printed in the United States of America

Library of Congress Cataloging-in-Publication Data

Main entry under title:

Empowerment evaluation: Knowledge and tools for self-assessment and
  accountability / edited by: David M. Fetterman, Shakeh J. Kaftarian,
  Abraham Wandersman.
      p.   cm.
  Includes bibliographical references and index.
  ISBN 0-7619-0024-1 (c: alk. paper).—ISBN 0-7619-0025-X (p:
alk. paper)
    1. Human services—United States—Citizen participation—
Evaluation.   2. Community health services—United States—Citizen
participation—Evaluation.   3. Community organization—United
States—Citizen participation—Evaluation.   4. Evaluation research
(Social action programs)—United States.   5. Present value analysis—
United States.   I. Fetterman, David M.   II. Kaftarian, Shakeh J.
III. Wandersman, Abraham.
  HV91.E54   1995
  361.8'0973—dc20   95-34986

This book is printed on acid-free paper.

96 97 98 99 10 9 8 7 6 5 4 3 2 1

Sage Production Editor: Gillian Dickens
Sage Typesetter: Andrea D. Swanson

Contents

Part IV: Theoretical and Philosophical Frameworks

Part V: Workshops, Technical Assistance, and Practice

Preface and Acknowledgments

Empowerment evaluation served as the theme of the 1993 American Evaluation Association annual meeting as well as the basis for Fetterman's presidential address. The topic stimulated conversations and arguments that spilled out into the hallways during the conference. Conference registration and attendance was at an all-time high and the conference chair attributed it directly to the theme. Clearly, it is a timely approach, reflecting the interests, needs, and practices of a significant number of evaluators. Empowerment evaluation has also touched a nerve among traditional evaluators, stimulating intense discussion in *Evaluation Practice*. This high level of intellectual and emotional engagement—both positive and negative—suggests that this approach speaks to issues at the very heart of the evaluation community. It acknowledges the significance of this new addition to the intellectual landscape—both as a contribution in its own right and as a tool in helping us refine and redefine evaluation use.

The conceptualization and crystallization of an idea are, however, only the first steps. This collection builds on this idea, generating both knowledge about empowerment evaluation and specific tools to build capacity in this area. In addition, self-assessment and accountability are pervasive concerns in society and around the world, in government, business, foundations, nonprofits, and academe. Empowerment evaluation has provided a philosophy, theoretical framework, and methods to address systematically these concerns. This collection brings us to an important stage in the evolution of empowerment evaluation, refining theory and practice.

This collection has been a team effort. Three editors brought their unique talents to this effort and worked diligently to make this dream a reality. Our contributors produced manuscripts of high quality under tight time lines. They responded to our critiques, incorporating our suggestions and in some cases using our comments as a catalyst for additional self-critique and improvement. In turn, they helped to shape our thinking about what empowerment evaluation is in theory and practice, often forcing us to rethink and revise our conception of the state of the art.

We have come together from very different starting points. We come from academe, government, nonprofits, and foundations; some of us are academics, others are practitioners. Our areas of focus are as diverse as our roles and affiliations, including public education, substance abuse prevention programs, battered women's shelters, and programs for individuals with disabilities, for HIV prevention, and for adolescent pregnancy prevention. We share a commitment to evaluation as a tool to build capacity and foster self-determination and program improvement.

We have benefited greatly from the support and critique of colleagues including Karen Kirkart, past president of the American Evaluation Association, and Daniel Stufflebeam, from the Evaluation Center at Western Michigan University. In their own way, each served to sharpen our conception of evaluation use and standards. The work of Jean Young, Allan Wicker, Wes Shera, Charles Usher, Nina Wallerstein, Rita O'Sullivan, and Arza Churchman was particularly informative and instructive, as they practice and theorize about empowerment evaluation in various settings. Many other colleagues also provided insights and assistance through their illuminating work and/or comments, including Will Shadish, Michael Scriven, Michael Patton, Eleanor Chelimsky, Ernest House, Chip Reichardt, Dianna Newman, Charles McClintock, Yvonna Lincoln and Egon Guba, John McLaughlin, Lois-ellin Datta, Bob Covert, Jim Sanders, Martha Ann Carey, and Jonathan Morell.

The hazard in acknowledging our intellectual debt is that valued colleagues will be omitted, as an exhaustive list is not possible. Nevertheless, we want to acknowledge a few colleagues who have been laboring in these same fields for years, well before this conception of evaluation had crystallized. Jennifer Greene, Jean King, Mark Jenness, and Zoe Barley have made enormous contributions to the area of collaborative and participatory evaluation. Their work and that of colleagues who trace back to action research activity laid the groundwork for empowerment evaluation.

The Center for Substance Abuse Prevention of the Department of Health and Human Services has been at the forefront of much of this work and should be recognized for its efforts and support of empowerment evaluation and this collection. The California Institute of Integral Studies has adopted this approach as part of its accreditation self-study and in so doing has created fertile ground for experimentation and knowledge development in this area. Stanford University, the Center for Substance Abuse Prevention, and the University of South Carolina have been focal points for training and development of empowerment evaluation knowledge and tools.

Program participants and community coalition members from across the United States and abroad have used empowerment evaluation to improve their own lives. As is the case in the adoption of any new approach, however, there has been some risk, and the dedication of all those involved in making this approach work is greatly appreciated. The Human Sciences Research Council and the Independent Development Trust in South Africa provided generous support to disseminate this approach and build capacity throughout their country. The University of Cape Town, the University of Natal, and Cape West in South Africa as well as the Evaluation Center at Western Michigan University graciously hosted presentations and opportunities for exchanges about empowerment evaluation.

David Fetterman would like to acknowledge the efforts of Deborah Waxman, his lifelong companion, who devoted many hours helping him refine his thinking, writing, and practice in this new field and in developing this collection. He also appreciates his entire family's support and encouragement. This personal odyssey has taken him in the United States from inner cities to Washington, D.C., and internationally from Canada to South Africa. He also appreciates their tolerance of an onerous traveling schedule, a seemingly endless series of long late-night telephone calls, and weekend workshops, training, and facilitation. Empowerment evaluation is personally rewarding, but it is also a time-consuming and often labor-intensive process. Throughout it all, their support has been invaluable.

Shakeh Kaftarian would like to acknowledge several people who have been instrumental in crystallizing, propelling, and supporting the developmental processes of her empowerment evaluation ideas in general and this book in particular. She would like to express appreciation to Elaine Johnson, the director of the Center for Substance Abuse Prevention, whose unwavering support of rigorous, adaptive, and empowering methods of evaluation has promoted evaluation and accountability in federal programs, as well as special appreciation for Mary Jensen's support of this book, which she recognized as a timely and valuable addition to the evaluation literature and a useful tool for academics, practitioners, and bureaucrats alike. Shakeh would like to thank her fellow Community Partnership evaluators across the country, who have helped sharpen her interest in empowerment evaluation. Last, but not least, she would like to thank her family for maintaining their invaluable presence in her life throughout her many empowering and consuming "journeys."

Abe Wandersman would like to thank his CRU colleagues at the University of South Carolina (Bob Goodman, Co-PI) and Pam Imm, Matt Chinman, Erin Morrissey, Maury Nation, David de la Cruz, Erica Adkins, Cindy Crusto, Katie Davino, Diana Seybolt, Pam Goodman, Simon Choi, and the members of the community coalitions for substance abuse prevention (project directors James Brown, Dian Crain, Kelli Kennison, Paul Pittman, Johnetta Davis, Greg Sparkman, Sheryl Taylor, and Kenneth Wright). They have inspired him to wrestle with the challenges of empowerment evaluation. The realities of the everyday challenges in the community, along with the needs and resources for planning and implementing programs and policies that work, have been a fertile ground for collaboratively developing knowledge and tools. Abe would also like to thank his family for their wonderful love and support and for constructive dialogues, all of which contribute so much to who he is.

Each of the authors in this collection believes that the time spent developing, cultivating, and refining this approach has been an investment in our communities and in the future. We realize that building capacity takes time and is a developmental process. Similarly, the development of empowerment evaluation will continue to take time as it evolves and adapts to new environments and new populations.

DAVID FETTERMAN
SHAKEH J. KAFTARIAN
ABRAHAM WANDERSMAN

INTRODUCTION
AND OVERVIEW

1

Empowerment Evaluation

An Introduction to Theory and Practice

DAVID M. FETTERMAN

Empowerment evaluation is an innovative approach to evaluation. It has been adopted in higher education, government, inner-city public education, nonprofit corporations, and foundations throughout the United States and abroad. A wide range of program and policy sectors use empowerment evaluation, including substance abuse prevention, accelerated schools, HIV prevention, crime prevention, welfare reform, battered women's shelters, agriculture and rural development, adult probation, adolescent pregnancy prevention, tribal partnership for substance abuse, self-determination for individuals with disabilities, and doctoral programs. Descriptions of these programs that use empowerment evaluation appear in this collection. *Empowerment Evaluation* (Fetterman, in press) provides additional insight into this new evaluation approach, including information about how to conduct workshops to train program staff members and participants to evaluate and improve program practice (see also Fetterman, 1994). In addition, this approach has been institutionalized within the American Evaluation Association[1] and is consistent with the spirit of the standards developed by the Joint Committee on Standards for Educational Evaluation (Fetterman, 1995; Joint Committee, 1994).[2]

Despite its increasingly wide use, empowerment evaluation is not a panacea; nor is it designed to replace other forms of evaluation. It meets a specific evaluation need: to help program participants evaluate themselves and their program to improve practice and foster self-determination. In this capacity, it may also influence other forms of evaluation and audit[3] to adopt a more collaborative and participatory tone. Empowerment evaluation is still evolving; there is much to learn, explore, refine, and improve. As with other forms of evaluation, we constantly learn more about the craft as we practice it.

Definition and Focus

Empowerment evaluation is the use of evaluation concepts, techniques, and findings to foster improvement and self-determination. It employs both qualitative and quantitative methodologies. Although it can be applied to individuals, organizations (at both intra- and extraorganizational levels),[4] communities, and societies or cultures, the focus is on programs. It is attentive to empowering processes and outcomes. Zimmerman's work on empowerment theory provides the theoretical framework for empowerment evaluation. According to Zimmerman (in press):

> A distinction between empowering processes and outcomes is critical in order to clearly define empowerment theory. Empowerment processes are ones in which attempts to gain control, obtain needed resources, and critically understand one's social environment are fundamental. The process is empowering if it helps people develop skills so they can become independent problem solvers and decision makers. Empowering processes will vary across levels of analysis. For example, empowering processes for individuals might include organizational or community involvement, empowering processes at the organizational level might include shared leadership and decision making, and empowering processes at the community level might include accessible government, media, and other community resources.
>
> Empowered outcomes refer to operationalization of empowerment so we can study the consequences of citizen attempts to gain greater control in their community or the effects of interventions designed to empower participants. Empowered outcomes also differ across levels of analysis. When we are concerned with individuals, outcomes might include situ-

ation specific perceived control, skills, and proactive behaviors. When we are studying organizations, outcomes might include organizational networks, effective resource acquisition, and policy leverage. When we are concerned with community level empowerment, outcomes might include evidence of pluralism, the existence of organizational coalitions, and accessible community resources.

Empowerment evaluation has an unambiguous value orientation—it is designed to help people help themselves and improve their programs using a form of self-evaluation and reflection. Program participants conduct their own evaluations and typically act as facilitators; an outside evaluator often serves as a coach or additional facilitator depending on internal program capabilities. Zimmerman's (in press) characterization of the community psychologist's role in empowering activities is easily adapted to the empowerment evaluator:

An empowerment approach to intervention design, implementation, and evaluation redefines the professional's role relationship with the target population. The professional's role becomes one of collaborator and facilitator rather than expert and counselor. As collaborators, professionals learn about the participants through their culture, their world view, and their life struggles. The professional works *with* participants instead of advocating *for* them. The professional's skills, interest, or plans are not imposed on the community; rather, professionals become a resource for a community. This role relationship suggests that what professionals do will depend on the particular place and people with whom they are working, rather than on the technologies that are predetermined to be applied in all situations. While interpersonal assessment and evaluation skills will be necessary, how, where, and with whom they are applied can not be automatically assumed as in the role of a psychotherapist with clients in a clinic.

Empowerment evaluation is necessarily a collaborative group activity, not an individual pursuit. An evaluator does not and cannot empower anyone; people empower themselves, often with assistance and coaching. This process is fundamentally democratic. It invites (if not demands) participation, examining issues of concern to the entire community in an open forum.

As a result, the context changes: The assessment of a program's value and worth is not the end point of the evaluation—as it often is in traditional evaluation—but part of an ongoing process of program improvement. This new context acknowledges a simple but often overlooked truth: that merit and worth are not static values. Populations shift, goals shift, knowledge about program practices and their value changes, and external forces are highly unstable. By internalizing and institutionalizing self-evaluation processes and practices, a dynamic and responsive approach to evaluation can be developed to accommodate these shifts. Both value assessments and corresponding plans for program improvement—developed by the group with the assistance of a trained evaluator—are subject to a cyclical process of reflection and self-evaluation. Program participants learn to continually assess their progress toward self-determined goals and to reshape their plans and strategies according to this assessment. In the process, self-determination is fostered, illumination generated, and liberation actualized.[5]

Value assessments are also highly sensitive to the life cycle of the program or organization. Goals and outcomes are geared toward the appropriate developmental level of implementation. Extraordinary improvements are not expected of a project that will not be fully implemented until the following year. Similarly, seemingly small gains or improvements in programs at an embryonic stage are recognized and appreciated in relation to their stage of development. In a fully operational and mature program, moderate improvements or declining outcomes are viewed more critically.

Despite its focus on self-determination and collaboration, empowerment evaluation and traditional external evaluation are not mutually exclusive. In fact, the empowerment evaluation process produces a rich data source that enables a more complete external examination.

Origins of the Idea

Empowerment evaluation has many sources. The idea first germinated during preparation of another book, *Speaking the Language of Power: Communication, Collaboration, and Advocacy* (Fetterman, 1993c). In developing that collection, I wanted to explore the many ways that evaluators and social scientists could give voice to the people they work with and bring their concerns to policybrokers. I found that, increasingly, socially concerned scholars in myriad fields are making their insights and findings available to decision makers. These scholars and practitioners address a host of significant issues, including conflict resolution, the drop-out problem, environmental health and safety, homelessness, educational reform, AIDS, American Indian concerns, and the education of gifted children. The aim of these scholars and practitioners was to explore successful strategies, share lessons learned, and enhance their ability to communicate with an educated citizenry and powerful policy-making bodies. Collaboration, participation, and empowerment emerged as common threads throughout the work and helped to crystallize the concept of empowerment evaluation.

Empowerment evaluation has roots in community psychology and action anthropology. Community psychology focuses on people, organizations, and communities working to establish control over their affairs. The literature about citizen participation and community development is extensive. Rappaport's (1987) "Terms of Empowerment/Exemplars of Prevention: Toward a Theory for Community Psychology" is a classic in this area. Sol Tax's (1958) work in action anthropology focuses on how anthropologists can facilitate the goals and objectives of self-determining groups, such as Native American tribes. Empowerment evaluation also derives from collaborative and participatory evaluation (Choudhary & Tandon, 1988; Oja & Smulyan, 1989; Papineau & Kiely, 1994; Reason, 1988; Shapiro, 1988; Stull & Schensul, 1987; Whitmore, 1990; Whyte, 1990).

A major influence was the national educational school reform movement with colleagues such as Henry Levin, whose Accelerated School Project (ASP) emphasizes the empowerment of parents, teachers, and administrators to improve educational settings. We worked to help design an appropriate evaluation plan for the Accelerated School Project that contributes to the empowerment of teachers, parents, students, and administrators (Fetterman & Haertel, 1990). The ASP team and I also mapped out detailed strategies for districtwide adoption of the project in an effort to help institutionalize the project in the school system (Stanford University and American Institutes for Research, 1992).

Dennis Mithaug's (1991, 1993) extensive work with individuals with disabilities to explore concepts of self-regulation and self-determination provided additional inspiration. We recently completed a 2-year Department of Education-funded grant on self-determination and individuals with disabilities. We conducted research designed to help both providers for students with disabilities and the students themselves become more empowered. We learned about self-determined behavior and attitudes and environmentally related features of self-determination by listening to self-determined children with disabilities and their providers. Using specific concepts and behaviors extracted from these case studies, we developed a behavioral checklist to assist providers as they work to recognize and foster self-determination.

Self-determination,[6] defined as the ability to chart one's own course in life, forms the theoretical foundation of empowerment evaluation. It consists of numerous interconnected capabilities, such as the ability to identify and express needs, establish goals or expectations and a plan of action to achieve them, identify resources, make rational choices from various alternative courses of action, take appropriate steps to pursue objectives, evaluate short- and long-term results (including reassessing plans and expectations and taking necessary detours), and persist in the pursuit of those goals. A breakdown at any juncture of this network of capabilities—as well as various environmental factors[7]—can reduce a person's likelihood of being self-determined. (See also Bandura, 1982, concerning the self-efficacy mechanism in human agency.)

A pragmatic influence on empowerment evaluation is the W. K. Kellogg Foundation's emphasis on empowerment in community settings. The foundation has taken a clear position concerning empowerment as a funding strategy:

> We've long been convinced that problems can best be solved at the local level by people who live with them on a daily basis. In other words, individuals and groups of people must be empowered to become change-makers and solve their own problems, through the organizations and institutions they devise. . . . Through our community-based programming, we are helping to empower various individuals, agencies, institutions, and organizations to work together to identify problems and to find quality, cost-effective solutions. In doing so, we find ourselves working more than ever with grantees with whom we have been less involved—smaller, newer organizations and their programs. (*Transitions,* 1992, p. 6)

Its work in the areas of youth, leadership, community-based health services, higher education, food systems, rural development, and families and neighborhoods exemplifies this spirit of putting "power in the hands of creative and committed individuals—power that will enable them to make important changes in the world" (p. 13). For example, one project—Kellogg's Empowering Farm Women to Reduce Hazards to Family Health and Safety on the Farm—involves a participatory evaluation component. The work of Sanders, Barley, and Jenness (1990) on cluster evaluations for the Kellogg Foundation also highlights the value of giving ownership of the evaluation to project directors and staff members of science education projects.

These influences, activities, and experiences form the background for this new evaluation approach. An eloquent literature on empowerment theory by Zimmerman (in press), Zimmerman, Israel, Schulz, and Checkoway (1992), Zimmerman and Rappaport (1988), and Dunst, Trivette, and LaPointe (1992), as discussed earlier, also informs this approach. A brief review of empowerment evaluation's many facets will illustrate its wide-ranging application.

Facets of Empowerment Evaluation

In this new context, training, facilitation, advocacy, illumination, and liberation are all facets—if not developmental stages—of empowerment evaluation. Rather than additional roles for an evaluator whose primary function is to assess worth (as defined by Stufflebeam, 1994, and Scriven, 1967), these facets are an integral part of the evaluation process. Cronbach's developmental focus is relevant and on target—the emphasis is on program development, improvement, and lifelong learning.

TRAINING

In one facet of empowerment evaluation, evaluators teach people to conduct their own evaluations and thus become more self-sufficient. This approach desensitizes and demystifies evaluation and ideally helps organizations internalize evaluation principles and practices, making evaluation an integral part of program planning. Too often, an external evaluation is an exercise in dependency rather than an empowering experience: In these instances, the process ends when the evaluator departs, leaving participants without the knowledge or experience to continue for themselves. In contrast, an evaluation conducted by program participants is designed to be ongoing and internalized in the system, creating the opportunity for capacity building. Jean Ann Linney and Abraham Wandersman's *Prevention Plus III* (1991), published by the U.S. Department of Health and Human Services' Center for Substance Abuse Prevention, illustrates how evaluators can teach people to conduct elementary evaluations of their own programs—in this case, primarily alcohol and other drug prevention programs (see Chapter 12). Similarly, Steven Mayer and associates at Rainbow Research highlight the educational value of this approach; they produced an evaluation "Toolbox" to help corporations document the effectiveness of their services for affordable housing residents and an Acts of Empowerment Evaluation Logbook for program use in documenting participants' achievements (see Chapter 15). The Charities Evaluation Services (CES), a United Kingdom-wide organization, provides training in self-evaluation and monitoring for members of nonprofit and community organizations. Libby Cooper, the CES director, emphasizes the need for participatory training to ensure that the experiences and expertise of those attending CES courses are acknowledged and developed. CES's work with a Belfast-based women's center focuses on the role women have played as a catalyst for the development of the wider community. CES provides training to help women monitor and evaluate their own work. In Western Ireland, they are evaluating the way an intermediary organization has sought to achieve the integration of people with disabilities. They are providing training to enable disabled people and their providers to participate with staff in the design and management of project services.

In empowerment evaluation, training is used to map out the terrain, highlighting categories and concerns. It is also used in making preliminary assessments of program components, while illustrating the need to establish goals, strategies to achieve goals, and documentation to indicate or substantiate progress. Training a group to conduct a self-evaluation can be considered equivalent to developing an evaluation or research design (as that is the core of the training), a standard part of any evaluation. This training is ongoing, as new skills are needed to respond to new levels of understanding. Training also becomes part of the self-reflective process of self-assessment (on a program level) in that participants must learn to recognize when more tools are required to continue ·and enhance the evaluation process. This self-assessment process is pervasive in an empowerment evaluation—built into every part of a program, even to the point of reflecting on how its own meetings are conducted and feeding that input into future practice.[8]

In essence, empowerment evaluation is the "give someone a fish and you feed her for a day; teach her to fish, and she will feed herself for the rest of her life" concept, as applied to evaluation. The primary difference is that in empowerment evaluation the evaluator and the individuals benefiting from the evaluation are typically on an even plane, learning from each other.

FACILITATION

Empowerment evaluators serve as coaches or facilitators to help others conduct a self-evaluation. For example, the Oakland Unified School District was in the process of self-evaluation to assess their progress in carrying out their 5-year plan. They had a district mission, a strategic approach, and a list of desired student outcomes. They adopted an empowerment evaluation approach. Superintendent Mesa, an enthusiastic supporter of the empowerment approach, and coordinators of the overall effort—Gary Yee, a former principal in the district, and Ed Ferran, a district staff member with extensive facilitation experience—recognized the value of the participatory process. Once staff members begin setting their own goals and identifying their own program performance indicators, program improvement becomes a powerful all-inclusive force. Staff members were asked to rate their performance, document that rating, and in some cases adjust their self-rating to accommodate group feedback. This process created a baseline against which to monitor future progress, established goals and milestones for the future, and highlighted the significance of documenting progress toward self-selected goals. It demystified the evaluation process, and helped staff members internalize evaluation as a way of thinking about what they were doing on a regular basis. It also put them in charge of their own destinies, as they selected the intermediate goals and objectives required to have an impact on the larger, long-term goals of improving student performance and reducing drop-out and crime rates.

In my role as coach, I provided general guidance and direction to the effort, attending sessions to monitor and facilitate as needed. It was critical to emphasize that the staff were in charge of this effort; otherwise, program participants initially tend to look to the empowerment evaluator as expert, which would make them dependent on an outside agent. In some instances, my task was to clear away obstacles and identify and clarify miscommunication patterns. I also participated in the first few cabinet-level meetings in the district, providing explanations, suggestions, and advice at various junctures to help ensure that the process had a fair chance.

An empowerment evaluation coach can also provide useful information about how to create facilitation teams (balancing analytical and social skills), work with resistant (but interested) units, develop refresher sessions to energize tired units, and resolve various protocol issues. Simple suggestions along these lines can keep an effort from backfiring or being seriously derailed.

A coach may also be asked to help create the evaluation design with minimal additional support. For example, the Hebrew Union College asked for assistance in designing an action, or empowerment-oriented, evaluation. After some consultation and a study of relevant literature,[9] the college reshaped its entire plan, choosing to have congregations throughout the country conduct their own self-evaluations. An empowerment evaluation coach was hired at a later date to help keep the process on track.

Whatever her contribution, the empowerment evaluation coach must ensure that the evaluation remains in the hands of program personnel. The coach's task is to provide useful information, based on her evaluator's training and past experience, to keep the effort on course.

ADVOCACY

In some instances, empowerment evaluators conduct an evaluation for a group after the goals and evaluation design have been collaboratively established. They may even serve as direct advocates—helping groups become empowered through evaluation. Evaluators often feel compelled to serve as advocates for groups that have no control over their own fates, such as the homeless or drop-out populations. In an empowerment setting, advocate evaluators allow participants to shape the direction of the evaluation, suggest ideal solutions to their problems, and then take an active role in making social change happen.

A common workplace practice provides a familiar illustration of self-evaluation and its link to advocacy on an individual level. Employees often collaborate with both supervisor and clients to establish goals, strategies for achieving those goals and documenting progress, and realistic time lines. Employees collect data on their own performance and present their case for their performance appraisal. Self-evaluation thus becomes a tool of advocacy. This individual self-evaluation process is easily transferable to the group or program level.

Evaluators have a moral responsibility to serve as advocates—after the evaluation has been conducted and if the findings merit it. In a national study of dropouts, my postevaluation activities included disseminating generally positive findings to appropriate policymakers and preparing a Joint Dissemination Review Panel Submission. A series of gifted and talented education evaluations culminated in a book recommending that the U.S. Department of Education establish a gifted and talented center. Based in part on this recommendation, the Department of Education appointed me to the panel that selected a consortium of universities to create the center.

Politically savvy evaluators often work with senators and representatives. For example, based on his evaluation findings in the Chicago School Reform effort, Fred Hess (1993) testified before a congressional committee in support of an act to establish a national Demonstration Project of Educational Performance Agreements for School Restructuring. This act would provide local schools with more flexibility in the use of federal funds, in exchange for commitments to improve student performance. Based on his work in program design and evaluation, Kim Hopper (1993) cofounded a local advocacy organization for the homeless in New York City. He has also served as an expert witness in public interest litigation involving the rights of homeless men and women. Margaret Weeks and Jean Schensul (1993) also demonstrated how ethnography and evaluation can be used as a tool of advocacy to empower people. Program staff in an AIDS prevention program used information about attitudes of injection drug users and prostitutes toward needle exchange to inform policy discussion and decision making. This same descriptive information was used to advocate for better access to services for HIV-positive people. Qualitative data were used to advocate for sustained funding for AIDS prevention programs on local, state, and national levels.

In another example, Linda Parker serves as an advocate for the Coushatta Tribe, in the role of economic development consultant (Parker & Langley, 1993). She combines her knowledge of the government grant system with a tribal officer's knowledge to help accomplish the tribe's objectives. Winning grants (with an evaluation) to serve tribal needs represents a concrete accomplishment in furthering the goals of self-determination.

Advocate evaluators write in public arenas to change public opinion, influence power brokers, and provide relevant information at opportune moments in the policy decision-making forum. An excellent editorial piece about school dropouts in Chicago highlighted evaluation findings concerning minority education and school failure. Hess's work was a catalyst for citywide educational and social change. Hopper writes newspaper editorials to respond critically to cultural "givens" or stereotypes about the homeless and to participate in social change on their behalf. In an op-ed piece in the *Chronicle of Higher Education,* I wrote about lessons learned in a controversial environmental health and safety evaluation at Stanford University. This editorial piece focused on classic organizational conflicts of interest existing within college campuses, and the benefits of empowering health and safety workers to ensure safer working environments in higher education (Fetterman, 1990a). Other op-ed pieces dispel myths about gifted and talented children and advocate on their behalf (Fetterman, 1990b). Evaluators thus used the media to build a case for the people we work with, attempting to inform a concerned and educated citizenry. These actions are in accord with Mills's (1959) position:

> There is no necessity for working social scientists to allow the potential meaning of their work to be shaped by the "accidents of its setting," or its use to be determined by the purposes of other men [or women]. It is quite within their powers to discuss its meaning and decide upon its uses as matters of their own policy. (p. 177)

ILLUMINATION   *like 4th Gen. Eval*

Illumination is an eye-opening, revealing, and enlightening experience. Typically, a new insight or understanding about roles, structures, and program dynamics is developed in the process of determining worth and striving for program improvement (see Parlett & Hamilton, 1976). Empowerment evaluation is illuminating on a number of levels. Two cases from the Oakland School District self-evaluation highlight the illuminating qualities of this process or approach. During one meeting, members of the district's early childhood group decided after a lengthy and somewhat circuitous discussion that they wanted to link their work with student academic outcomes or test data—something they'd never done before in assessing students' performance. Working with various district administrators, they extracted the CTBS test data for children in the early childhood program from the district management information system and compared them with data for similar students in the district but not in the program. The data documented significantly higher educational achievement by students in the early childhood program. Program staff members found this to be an eye-opening or illuminating experience. The next tasks were to determine whether these findings held up with additional comparison, and to dig deeper to identify the specific reasons for the difference. This led to a detailed critical review of the entire program. It also opened doors that they did not know existed, such as access to an existing student database within the district bureaucracy to help them understand the differences, measure the impact, and improve their program.

A meeting with one of the largest and most powerful units in the district resulted in a research epiphany. Unit members thought of themselves as a very successful group, in spite of the district's overall poor performance. When one facilitator asked them to provide some evidence of their effectiveness, they pointed to their work in the area of school climate. After some discussion, they suggested that leadership training was the most significant variable affecting school climate (of the variables they had control over). They claimed to have five leadership teams operating at a high level of effectiveness. After requesting and receiving documentation to support this rating, the facilitator asked if they would have more impact if they had more teams. One member of the unit said, "We could have a dramatic effect if we had more teams and we worked at more schools." She then proceeded, with the assistance of the facilitator, to chart a growth curve with an x and y axis and a dotted line running through it at a 45-degree angle predicting the type of positive impact anticipated from this increased effort. They agreed to set this new goal for the unit, rearrange their schedules and workloads to accommodate the expanded number of schools, and work toward this goal over the academic year—collecting documentation about their progress throughout the year. This administrator, with little or no research background, developed a testable, researchable hypothesis in the middle of a discussion about indicators and self-evaluation. It was not only illuminating to the group (and to her), it revealed what they could do as a group when given the opportunity to think about problems and come up with workable options, hypotheses, and tests.

This experience of illumination holds the same intellectual intoxication each of us experienced the first time we came up with a researchable question. The process creates a dynamic community of learners as people engage in the art and science of evaluating themselves.

LIBERATION

Illumination often sets the stage for liberation. It can unleash powerful, emancipatory forces for self-determination. Liberation is the act of being freed or freeing oneself from preexisting roles and constraints. It often involves new conceptualizations of oneself and others. Empowerment evaluation can also be liberating. Many of the examples in this discussion demonstrate how helping individuals take charge of their lives—and find useful ways to evaluate themselves— liberates them from traditional expectations and roles. They also demonstrate how empowerment evaluation enables participants to find new opportunities, see existing resources in a new light, and redefine their identities and future roles.

For example, school nurses in the Oakland Public School system are using this approach to help them understand their own evolving role in the school district. Nurses are becoming more involved in assessing the life circumstances of the entire student population, rather than simply meeting individual student needs. They view the empowerment evaluation meeting activity as an opportunity to help define what their role will be in the future. In the process of redefining their role, they have designed specific tasks that will help them emerge as life circumstance-oriented health care providers, including conducting a school-wide assessment of the health conditions at the various sites, such as the percentage of students with asthma at each school site.

Empowerment evaluation can also be liberating on a larger sociopolitical level. Johann Mouton, executive director of the Centre for Science Development at the Human Sciences Research Council,[10] and Johann Louw from the Department of Psychology at the University of Cape Town invited me to speak about empowerment evaluation and conduct workshops throughout South Africa after apartheid had ended but before the elections. These two individuals and the institutions they represent "reject racism and racial segregation and strive to maintain a strong tradition of non-discrimination with regard to race, religion, and gender."[11] The Centre for Science Development was the national funding agency for the human sciences in South Africa, and my empowerment evaluation workshops were conducted under the auspices of a then-new Directorate: Research Capacity Building, which focused primarily on building research capacity among black scholars in the country. Over a third of the participants in the workshops were black. This was a historic achievement by South African standards.

When Johann Louw and I first met, he said he was "intrigued and interested [in the approach, since] as you can imagine, empowerment is very much on the social agenda in this country" (J. Louw, 1993, personal communication). He invited me to work with him, assisting in the evaluation of various programs administered in and by an impoverished black community near Cape Town (personal communication, 1993; Fetterman, 1994). These community members were implementing and evaluating a broad range of community participation health care programs. They used self-evaluation to monitor and build on their successes and failures. This commendable work took place despite a context of disenfranchisement, high rates of unemployment, and disease. Acts of violence were a part of daily life.[12] (See Fetterman, 1993a, for a discussion of the culture of violence and the balance between hope and fear in South Africa.) This progressive, self-reflective, impoverished black community reflects the real spirit of hope persisting despite South Africa's culture of violence.[13] In another example, the Independent Development Trust, under the guidance of its director Professor Merlyn Mehl from the University of the Western Cape, is building empowerment evaluation into the process of reformulating national educational goals, including one of the country's most ambitious educational undertakings—the whole-school improvement of 1,000 primary and secondary schools across the country (Mehl, Gillespie, Foale, & Ashley, 1995). This new nation is rethinking and reevaluating everything—from social attitudes to land distribution. The issue of empowerment speaks to the heart of the reconstruction of South Africa.

Steps of Empowerment Evaluation

There are several pragmatic steps involved in helping others learn to evaluate their own programs: (a) taking stock or determining where the program stands, including strengths and weaknesses; (b) focusing on establishing goals (determining where you want to go in the future with an explicit emphasis on program improvement); (c) developing strategies and helping participants determine their own strategies to accomplish program goals and objectives; and (d) helping program participants determine the type of evidence required to document credibly progress toward their goals.

STEP 1: TAKING STOCK

One of the first steps in empowerment evaluation is taking stock. Program participants are asked to rate their program on a 1 to 10 scale, with 10 being the highest level. They are asked to make the rating as accurate as possible. Many participants find it less threatening or overwhelming to begin by listing, describing, and then rating individual activities in their program, before attempting a gestalt or overall unit rating. Specific program activities might include recruitment, admissions, pedagogy, curriculum, graduation, and alumni tracking in a school setting. The potential list of components to rate is endless, and each participant must prioritize the list of items—typically limiting the rating to the top 10 activities. Program participants are also asked to document their ratings (both the ratings of specific program components and the overall program rating). Typically, some participants give their programs an unrealistically high rating. The absence of appropriate documentation, peer ratings, and a reminder about the realities of their environment—such as a high drop-out rate, students bringing guns to school, and racial violence in a high school—help participants recalibrate their rating, however. In some cases, ratings stay higher than peers consider appropriate. The significance of this process, however, is not the actual rating so much as it is the creation of a baseline from which future progress can be measured. In addition, it sensitizes program participants to the necessity of collecting data to support assessments or appraisals.

STEP 2: SETTING GOALS

After rating their program's performance and providing documentation to support that rating, program participants are asked how highly they would like to rate their program in the future. Then they are asked what goals they want to set to warrant that future rating. These goals should be established in conjunction with supervisors and clients to ensure relevance from both perspectives. In addition, goals should be realistic, taking into consideration such factors as initial conditions, motivation, resources, and program dynamics.

It is important that goals be related to the program's activities, talents, resources, and scope of capability. One problem with traditional external evaluation is that programs have been given grandiose goals or long-term goals that participants could only contribute to in some indirect manner. There was no link between their daily activities and ultimate long-term program outcomes in terms of these goals. In empowerment evaluation, program participants are encouraged to select intermediate goals that are directly linked to their daily activities. These activities can then be linked to larger, more diffuse goals, creating a clear chain of outcomes.

Program participants are encouraged to be creative in establishing their goals. A brainstorming approach is often used to generate a new set of goals. Individuals are asked to state what they think the program should be doing. The list generated from this activity is refined, reduced, and made realistic after the brainstorming phase, through a critical review and consensual agreement process.

There are also a bewildering number of goals to strive for at any given time. As a group begins to establish goals based on this initial review of their program, they realize quickly that a consensus is required to determine the most significant issues to focus on. These are chosen according to significance to the operation of the program, such as teaching; timing or urgency, such as recruitment or budget issues; and vision, including community building and learning processes.

Goal setting can be a slow process when program participants have a heavy work schedule. Sensitivity to the pacing of this effort is essential. Additional tasks of any kind and for any purpose may be perceived as simply another burden when everyone is fighting to keep their heads above water.

STEP 3: DEVELOPING STRATEGIES

Program participants are also responsible for selecting and developing strategies to accomplish program objectives. The same process of brainstorming, critical review, and consensual agreement is used to establish a set of strategies. These strategies are routinely reviewed to determine their effectiveness and appropriateness. Determining appropriate strategies, in consultation with sponsors and clients, is an essential part of the empowering process. Program participants are typically the most knowledgeable about their own jobs, and this approach acknowledges and uses that expertise—and in the process, puts them back in the "driver's seat."

STEP 4: DOCUMENTING PROGRESS

In Step 4, program participants are asked what type of documentation is required to monitor progress toward their goals. This is a critical step. Each form of documentation is scrutinized for relevance to avoid devoting time to collecting information that will not be useful or relevant. Program participants are asked to explain how a given form of documentation is related to specific program goals. This review process is difficult and time-consuming but prevents wasted time and disillusionment at the end of the process. In addition, documentation must be credible and rigorous if it is to withstand the criticism that this evaluation is self-serving. (See Fetterman, 1994, for a detailed discussion of these steps and case examples.)

Caveats and Concerns

Is Research Rigor Maintained? Empowerment evaluation is one approach among many being used to address social, educational, industrial, health care, and many other problems. As with the exploration and development of any new frontier, this approach requires adaptations, alterations, and innovations. This does not mean that significant compromises must be made in the rigor required to conduct evaluations. Although I am a major proponent of individuals taking evaluation into their own hands and conducting self-evaluations, I recognize the need for adequate research, preparation, and planning. These first discussions need to be supplemented with reports, texts, workshops, classroom instruction, and apprenticeship experiences if possible. Program personnel new to evaluation should seek the assistance of an evaluator to act as coach, assisting in the design and execution of an evaluation. Further, an evaluator must be judicious in determining when it is appropriate to function as an empowerment evaluator or in any other evaluative role.

Does This Abolish Traditional Evaluation? New approaches require a balanced assessment. A strict constructionist perspective may strangle a young enterprise; too liberal a stance is certain to transform a novel tool into another fad. Colleagues who fear that we are giving evaluation away are right in one respect—we are sharing it with a much broader population. But those who fear that we are educating ourselves out of a job are only partially correct. Like any tool, empowerment evaluation is designed to address a specific evaluative need. It is not a substitute for other forms of evaluative inquiry or appraisal. We are educating others to manage their own affairs in areas they know (or should know) better than we do. At the same time, we are creating new roles for evaluators to help others help themselves.

How Objective Can a Self-Evaluation Be? Objectivity is a relevant concern. We needn't belabor the obvious point that science and specifically evaluation have never been neutral. Anyone who has had to roll up her sleeves and get her hands dirty in program evaluation or policy arenas is aware that evaluation, like any other dimension of life, is political, social, cultural, and economic. It rarely produces a single truth or conclusion. In the context of a discussion about self-referent evaluation, Stufflebeam (1994) states,

> As a practical example of this, in the coming years U.S. teachers will have the opportunity to have their competence and effectiveness examined against the standards of the National Board for Professional Teaching Standards and if they pass to become nationally certified. (p. 331)

Regardless of one's position on this issue, evaluation in this context is a political act. What Stufflebeam considers an opportunity, some teachers consider a threat to their livelihood, status, and role in the community. This can be a screening device in which social class, race, and ethnicity are significant variables. The goal is "improvement," but the questions of for whom and at what price remain valid.

To assume that evaluation is all in the name of science or that it is separate, above politics, or "mere human feelings"—indeed, that evaluation is objective—is to deceive oneself and to do an injustice to others. Objectivity functions along a continuum—it is not an absolute or dichotomous condition of all or none. Fortunately, such objectivity is not essential to being critical. For example, I support programs designed to help dropouts pursue their education and prepare for a career; however, I am highly critical of program implementation efforts. If the program is operating poorly, it is doing a disservice both to former dropouts and to taxpayers.

One needs only to scratch the surface of the "objective" world to see that it is shaped by values, interpretations, and culture. Whose ethical principles are evaluators grounded in? Do we all come from the same cultural, religious, or even academic tradition? Such an ethnocentric assumption or assertion flies in the face of our accumulated knowledge about social systems and evaluation. Similarly, assuming that we can "strictly control bias or prejudice" is naive, given the wealth of literature available on the subject, ranging from discussions about cultural interpretation to reactivity in experimental design.[14]

What About Participant or Program Bias? The process of conducting an empowerment evaluation requires the appropriate involvement of stakeholders. The entire group—not a single individual, not the external evaluator or an internal manager—is responsible for conducting the evaluation. The group thus can serve as a check on individual members, moderating their various biases and agendas.

No individual operates in a vacuum. Everyone is accountable in one fashion or another and thus has an interest or agenda to protect. A school district may have a 5-year plan designed by the superintendent; a graduate school may have to satisfy requirements of an accreditation association; an outside evaluator may have an important but demanding sponsor pushing either time lines or results, or may be influenced by training to use one theoretical approach rather than another.

In a sense, empowerment evaluation minimizes the effect of these biases by making them an explicit part of the process. The example of a self-evaluation in a performance appraisal is useful again here. An employee negotiates with his or her supervisor about job goals, strategies for accomplishing them, documentation of progress, and even the time line. In turn, the employee works with clients to come to an agreement about acceptable goals, strategies, documentation, and time lines. All of this activity takes place within corporate, institutional, and/or community goals, objectives, and aspirations. The larger context, like theory, provides a lens in which to design a self-evaluation. Supervisors and clients are not easily persuaded by self-serving forms of documentation. Once an employee loses credibility with a supervisor, it is difficult to regain it. The employee thus has a vested interest in providing authentic and credible documentation. Credible data (as agreed on by supervisor and client in negotiation with the employee) serve both the employee and the supervisor during the performance appraisal process.

Applying this approach to the program or community level, superintendents, accreditation agencies, and other "clients" require credible data. Participants in an empowerment evaluation thus negotiate goals, strategies, documentation, and time lines. Credible data can be used to advocate for program expansion, redesign, and/or improvement. This process is an open one, placing a check on self-serving reports. It provides an infrastructure and network to combat institutional injustices. It is a highly (often brutally) self-critical process. Empowerment evaluation is successful because it adapts and responds to existing decision-making and authority structures on their own terms (see Fetterman, 1993c). It also provides an opportunity and a forum to challenge authority and managerial facades by providing data about actual program operations—from the ground up. The approach is particularly valuable for disenfranchised people and programs to ensure that their voices are heard and that real problems are addressed.

POSITIONS OF PRIVILEGE

Empowerment evaluation is grounded in my work with the most marginalized and disenfranchised populations, ranging from urban school systems to community health programs in South African townships, who have educated me about what is possible in communities overwhelmed by violence, poverty, disease, and neglect. They have also repeatedly sensitized me to the power of positions of privilege. One dominant group has the vision, makes and changes the rules, enforces the standards, and need never question its own position or seriously consider any other. In such a view, differences become deficits rather than additive elements of culture. People in positions of privilege dismiss the contributions of a multicultural world. They create rational policies and procedures that systematically deny full participation in their community to people who think and behave differently.

Evaluators cannot afford to be unreflective about the culturally embedded nature of our profession. There are many tacit prejudgments and omissions embedded in our primarily Western thought and behavior. These values, often assumed to be superior, are considered natural. Western philosophies, however, have privileged their own traditions and used them to judge others who may not share them, disparaging such factors as ethnicity and gender. In addition, they systematically exclude other ways of knowing. Some evaluators are convinced that there is only one position and one sacred text in evaluation, justifying exclusion or excommunication for any "violations" or wrong thinking (see Stufflebeam, 1994). Scriven's (1991, p. 260) discussion about perspectival evaluation is instructive in this context, highlighting the significance of adopting multiple perspectives, including new perspectives.

We need to keep open minds, including alternative ways of knowing—but not empty heads. Skepticism is healthy; cynicism, blindness, and condemnation are not, particularly for emerging evaluative forms and adaptations. New approaches in evaluation and even new ways of knowing are needed if we are to expand our knowledge base and respond to pressing needs. As Campbell (1994) states, we should not "reject the new epistemologies out of hand. . . . Any specific challenge to an unexamined presumption of ours should be taken seriously" (p. 293). Patton (1994) might be right "that the world will not end in a subjective bang, but in a boring whimper as voices of objectivity [drift] off into the chaos" (p. 312).

Evaluation must change and adapt as the environment changes, or it will either be overshadowed by new developments or—as a result of its unresponsiveness and irrelevance—follow the path of the dinosaurs to extinction. People are demanding much more of evaluation and are not tolerant of the limited role of the outside expert who has no knowledge of or vested interest in their program or community. Participation, collaboration, and empowerment are becoming requirements in many community-based evaluations, not recommendations. Program participants are conducting empowerment and other forms of self- or participatory evaluations with or without us (the evaluation community). I think it is healthier for all parties concerned to work together to improve practice rather than ignore, dismiss, and condemn evaluation practice; otherwise, we foster the development of separate worlds operating and unfolding in isolation from each other.

DYNAMIC COMMUNITY OF LEARNERS

Many elements must be in place for empowerment evaluation to be effective and credible. Participants must have the latitude to experiment, taking both risks and responsibility for their actions. An environment conducive to sharing successes and failures is also essential. In addition, an honest, self-critical, trusting, and supportive atmosphere is required. Conditions need not be perfect to initiate this process. The accuracy and usefulness of self-ratings, however, improve dramatically in this context. An outside evaluator who is charged with monitoring the process can help keep the effort credible, useful, and on track, providing additional rigor, reality checks, and quality controls throughout the evaluation. Without any of these elements in place, the exercise may be of limited utility and potentially self-serving. With many of these elements in place, the exercise can create a dynamic community of transformative learning.

SPREADING THE WORD

Empowerment evaluation is drawing a great deal of attention. It was the theme of the annual meeting of the 1993 American Evaluation Association as well as the subject of my presidential address. This collection about empowerment evaluation builds on the foundation established for this approach and provides case examples and recommendations about the diverse applications of this approach. Evaluators throughout the world, ranging from OXFAM[15] in England to scholars in Israel[16] and auditors in Canada and Texas,[17] have expressed their interest in this new approach. It crystallizes what many of these evaluators are already doing—serving as change agents to help others help themselves. Notable examples include work under way at Victoria University of Technology in Australia,[18] the School of Social Work at the University of Hawaii at Manoa,[19] the Minority Affairs office at the University of Wisconsin—Madison System,[20] the University of North Carolina at Chapel Hill,[21] the College of Education at the University of Arizona,[22] the Psychology Department at the University of Rhode Island,[23] Université Laval in Quebec,[24] Keystone University Research Corporation,[25] the College of Human Ecology at Cornell University,[26] Modesto Junior College,[27] and the Virginia Commonwealth University.[28] Numerous organizations are working in precisely the same direction at the same time, including such diverse organizations as the Knowledge Utilization Society,[29] the Transition Research Institute (funded by the Office of Special Education and Rehabilitative Services),[30] the National Research Center on the Gifted and Talented,[31] the Evaluation Center at Western Michigan University,[32] the Independent Sector,[33] the Wisconsin School Evaluation Consortium,[34] the California Institute of Integral Studies,[35] the Center for Substance Abuse Prevention,[36] the Multicultural Resource Center,[37] Full Citizenship, Inc., and the National Institute on Disability Rehabilitation Research (of the U.S. Department of Education),[38] the National Center for Improving Science Education,[39] the General Accounting Office,[40] and as discussed earlier universities, foundations, and impoverished black communities in South Africa.

Empowerment evaluation is creating a new niche in the intellectual landscape of evaluation. This approach is political in that it has an agenda—empowerment. It is not, however, liberal or conservative ideologically, or positivist or phenomenological per se. It knows no political or geographic boundaries. It has a bias for the disenfranchised, including minorities, disabled individuals, and women. Empowerment evaluation can be used to help anyone with a desire for self-determination, however. The ultimate test of any new approach is that, as it becomes more clearly defined, useful, and acceptable, it becomes absorbed into the mainstream of evaluation. I look forward to the day when it will be simply one more tool in the evaluator's toolbox.

Overview

This collection contains a wealth of philosophical, theoretical, and practical material concerning empowerment evaluation. It is divided into six parts: introduction; breadth and scope; context; theoretical and philosophical frameworks; workshops, technical assistance, and practice; and conclusion.

The second part highlights the breadth and scope of empowerment evaluation as it is adopted in academic and foundation settings. Part II includes chapters by Henry Levin, Director of the Accelerated Schools Project, and Ricardo Millett, Director of Evaluation at the W. K. Kellogg Foundation. Empowerment evaluation also has been adopted in government settings, such as the Center for Substance Abuse Prevention (see Chapter 9, by Robert Yin, Shakeh Kaftarian, and Nancy Jacobs), and Texas state agencies (see Chapter 4, by Joyce Keller).

Part III includes chapters by Joyce Keller, Cynthia Gómez and Ellen Goldstein, Cheryl Grills et al., and Arlene Andrews. This part highlights the various environments or contexts in which empowerment evaluation is conducted, ranging from resistant to responsive. In some cases, a significant effort is required to move from passive-compliance orientations to a self-determining atmosphere and orientation. Other environments are particularly conducive to the pursuit of empowerment evaluation because they already have a tradition of self-determination, community organizing, and activism.

Part IV, focusing on theoretical, philosophical, and organizing frameworks, includes the work of Stephen Fawcett et al.; Robert Yin, Shakeh Kaftarian, and Nancy Jacobs; John Stevenson, Roger Mitchell, and Paul Florin; and Dennis Mithaug. They address a variety of critical concerns in empowerment evaluation, including the role of empowerment theory, processes and outcomes, multiple levels of empowerment—individual, organizational, and societal and/or cultural—and capacity building.

Part V provides concrete discussions about workbooks, workshops, technical assistance, instruments, planning, and advice to strengthen the links between evaluation and community capacity building. Authors include Jean Ann Linney and Abraham Wandersman, Margret Dugan, Frances Butterfoss et al., and Steven Mayer. Although these areas necessarily overlap in practice, they are separated here to provide a clearer view of key elements in the theory and practice of empowerment evaluation.

BREADTH AND SCOPE

Henry Levin's chapter, "Empowerment Evaluation and Accelerated Schools," demonstrates the scope and value of empowerment evaluation. The Accelerated Schools Project (ASP), established in 1986-1987, is part of the national educational reform movement in the United States. It encompasses more than 700 elementary and middle schools in 37 states. ASP embodies the concept of empowerment evaluation—fostering self-determination and ongoing program improvement. Levin highlights the congruence between empowerment and empowerment evaluation in Accelerated Schools and the overall constructs of community empowerment developed by Fetterman, Rappaport, and Zimmerman. For example, ASP principles include unity of purpose, empowerment with responsibility, and building on strengths. Values include viewing the school as the center of expertise, equity, community, risk taking, experimentation, reflection, participation, trust, and communication. In addition, ASP staff members provide training to build capacity and serve as coaches (rather than experts) to work with schools. Levin also speaks in terms of the transformation of an entire school culture, which parallels the organizational transformations associated with many empowerment evaluation activities. One of ASP's primary goals is to replace academic remediation with acceleration—using teaching and learning approaches associated with gifted and talented education to advance the development of all children. This goal represents a profound respect for individual potential and capacity, a democratization of knowledge, and an appreciation of existing tools and technologies—adapting instead of reinventing the wheel. (See Fetterman, 1988, for a detailed discussion about how gifted and talented education can be used as a model of excellence for the American educational system.) ASP requires school staff, parents, and students to take responsibility for major decisions that will determine educational outcomes. In addition, the ASP approach requires inclusive governance and problem-solving systems. Fetterman and Levin have worked together to design planning and self-evaluation models that match the ASP context of collaboration, participation, and empowerment. All members of the school community are involved in the process of taking stock through systematic inquiry, which includes identifying significant issues and activities, developing evaluative questions, collecting and analyzing data, and reporting findings. Baseline data from this process are compared with the school's vision. Differences between the baseline and the vision are prioritized and form the basis of plans for the future. Members of the school community develop criteria to determine whether they are approximating their goals, including specific observable outcomes. Moreover, ASP requires routine critical reflection about its own processes and practices, highlighting in practice the concerns of Yin, Kaftarian, and Jacobs, as well as Butterfoss et al., about assessing the quality of the process. ASP is a reflective, self-evaluating entity by its nature, examining its fidelity to the process and to self-determined goals in its quest for continuous improvement.

Ricardo Millett's chapter, "Empowerment Evaluation and the W. K. Kellogg Foundation," shifts the focus to foundations. The Kellogg Foundation is one of the largest in the world, with expenditures exceeding $260 million per year, including $20 to $25 million to support evaluation activities. The Kellogg Foundation, like the Accelerated Schools Project, has both influenced and been shaped by empowerment evaluation. The foundation's approach to program evaluation from its earliest conception was designed to "help others help themselves." Evaluation at the foundation is considered part of programming rather than a separate function, viewed as a tool for organizational learning, a developmental ongoing process (not a single summative report card), and a collaborative relationship between grantmaker and -seeker. Evaluation must also be sensitive to the human, political, cultural, and contextual elements affecting the programs. Traditional evaluation methodologies have not always been compatible with building capacity or active participation in the evaluation process—in some cases, proving to be insensitive and disempowering. Empowerment evaluation at the Kellogg Foundation is designed to "improve, not prove." The foundation recognizes the tension in its concurrent interest in judging whether a program made a difference—basic accountability data—and attempts to balance the need for valid and reliable data (and accountability) against the primary commitment of building capacity and improving programs. By placing evaluation responsibility in the hands of grantees, an imperfect but healthier balance is maintained. The foundation has systematically attempted to increase grantees' capacity to control their own evaluations. For example, clusters of grantees working on similar projects are assembled to share lessons learned and, in the process, build their capacity to improve program practice. The aim is to help institutionalize evaluation within each organization, so that it has the tools it needs long after foundation funding ceases.

Two other chapters, appearing in later sections, illustrate the use of empowerment evaluation in federal, state, and local government settings. The Center for Substance Abuse Prevention (CSAP) community partnerships programs provide an excellent example of the use of empowerment evaluation to build capacity and improve programs throughout the country (Chapter 9). CSAP has taken the lead in a collaborative model that requires local-level self-evaluations designed to improve problem solving and program practice in more than 250 comprehensive community programs, devoting 15% of CSAP's budget to these evaluation efforts. CSAP's efforts add another dimension to empowerment evaluation, highlighting its role in government settings. Texas state agencies also provide an excellent example of how empowerment evaluation is being used to build capacity and provide a measure of accountability within state government (Chapter 4).

CONTEXT

Joyce Keller's chapter, "Empowerment Evaluation and State Government: Moving From Resistance to Adoption," highlights the importance of a receptive environment for empowerment evaluation to take root, focusing on management characteristics such as risk taking and willingness to share responsibility. This chapter is particularly persuasive because it is a personal odyssey that began in an audit office where resistance to this type of approach was high. Despite this resistance, Keller developed a receptive environment. Her success demonstrates the power of this approach and the range or scope of settings in which it can be used. The chapter also provides a close look at the process of empowerment evaluation. It illustrates the personal immersion necessary for this work as well as the importance of continuous self-evaluation. This close, daily, collaborative effort often means that evaluators undergo a personal and professional metamorphosis, just as program participants do. The chapter is also valuable in illustrating the four stages of empowerment evaluation—taking stock, setting goals, developing strategies, and documenting progress toward goals—and the various facets of empowerment evaluation, including training, facilitation, advocacy, illumination, and liberation.

"The HIV Prevention Evaluation Initiative: A Model for Collaborative and Empowerment Evaluation," by Cynthia Gómez and Ellen Goldstein, builds on Keller's chapter by demonstrating how the authors overcame historical barriers to cooperation and understanding between researchers and community practitioners. Gómez and Goldstein used an empowerment evaluation approach, including training, facilitation, advocacy, and illumination, to forge a collaborative working relationship between service providers, academics, and private funders in the fight against AIDS—the leading cause of death for adults aged 22 to 45 in the United States. Despite long-standing resistance, they succeeded in creating an atmosphere in which group participation and support emerged and empowerment evaluation principles were embraced. This approach was designed to empower rather than judge, to share skills and knowledge rather than to find fault, and to improve program practice. Evaluators and researchers served as facilitators rather than implementors of the evaluation effort. Workshops were designed for capacity building—helping community-based organizations learn how to write proposals for support and to evaluate their own services. The authors also provide a list of lessons learned for others interested in developing collaborative and empowering evaluation-based models. Consistent with the Joint Committee on Standards for Educational

Evaluation's meta-evaluation standard, a meta-evaluation of this effort is being conducted by an external independent agency to determine if this collaborative effort resulted in changes in prevention programs, prevention research priorities, funding priorities, and the community-based organization's self-evaluation capabilities.

In contrast to Keller's chapter, Cheryl Grills, Karen Bass, Didra Brown, and Aletha Akers discuss the role of empowerment evaluation within a community that has a tradition of self-determination. In their chapter, "Empowerment Evaluation: Building Upon a Tradition of Activism in the African American Community," they demonstrate how evaluation concepts and strategies are being used by a community coalition for community organizing, public policy work, and planning and prevention strategies for substance abuse prevention and treatment. Civil rights and community activists constantly debrief, critique, evaluate, and refine their work to address economic problems that contribute to cocaine trafficking, educational inequalities that result in unemployment, and the proliferation of liquor stores that contribute to a host of destructive and often violent behaviors. Traditional evaluation approaches were viewed with skepticism because they stood apart from the community and because they were not in touch with the cultural or political context of the community. The values of empowerment evaluation are consistent with community activism, and the evaluation rigor enhanced and refined an existing self-reflective cultural practice within the coalition. The merger of community activism and social science evaluation technology served to enfranchise the community. As in most of the works in this collection, the communities' efforts built on strengths (rather than using a deficit-based model), focused on evaluation capacity building using both qualitative and quantitative data, and was a collaborative partnership. They also adopted an empowerment evaluation perspective that recognized contextual factors that mitigate against self-determining behavior—social, political, economic, and historical. Attention is given to how problems are defined because this shapes the measures used to determine both outcome and impact. Accomplishments are placed in perspective, often valuing and reinforcing small increments in the process of improving program practice.

Arlene Andrews's chapter, "Realizing Participant Empowerment in the Evaluation of Nonprofit Women's Services Organizations: Notes

From the Front Line," provides a compelling picture of people searching for a place they can call home and the role of evaluation and service organizations, such as a United Way-affiliated battered women's emergency shelter and Habitat for Humanity, in helping them pursue this goal. This portrait, like that in Grills et al.'s chapter, describes the congruence between local nonprofit women's services organizations with a tradition of empowerment and empowerment evaluation. Andrews was a founding member of the battered women's agency, which facilitated her role as an empowerment evaluation coach. Andrews's chapter describes how empowerment evaluation principles have been applied in this context, focusing on such aspects as self-determination, primacy of and respect for program participants, reciprocity, and feminist principles (highlighting participatory and helping relationships), perpetual innovation and assessment, and ecological context—community resources, public policy, market demand, and advocacy. Her case study describes how the interorganizational collaboration is developing and attempting to institutionalize evaluation. Empowerment evaluation was adopted precisely because it is consistent with the ideals of the service organizations, responsive to and dependent on the wisdom of program participants, and focused on program improvement and accountability.

THEORETICAL AND PHILOSOPHICAL FRAMEWORKS

A framework for empowerment evaluation is presented articulately in "Empowering Community Health Initiatives Through Evaluation" by Stephen Fawcett, Adrienne Paine-Andrews, Vincent Francisco, Jerry Schultz, Kimber Richter, Rhonda Lewis, Kari Harris, Ella Williams, Jannette Berkley, Christine Lopez, and Jacqueline Fisher. They recognize the research tradition in which empowerment evaluation has been constructed, a tradition aimed at legitimizing community members' experiential knowledge, acknowledging the role of values in research, empowering community members, democratizing research inquiry, and enhancing the relevance of evaluation data for communities. Building on the basic steps to empowerment evaluation outlined by Fetterman, they present an evaluative sequence of steps used to contribute to empowerment in community coalitions for the prevention of adolescent pregnancy and substance abuse and in a tribal partnership for prevention of substance abuse. These steps include assessing community concerns and resources, setting a mission and objectives, developing strategies and action plans, monitoring process and outcome, communicating information to relevant audiences, and promoting adaptation, renewal, and institutionalization. They also highlight the multilevel character of a key empowerment concept: the importance of gaining influence over events and outcomes crucial to an individual, group, or community. Fawcett and colleagues, like many authors in this collection, emphasize the value of empowerment evaluation as a capacity building process grounded in participatory and other forms of collaborative research. Moreover, as is common in empowerment evaluation, they explicitly join technical assistance and evaluation in an integrated support system—viewing it as one conceptual package. They, more than any other contributors in the collection, clearly capture the tension between the traditional goal of evaluation—the systematic investigation of the worth or merit of an object—and the multiple goals of empowerment evaluation, which involve self-assessment of merit as a tool to foster improvement and self-determination at every stage.

Robert Yin, Shakeh Kaftarian, and Nancy Jacobs make a number of contributions to the evolving concept and practice of empowerment evaluation in their chapter, "Empowerment Evaluation at Federal and Local Levels: Dealing With Quality." Building on empowerment theory and values, they reinforce one of the fundamental elements of the empowerment process: groups becoming empowered together. Technical assistance and training are conducted within the context of an ongoing evaluative process, not as ends in themselves. In addition, they emphasize the pivotal role federal staff members and community program participants play in designing the framework for evaluating community partnerships. Their concern for quality and validity focuses on formal evaluation standards and empowerment evaluation outcomes that can be tracked. They also offer a unique vantage point in that their empowerment evaluation activity involved a federal agency, local evaluators, and a cross-site evaluation team. This experience has contributed to the development and refinement of a conceptual framework as they explicate the dimensions of empowerment evaluation.

John Stevenson, Roger Mitchell, and Paul Florin's intellectually engaging chapter, "Evaluation and Self-Direction in Community Prevention Coalitions," is grounded in the context of community coalitions, focusing in large part on the reduction of alcohol and other drug abuse. They rely heavily on utilization-oriented methods and recognize the critical role evaluators play in building the learning capacity of organizations as part of the evaluation process—fundamental aspects of empowerment evaluation practice. These authors also adopt a multilevel approach to empowerment, including individual, intraorganizational, and extraorganizational levels of analysis. They pose a series of self-critical questions, dealing with topics ranging from role responsibilities to the expertise of program participants to conduct self-evaluations. Workshops and individual technical assistance are critical components of capacity building for them as they share the evaluation effort. The authors conclude with a level-headed assessment that empowerment evaluation successes "can occur at different levels of expertise," and point to specific limitations of this approach. This is an important refinement of the empowerment evaluation framework.

Dennis Mithaug's "Fairness, Liberty, and Empowerment Evaluation" contributes in several important ways to the field. In concert with other contributors, he believes that empowerment evaluation should be used to enhance capacity in the areas of personal improvement, organizational restructuring, and societal reform. Unlike most other evaluation efforts described here, however, his focus is on individual and societal levels of analysis. Using the individual as a metaphor, he explains how empowerment evaluation is the logical method for monitoring fairness and liberty for all, across cultures. Mithaug expands the boundaries of empowerment evaluation use, employing it for the disenfranchised on all levels, ranging from individuals with disabilities to desegregated schools for African American students. He also transcends the societal level, invoking the International Covenant on Civil and Political rights adopted by the General Assembly of the United Nations, which states: "All peoples have the right to self-determination." Empowerment evaluation, which is anchored in self-determination, is an appropriate tool to measure our progress toward social justice on a global scale.

WORKSHOPS, TECHNICAL ASSISTANCE, AND PRACTICE

The chapter "Empowering Community Groups With Evaluation Skills: The Prevention Plus III Model," by Jean Ann Linney and Abraham Wandersman, serves as a transition from theoretical and philosophical concerns to practice. It presents a four-step approach to assessing school and community prevention programs for nonprofessionals. This approach has been used throughout the United States and internationally; more than 50,000 copies of their self-help workbook *Prevention Plus III* have been distributed. The widespread use of this approach is evident in this collection: Fetterman highlights the value of this tool and has shared it with American and South African academics and service providers; Stevenson and colleagues use it in their workshops; and Dugan describes how she used and adapted the approach to the local context as part of training and technical assistance. The popularity of this self-help book is a function of many interwoven factors: a shift from national to local responsibility and choice concerning community problems, a growing desire on the part of individuals to become more involved in the community to improve the quality of family life, and fiscal realities such as an increased concern with accountability and documentation paired with no financial support. This chapter distills evaluation to four critical components: (a) question formation, including the identification of goals and desired outcomes; (b) implementation/process evaluation; (c) assessment of outcomes; and (d) assessment of impact. It provides a logic of evaluation, highlighting the importance of thinking about the interconnectedness of planning, implementation, and outcomes. It also discusses the value of thinking causally—for example, understanding the relationship of problem identification to intervention and intervening variables, monitoring implementation and outcomes, making assessments, and questioning the underlying causal model. This chapter discusses some of the most common needs of nonprofessional evaluators, including knowing where to start (formulating evaluation questions and translating them into project-specific indices and evaluation activities), identifying evaluation instruments and measures, and recognizing the value of ongoing formative evaluation. This guide to evaluation reflects important concepts of empowerment evaluation: using evaluation to build capacity and foster improvement and self-determination.

"Participatory and Empowerment Evaluation: Lessons Learned in Training and Technical Assistance," by Margret Dugan, focuses on a nonprofit organization involved in assessing the effectiveness of its curriculum-based program to prevent alcohol, tobacco, and other drug use by at-risk youth. The state of Texas mandated that all its grantees have a program evaluation component. The first attempt using a traditional evaluation approach was a crushing failure. As the authors analyzed what went wrong, however, they learned that their clients wanted to be a part of the process. Clients also wanted training in small groups and ongoing technical assistance. Clients wanted to be scientific but not detached, and to determine the course of their own development. Dugan developed a schema contrasting the needs of participants as compared with those of the funding agency, demonstrating that initially evaluators had tried to conduct a funder-oriented evaluation in a context demanding participation. These episodes led to a more collaborative arrangement that evolved into an empowerment evaluation. The evaluators became facilitators, advocates, trainers, coaches, mentors, and occasionally experts. They recognized the pluralism of values in each program and the need to illuminate multiple perspectives. They also understood that there was no single truth acceptable to all members of a program. At the same time, they were concerned about validity and participant objectivity. With funder assurance that the focus was on program improvement, not perfection, however, program participants provided an objective, often brutally self-critical, and analytical review of their program. This transformation didn't happen all by itself—it required training and technical assistance. Dugan provides a useful evaluation framework to facilitate this process, identifying stages, participants' tasks, evaluators' tasks, and evaluators' roles. In addition, she highlights the role of working in small groups; using participants' materials, slides, role-plays, and a fair amount of humor; and organizing support groups or fax buddies. Dugan also introduced theory-driven evaluation into the training, recognizing that understanding how the program works is the single most important evaluation tool for the novice. State-level grant monitors, program and evaluation specialists, and other funding stakeholders also participated in the training to help them understand the process and thus more effectively serve the grantees. The entire effort was guided by simplicity and validity at a low cost, using tested public domain materials—specifically the *Prevention Plus III* guide. Training was a part of the evaluation process, resulting in an evaluation plan with time lines.

Planning is a critical facet of program operation and evaluation, as evidenced in every chapter of this volume. Frances Butterfoss, Robert Goodman, Abraham Wandersman, Robert Valois, and Matthew Chinman underscore this aspect of community coalition work in their chapter, "The Plan Quality Index: An Empowerment Evaluation Tool for Measuring and Improving the Quality of Plans." Like Yin, Kaftarian, and Jacobs, Butterfoss and colleagues focus on quality—in this case, the quality of plans. Their tool was designed to help evaluate and understand the planning component of coalitions. They stress that careful, participatory planning is required to achieve community-based goals. In addition, plans are important intermediate outcomes of their work. Their Plan Quality Index was initially a quantitative instrument designed to provide ratings used for evaluation research purposes. It evolved into a narrative feedback format supplemented with consultation and training, based on community coalition reactions to each iteration of the index. In concert with the principles and practice of empowerment evaluation, the development and refinement of tools themselves are a product of participation. In this case, the various iterations of the tool marked the transition of the evaluators from traditional evaluators—training raters to review plans and report numerical score ratings, to technical assistants and consultants analyzing the scores and providing feedback, to problem-solving collaborative colleagues and coaches helping to improve implementation plans. Evaluating prevention plans proved to be a challenging task, as evidenced by the need to respond to each critique and revise accordingly. In addition to evaluative feedback, they learned how important it was to build on strengths, enter into a collaborative problem-solving mode, and provide anticipatory guidance to help community coalition members produce a quality plan. The final product—the revised index and the plan—was more useful and meaningful to community staff members and participants as a function of this empowering process.

Steven Mayer's chapter, "Building Community Capacity With Evaluation Activities That Empower," like those of Grills and colleagues and others, highlights the link between a community activist's bias and empowerment evaluation. He emphasizes common characteristics including building community capacity, building on strengths, participant participation, utility, and empowerment. In addition, he suggests three strategies that help align evaluation with community capacity building: (a) Create a constructive environment for the evaluation; (b) actively include the voices of intended beneficiaries; and (c) help communities use evaluation findings to strengthen community responses. Like many of the authors in this collection, Mayer and colleagues did not begin with empowerment evaluation, they discovered it only after finding traditional approaches wanting. The chapter concludes with a brief discussion of a few examples, including women's empowerment logs, an affordable housing evaluation toolbox, a leadership program for community foundations, a crime prevention assessment, and a drug abuse prevention assessment.

In the final part, David Fetterman's "Conclusion: Reflections on Emergent Themes and the Next Steps" brings this phase of the discussion to a close. He uses themes drawn from this volume to help characterize empowerment evaluation. In addition, Fetterman provides a brief list of areas that need further exploration and development to refine empowerment evaluation theory and practice in the future.

Notes

1. It has been institutionalized as part of the Collaborative, Participatory, and Empowerment Evaluation Topical Interest Group (TIG). TIG chairs are David Fetterman and Jean King. All interested evaluators are invited to join the TIG and attend our business meetings, which are open to any member of the association.

2. Although there are many problems with the standards and the application of the standards to empowerment evaluation, and the fact that they have not been formally adopted by any professional organization, they represent a useful tool for self-reflection and examination. Empowerment evaluation meets or exceeds the spirit of the standards in terms of utility, feasibility, propriety, and accuracy (see Fetterman, 1995, for a detailed examination).

3. The Texas Office of the State Auditor is already successfully using empowerment evaluation (Keller, Chapter 4, this volume).

4. See Stevenson, Mitchell, and Florin (Chapter 10) for a detailed explanation of these distinctions. See also Zimmerman (in press) for more detail about empowerment theory focusing on psychological, organizational, and community levels of analysis.

5. Using traditional evaluation language, the investigation of worth or merit and plans for program improvement become the means by which self-determination is fostered, illumination generated, and liberation actualized.

6. One of the many heartwarming stories that emerged from the case study section of the individuals with disabilities and self-determination study highlights what self-determination is all about. This story involves a young high school girl who has cerebral palsy and is quadriplegic. In elementary school, she was classified as a special education, mentally retarded student and grouped accordingly. She knew she did not belong in this special education class. One day during recess, she hid behind some advanced students she had been speaking with and followed them right into their classroom in her motorized wheelchair. She made sure there were plenty of students in front of her to camouflage her entrance. She knew she belonged with them, and she gambled (successfully) that no one would have the nerve to kick her out. No one did, and the teachers quickly learned that she was a gifted and talented student—not a special education, mentally retarded, or remedial education student. This is an example of gutsy self-determination.

7. Concerning individuals with disabilities, self-determination exists in varying degrees and is enhanced or diluted by developmental factors (including age and maturity), type or degree of disability, and environmental conditions. For example, a supportive provider and a supportive school environment generate opportunities and encourage risk taking, exploration, and the development of abilities. The absence of these supportive environmental features limits opportunities, creates obstacles, and fosters dependency and/or despondent behavior.

8. In anthropology and folklore, this is called a folk culture or more specifically an evaluation folk culture.

9. One of the most significant recommendations included reading John Watkins's (1992) chapter "Critical Friends in the Fray: An Experiment in Applying Critical Ethnography to School Restructuring," as well as various books about evaluation methodology.

10. The Centre for Science Development provided complete support for these activities, including the keynote presentation at the national Symposium on Program Evaluation and the Empowerment Evaluation and Qualitative Workshops in Johannesburg, Pretoria, Durban, and Cape Town.

11. This phrase also represents a self-rating. The key to understanding empowerment evaluation is precisely in the interpretation of this self-rating. It is a form of cultural interpretation. Individuals who read this sentence and conclude "I don't believe them" are viewing the present through the lens of the past. Interpreting this statement as the place to begin, rather than a place to conclude, allows you to ask: What's next? What will I do to accomplish this? How will I monitor and document it? What do you plan to do next year to build on successes and failures?

12. Violence and fear permeate the consciousness of every South African. The newspapers have become a daily record of stonings, stabbings, and shootings. My drive to this community passed directly by Guguletu; Amy Biehl, a Fulbright scholar and Stanford graduate, was stabbed and beaten to death only a few miles from where I worked.

13. See the *Training for Self-Evaluation at Ithusheng Health Centre* report (1993) by Hester van der Walt and Lies Hoogendoorn for an excellent example of how to train nonliterate community members and program participants to conduct a self-evaluation.

14. See Fetterman (1982) and Conrad (1994) for additional discussion about reactivity.

15. OXFAM was founded in 1942. They work "with poor people regardless of race or religion in their struggle against hunger, disease, exploitation and poverty in Africa, Asia, Latin America and the Middle East through relief, development, research overseas and public education at home." OXFAM contacted AEA in December 1992, through the president-elect, communicating their clear interest in empowerment evaluation.

16. Professor Arza Churchman and her doctoral students, from Technion—Israel Institute of Technology, are working on the development of a theory of empowerment within the context of community planning.

17. The Texas Office of the State Auditor found that virtually no evaluation had been performed of the effectiveness of probation itself or of individual rehabilitation programs in the Texas adult probation system. In an effort to shift this mentality of compliance (or noncompliance) to one of effectiveness, the state auditor's office set up an evaluation model of probation programs statewide and of specific probation intervention programs, emphasizing the responsibility of entities to perform their own ongoing effectiveness evaluations.

18. Delwyn Goodrick's work in the areas of AIDS, the evaluation of homelessness prevention, birthing needs, eating disorders, and participatory evaluation for the Commonwealth Department of Finance highlight the utility of the empowerment evaluation approach. See also Wadsworth's self-evaluation and research work as represented by the Action Research Issues Association in Melbourne.

19. Charles Rapp, Wes Shera, and Walter Kisthardt's work in the area of consumer empowerment highlights the role of ethnography in empowerment research and evaluation.

20. Hazel Symonett's work in the Minority Affairs office at the University of Wisconsin System highlights the power of empowerment evaluation and self-evaluation throughout a university system, as the university designs for diversity in a multicultural environment.

21. Charles Usher's child welfare reform initiatives work at the University of North Carolina at Chapel Hill with the Center for the Study of Social Policy is quite consistent with empowerment evaluation. His 1993 report titled *Self-Evaluation in the Prince George's County Services Reform Initiative* is an instructive and useful example of empowerment evaluation (see Usher, 1995).

22. Amy Schlessman-Frost's work in the area of democratic models and multicultural educational evaluation has clear implications and applications for empowerment evaluation.

23. John Stevenson's efforts with the Community Research and Services Team at the University of Rhode Island have aspired for several years to play the kind of role required to conduct empowerment evaluations. They have been influenced by many of the same sources of inspiration described in this text. They also identify with the action research tradition initiated by Kurt Lewin. They are attempting to build the capacity of local prevention efforts with evaluation skills. They discuss some of the obstacles associated with such efforts, including problems with single training sessions with little or no follow-through.

24. Helene Johnson is an evaluation consultant at Universite Laval, Direction genérale du premier cycle. Building on a stakeholder evaluation approach, she conducts periodic evaluations of university programs in a manner that empowers participants—often providing a voice for students in their communication with faculty and administrators.

25. Joyce Miller Iutcovich's work in assessing the needs of rural elderly is based on an empowerment model.

26. William Trochim's application of concept mapping to school districts and supported employment programs for persons with severe mental illness highlights the participatory component of empowerment evaluation, as the content of the map is entirely determined by the group. In addition, see Elizabeth Whitmore, Willem van der Eyken, Barbara Clinton, Jennifer Greene, Doreen Greenstein, and Daniel Selener's views on this subject as presented in the Cornell Empowerment Project.

27. David Baggett (1994) is using empowerment evaluation to evaluate model demonstration projects. He advises projects to adopt a portfolio process to collect information about their activities, ranging from planning to research and dissemination activities. He has also used empowerment evaluation to guide the development of an employment portfolio.

28. Sally Schumacher and Wendy Wood use empowerment evaluation as they explore strategies to generate national policy options for adults with traumatic brain injury.

29. The president of the Knowledge Utilization Society invited me to make a plenary presentation about empowerment evaluation at their Seventh Annual Meeting on April 21, 1993. The theme of the conference was "Using Knowledge to Empower Organizational Change: Working Smarter and Targeting for Results."

30. The Director of the Transition Research Institute at the University of Illinois at Urbana-Champaign invited me to conduct an empowerment evaluation workshop for all the directors of Office of Special Education and Rehabilitative Services-funded model transition demonstration projects as well as directors of State Systems for Transition Services for Youth With Disabilities Programs and directors of Regional Resource Centers and project officers, including Michael Ward. The focus of their evaluation technical assistance matches the empowerment evaluation approach—helping people help themselves through evaluation. "The workshop goal is to increase the capacity of directors of model transition demonstration projects to discover, understand, and believe that evaluation activities can lead to self-determination, that is, evaluation practice can and should be integral to program planning and implementation." The evaluation of the workshop documented significant success in each of these areas.

31. The center has issued and widely disseminated a report about self-evaluation titled *Evaluate Yourself* (Fetterman, 1993b). In addition, I provided an empowerment evaluation workshop for teachers and researchers at the center's Building a Bridge Between Research and Classroom Practices in Gifted Education Conference (1995).

32. James Sanders's evaluation work with grassroots community groups while at the Kellogg Foundation highlighted the "concept of evaluation as a human activity that is the responsibility of all who are involved in the project." In addition, he focuses on the internalization of evaluation concepts and practices for self-improvement and capacity building. Zoe Barley and Mark Jenness's multisite evaluation work of community-based programs with the Kellogg Foundation also represents a form of empowerment evaluation. They are conducting cluster evaluations for Kellogg with an emphasis on empowering the science education cluster projects. In addition, I was an invited visiting scholar at Western Michigan University for the express purpose of presenting and exploring empowerment evaluation.

33. Sandra Trice Gray, Vice President, Leadership and Management and International Initiatives, from the Independent Sector, has developed under her leadership an elaborate vision of evaluation "as a means of achieving organizational effectiveness and renewal." Their approach follows the empowerment evaluation model, ranging from asking groups to identify their own goals to linking evaluation to strategic planning and

achievement of a program's mission (see Gray, 1993, *Leadership Is: A Vision of Evaluation.*)

34. Jake Blasczyk, Director of the Wisconsin School Evaluation Consortium and the Wisconsin North Central Association, is helping school districts put in place long-range plans to reform education. Moreover, he has developed an excellent self-study guide for program evaluation that he is using in 40% of Wisconsin's K-12 school districts.

35. As the director of research and evaluation at the California Institute of Integral Studies, I have been provided with the opportunity to initiate an empowerment evaluation approach to self-assessment and improvement in the School for Transformative Learning. This approach has been adopted as part of the accreditation self-study process to improve teaching, research, governance, and administration. In addition, it is designed to be highly interactive in both face-to-face communication at the institute and through synchronous and asynchronous electronic communication throughout the United States using America Online and the Electronic University Network.

36. Darlind Davis's plenary presentation at the 1993 American Evaluation Association annual meeting in Dallas, Texas, highlighted the center's commitment to empowerment evaluation in their work. In addition, this organization was instrumental in publishing *Prevention Plus III.*

37. The center has asked me to provide empowerment evaluation workshops to help parents evaluate gifted and talented programs in Los Angeles—focusing on the needs of minority gifted and talented students.

38. R. M. Stineman from Full Citizen, Inc., is developing an empowerment evaluation tool for community compliance with the Americans With Disabilities Act of 1990. They are using this tool to ensure involvement of city government, the business community, people with disabilities and their families, disability organizations, and other community leaders. This project is funded by NIDRR. I am serving as a coach and consultant on the project.

39. M. Jean Young and Susan Loucks-Horsley use empowerment evaluation through technical assistance to build the capacity of Department of Energy Precollege Education staff members.

40. Eleanor Chelimsky, Assistant Comptroller General for Program Evaluation and Methodology, has collected evaluative data from program beneficiaries to illuminate a social situation in a way that also assists decision makers to understand the particular impacts of a program on relevant parties. In essence, GAO uses empowerment evaluation by giving voice to patients, the disabled, businesspersons, and immigrants, all of whom may be the intended beneficiaries of government programs.

References

Baggett, D. (1994). Using a project portfolio: Empowerment evaluation for model demonstration projects. *Interchange, 13*(1), 5-7. (Transition Research Institute, Champaign: University of Illinois at Urbana-Champaign)

Bandura, A. (1982). Self-efficacy mechanism in human agency. *American Psychologist, 37,* 122-147.

Campbell, D. T. (1994). Retrospective and prospective on program impact assessment. *Evaluation Practice, 15*(3), 291-298.

Choudhary, A., & Tandon, R. (1988). *Participatory evaluation*. New Delhi, India: Society for Participatory Research in Asia.

Conrad, K. J. (1994). *Critically evaluating the role of experiments: New directions for program evaluation* (No. 63). San Francisco: Jossey-Bass.

Dunst, C. J., Trivette, C. M., & LaPointe, N. (1992). Toward clarification of the meaning and key elements of empowerment. *Family Science Review, 5*(1-2), 111-130.

Fetterman, D. M. (1982). Ibsen's baths: Reactivity and insensitivity (A misapplication of the treatment-control design in a national evaluation). *Educational Evaluation and Policy Analysis, 4*(3), 261-279.

Fetterman, D. M. (1988). *Excellence and equality: A qualitatively different perspective on gifted and talented education*. Albany: State University of New York Press.

Fetterman, D. M. (1990a, March 21). Health and safety issues: Colleges must take steps to avert serious problems. *Chronicle of Higher Education*, p. A48.

Fetterman, D. M. (1990b). Wasted genius. *Stanford Magazine, 18*(2), 30-33. (Reprinted in the *San Jose Mercury*, July 22, 1990)

Fetterman, D. M. (1993a, October 3). Confronting a culture of violence: South Africa nears a critical juncture. *San Jose Mercury*, pp. 1C, 4C.

Fetterman, D. M. (1993b). *Evaluate yourself* (National Research Center on the Gifted and Talented). Storrs: University of Connecticut.

Fetterman, D. M. (1993c). *Speaking the language of power: Communication, collaboration, and advocacy (Translating ethnography into action)*. London: Falmer.

Fetterman, D. M. (1994). Steps of empowerment evaluation: From California to Cape Town. *Evaluation and Program Planning, 17*(3), 305-313.

Fetterman, D. M. (1995). In response to Dr. Daniel Stufflebeam's: "Empowerment evaluation, objectivist evaluation, and evaluation standards: Where the future of evaluation should not go and where it needs to go." *Evaluation Practice, 16*(2), 177-197.

Fetterman, D. M. (in press). *Empowerment evaluation*. Thousand Oaks, CA: Sage.

Fetterman, D. M., & Haertel, E. H. (1990). *A school-based evaluation model for accelerating the education of students at-risk*. (Clearinghouse on Urban Education, ERIC Document Reproduction Service No. ED 313 495)

Gray, S. T. (Ed.). (1993). *Leadership is: A vision of evaluation*. Washington, DC: Independent Sector.

Hess, F. A., Jr. (1993). Testifying on the Hill: Using ethnographic data to shape public policy. In D. M. Fetterman (Ed.), *Speaking the language of power: Communication, collaboration, and advocacy (Translating ethnography into action)* (pp. 38-49). London: Falmer.

Hopper, K. (1993). On keeping an edge: Translating ethnographic findings and putting them to use: NYC's homeless policy. In D. M. Fetterman (Ed.), *Speaking the language of power: Communication, collaboration, and advocacy (Translating ethnography into action)* (pp. 19-37). London: Falmer.

Joint Committee on Standards for Educational Evaluation. (1994). *The program evaluation standards*. Thousand Oaks, CA: Sage.

Linney, J. A., & Wandersman, A. (1991). *Prevention Plus III: Assessing alcohol and other drug prevention programs at the school and community level: A four-step guide to useful program assessment*. Rockville, MD: U.S. Department of Health and Human Services, Office of Substance Abuse Prevention.

Mehl, M., Gillespie, G., Foale, S., & Ashley, M. (1995). Project in progress: The first year and a half of the Thousand Schools Project (May 1993-December 1994). Cape Town, South Africa: Independent Development Trust.

Mills, C. W. (1959). *The sociological imagination*. New York: Oxford University Press.

Mithaug, D. E. (1991). *Self-determined kids: Raising satisfied and successful children*. New York: Macmillan (Lexington imprint).

Mithaug, D. E. (1993). *Self-regulation theory: How optimal adjustment maximizes gain*. New York: Praeger.

Oja, S. N., & Smulyan, L. (1989). *Collaborative action research*. Philadelphia: Falmer.

Papineau, D., & Kiely, M. C. (1994). Participatory evaluation: Empowering stakeholders in a community economic development organization. *Community Psychologist, 27*(2), 56-57.

Parker, L., & Langley, B. (1993). Protocol and policy-making systems in American Indian tribes. In D. M. Fetterman (Ed.), *Speaking the language of power: Communication, collaboration, and advocacy (Translating ethnography into action)* (pp. 70-75). London: Falmer.

Parlett, M., & Hamilton, D. (1976). Evaluation as illumination: A new approach to the study of innovatory programmes. In D. Hamilton (Ed.), *Beyond the numbers game*. London: Macmillan.

Patton, M. (1994). Developmental evaluation. *Evaluation Practice, 15*(3), 311-319.

Rappaport, J. (1987). Terms of empowerment/exemplars of prevention: Toward a theory for community psychology. *American Journal of Community Psychology, 15,* 121-148.

Reason, P. (Ed.). (1988). *Human inquiry in action: Developments in new paradigm research*. Newbury Park, CA: Sage.

Sanders, J. R., Barley, Z. A., & Jenness, M. R. (1990). *Annual report: Cluster evaluation in science education*. Unpublished report.

Scriven, M. S. (1967). The methodology of evaluation. In R. E. Stake (Ed.), *Curriculum evaluation* (AERA Monograph Series on Curriculum Evaluation, Vol. 1). Chicago: Rand McNally.

Scriven, M. S. (1991). *Evaluation thesaurus* (4th ed.). Newbury Park, CA: Sage.

Shapiro, J. P. (1988). Participatory evaluation: Toward a transformation of assessment for women's studies programs and projects. *Educational Evaluation and Policy Analysis, 10*(3), 191-199.

Stanford University and American Institutes for Research. (1992). *A design for systematic support for accelerated schools: In response to the New American Schools Development Corporation RFP for designs for a new generation of American schools*. Palo Alto, CA: Author.

Stufflebeam, D. L. (1994). Empowerment evaluation, objectivist evaluation, and evaluation standards: Where the future of evaluation should not go and where it needs to go. *Evaluation Practice, 15*(3), 321-338.

Stull, D., & Schensul, J. (1987). *Collaborative research and social change: Applied anthropology in action*. Boulder, CO: Westview.

Tax, S. (1958). The Fox Project. *Human Organization, 17,* 17-19.

Transitions. (1992). Battle Creek, MI: W. K. Kellogg Foundation.

Usher, C. (1993). *Self-evaluation in the Prince George's County Services Reform Initiative*. Chapel Hill: University of North Carolina, School of Social Work, Center for the Study of Social Policy.

Usher, C. (1995). Improving evaluability through self-evaluation. *Evaluation Practice, 16*(1), 59-68.

van der Walt, H., & Hoogendoorn, L. (1993). *Training for self-evaluation at Ithusheng Health Centre.* Tygerberg, South Africa: Centre for Epidemiological Research in Southern Africa, Medical Research Council.

Watkins, J. M. (1992). Critical friends in the fray: An experiment in applying critical ethnography to school restructuring. In F. Hess (Ed.), *Empowering teachers and parents: School restructuring through the eyes of anthropologists* (pp. 207-228). Westport, CT: Bergin & Garvey.

Weeks, M. R., & Schensul, J. J. (1993). Ethnographic research on AIDS risk behavior and the making of policy. In D. M. Fetterman (Ed.), *Speaking the language of power: Communication, collaboration, and advocacy (Translating ethnography into action)* (pp. 50-69). London: Falmer.

Whitmore, E. (1990). Empowerment in program evaluation: A case example. *Canadian Social Work Review, 7*(2), 215-229.

Whyte, W. F. (Ed.). (1990). *Participatory action research.* Newbury Park, CA: Sage.

Zimmerman, M. A. (in press). Empowerment theory: Psychological, organizational, and community levels of analysis. In J. Rappaport & E. Seldman (Eds.), *Handbook of community psychology.* New York: Plenum.

Zimmerman, M. A., Israel, B. A., Schulz, A., & Checkoway, B. (1992). Further explorations in empowerment theory: An empirical analysis of psychological empowerment. *American Journal of Community Psychology, 20*(6), 707-727.

Zimmerman, M. A., & Rappaport, J. (1988). Citizen participation, perceived control, and psychological empowerment. *American Journal of Community Psychology, 16*(5), 725-750.

BREADTH AND SCOPE

Empowerment Evaluation
and Accelerated Schools

HENRY M. LEVIN

David Fetterman (1994) has defined *empowerment evaluation* as "the use of evaluation concepts and techniques to foster self-determination" (p. 1). This definition is compelling and parsimonious. Much of the evaluation literature focuses on the construction and use of evaluations by those who govern the lives of others. When the state sets out a testing program for schools, it defines what is to be evaluated and how it is to be evaluated. Even if it shares the data with schools, it has not sought their participation in defining what is to be evaluated, how it is to be evaluated, and the best form for reporting results that will foster good decision making. Embedded in Fetterman's definition are two constructs. First, empowerment requires that those whose lives or activities are affected by decisions participate in meaningful

AUTHOR'S NOTE: The author is Director of the National Center for the Accelerated Schools Project, Stanford University. He wishes to thank Pia Wong for her comments and David Fetterman for his encouragement and continuing interest in the Accelerated Schools Project. An earlier version was presented as an invited keynote address at the 1993 annual meetings of the American Evaluation Association, Dallas (November 1, 1993).

49

ways in making those decisions. Second, good decisions are informed decisions that can be enhanced by information that can be provided through systematic evaluations. As a corollary of these two constructs, empowerment evaluation must necessarily require that the design, implementation, and use of evaluations incorporate the meaningful input of those who will be affected by the consequences of their use. This is a perspective that can be found in a broader literature on empowerment theory and especially represented in the writings of its leading advocates (Rappaport, 1981, 1985; Zimmerman, in press).

In this chapter, I will address the Accelerated Schools movement, a movement that Fetterman (1994, p. 2) has cited as a good example of empowerment evaluation. First, I will provide a brief description of the Accelerated Schools Project, its philosophy, process, and practices. Second, I will focus on the mechanics of empowerment in the Accelerated Schools Project. Finally, I will show how evaluation is foundational to and permeates the approach. I believe that there is a high congruity between empowerment and empowerment evaluation in Accelerated Schools and the overall constructs and concepts of community empowerment developed by Fetterman, Rappaport, and Zimmerman.

Accelerated Schools

The Accelerated Schools Project (ASP) was established in 1986-1987 with the launching of two pilot elementary schools serving high concentrations of at-risk students. Its goal was and continues to be the elimination of academic remediation by using all of the schools' resources to accelerate the growth and development of all students to bring them into the academic mainstream by the end of elementary school (Levin, 1987a, 1987b). That is, the focus of Accelerated Schools is to advance the development of all children through using teaching and learning approaches that are usually reserved for gifted and talented students.

ASP has focused especially on schools with high concentrations of at-risk students of whom large numbers had previously been relegated to remedial programs and special education as learning disabled children. The project grew out of earlier research that had examined the demographic characteristics of at-risk students and their educational prospects in conventional schools (Levin, 1986, 1987b, 1988). It found that the predominant policy of tracking these students into remedial instruction characterized by drill and practice and associative learning had extremely deleterious consequences. At-risk children got farther and farther behind the educational mainstream the longer that they were in school, and many began to view school as punishing and arduous, even in the early elementary years.

Out of this research came a quest for a different kind of school that would accelerate rather than remediate. Acceleration necessitates the remaking of the school to advance the academic and social development of children in at-risk situations, not slow it down. This means creating a school in which all children are viewed as capable of benefiting from a rich instructional experience rather than relegating them to a watered-down one. It means a school that creates powerful learning situations for all children, ones that integrate curriculum, instructional strategies, and context (climate and organization) rather than providing piecemeal changes through new textbooks or instructional packages. It means a school whose culture is transformed internally to encompass the needs of all students through creating a stimulating educational experience that builds on their identities and strengths.

Such transformation is neither simple nor swift. Schools are provided with training and follow-up experiences as part of a systematic process (Hopfenberg et al., 1993). The training and follow-up activities require the participation of the full school staff, parents, and students. A coach is trained to work with the school using constructivist activities that engage the members of the school community in problem-solving experiences that lead to a sequence of major activities that the school undertakes over subsequent months. The focus is on internalizing the Accelerated School philosophy and values throughout the school by employing constructivist activities and school processes that lead to school change and the transformation of school culture (Finnan, 1994). Ultimately, these emerge in a school governance and decision-making process that focuses on the creation of powerful learning situations for all children. (The details can be found in *The Accelerated Schools Resource Guide,* Hopfenberg et al., 1993.) Through a particular governance structure and an inquiry approach to decision making, the school addresses its major problem areas in a way that will create powerful learning throughout the school.

Over the 8 years since the initiation of the two pilot schools, the Accelerated Schools Project has expanded considerably in terms of the number of schools, coaches, and regional centers and in the depth and sophistication of the transformation process. In 1994-1995, the project encompassed over 700 elementary and middle schools in 37 states with a typical cost of about $30 to $40 per student. Some 200 coaches had been trained and were being mentored by the National Center through communication, site visits, and retreats. In addition, 10 regional centers had been established to work with schools and train or cotrain coaches with the national center. The project had also initiated its first collaborative effort to transform entire school districts into accelerated entities that would support their constituent Accelerated Schools.

Evaluations of Accelerated Schools have shown substantial gains in student achievement and attendance, full inclusion of special needs children in the mainstream, parental participation, and numbers of students meeting traditional gifted and talented criteria. They have also shown reductions in the numbers of students repeating grades, in student suspensions, and in school vandalism (see, e.g., English, 1992; Knight & Stallings, 1995; Chasin & Levin, 1995; McCarthy & Still, 1993; Wong, 1994). These evaluations have included multiyear assessments in which Accelerated Schools have been compared with control schools (Knight & Stallings, 1995; McCarthy & Still, 1993).

Accelerated Principles and Values

The Accelerated School is not just a collection of programs or an attempt to put together a school through piecemeal accumulation of different policies and practices. It is a set of practices based upon a coherent philosophy and principles. The goal of the Accelerated School is to bring all students into a meaningful educational mainstream, to create for all children the dream school we would want for our own children. This is the guiding sentiment for the transformation of an Accelerated School, one that is embodied in its three central principles: (a) unity of purpose, (b) empowerment with responsibility, and (c) building on strengths.

UNITY OF PURPOSE

Unity of purpose refers to the common purpose and practices of the school on behalf of all its children. Traditional schools separate children according to abilities, learning challenges, and other distinctions; staff are divided according to their narrow teaching, support, or administrative functions; and parents are usually relegated to only the most marginal of roles by the school in the education of their children. Accelerated Schools require that the schools forge a unity of purpose around the education of all students and all of the members of the school community, a living vision, and a culture of working together on behalf of all of the children. Strict separation of either teaching or learning roles works against this unity and results in different expectations for different groups of children. Accelerated Schools formulate and work toward high expectations for all children, and children internalize these high expectations for themselves. Unity of purpose means widespread empowerment to set and implement goals, standards, and dreams for the school community that will become the subject of assessment and evaluation by that community.

EMPOWERMENT WITH RESPONSIBILITY

Empowerment with responsibility refers to who makes the educational decisions and takes responsibility for their consequences. Traditional schools rely on higher authorities at school district and state levels as well as on the content of textbooks and instructional packages formulated by publishers who are far removed from schools. Staff at the school site have little discretionary power over most of the major curriculum and instructional practices of the school, and students and parents have almost no meaningful input into school decisions. In this respect, the powerlessness leads to a feeling of exclusion in terms of the ability to influence the major dimensions of school life.

An Accelerated School requires that school staff, parents, and students take responsibility for the major decisions that will determine educational outcomes. The school is no longer a place in which roles, responsibilities, practices, and curriculum content are determined by forces beyond the control of its members. In its daily operations, the school community hones its unity of purpose through making and implementing the decisions that will determine its destiny. At the same time, the school takes responsibility for the consequences of its decisions through continuous assessment and accountability, holding as its ultimate purpose its vision of what the school will become. This is accomplished through a parsimonious, but highly effective, system of governance and problem solving that ensures inclusion of students, staff, and parents in the daily life of the school.

BUILDING ON STRENGTHS

Traditionally, schools have been far more assiduous about identifying the weaknesses of their students than looking for their strengths. A focus on weaknesses or deficiencies leads naturally to organizational and instructional practices in which children are tracked according to common deficiencies. The logic is that "lower" groups cannot keep up with a curricular pace that is appropriate for "higher" groups. But Accelerated Schools begin by identifying strengths of participants and building on those strengths to overcome areas of weakness.

In this respect, all students are treated as gifted and talented students, because the gifts and talents of each child are sought out and recognized. Such strengths are used as a basis for providing enrichment and acceleration. As soon as one recognizes that all students have strengths and weaknesses, a simple stratification of students no longer makes sense. Strengths include not only the various areas of intelligence identified by Gardner (1983) but also areas of interest, curiosity, motivation, and knowledge that grow out of the culture, experiences, and personalities of all children. Classroom themes can be those in which children show interest and curiosity and in which reciprocal teaching, cooperative learning, peer and cross-age tutoring, and individual and group projects can highlight the unique talents of each child in classroom and school activities. These group processes and the use of specialized staff can both recognize and build on the particular strengths and contributions of each child while providing assistance in areas of need within the context of meaningful academic work.

Accelerated Schools require that each child be fully included in the activities of the school while validating the child's strengths and addressing her areas of special need. This can be done in regular classrooms employing classroomwide and schoolwide curricular approaches that are based upon inclusion of every child in the central life of the school. It can be done not only with multiability grouping and multiage grouping but by recognizing that all children have different profiles of strengths that can be used to complement each other and to create strong teams that provide internal reinforcement among students.

It should also be noted that the process of building on strengths is not just limited to students. Accelerated Schools also build on the strengths of parents, teachers, and other school staff. Parents can be powerful allies if they are placed in productive roles and provided with the skills to work with their own children. Teachers bring gifts of insight, intuition, and organizational acumen to the instructional process that are often untapped by the mechanical curricula that are so typical of remedial programs. By acknowledging the strengths found among participants within the entire school community, all of the participants are expected to contribute to success.

ACCELERATED SCHOOLS VALUES

Accelerated Schools also exemplify a set of values that permeate relationships and activities. These include the school as a center of expertise, equity, community, risk taking, experimentation, reflection, participation, trust, and communication. These values focus the school on the inner power, vision, capabilities, and solidarity of the school community. But, especially important are such values as equity, the view that the school has an obligation to all children to create for them the dream school that we would want for our own children. Such a school must treat children equitably and must address equitable participation and outcomes. The school is viewed as an overall community rather than as a building with many separate communities represented, although the cultures and experiences of different students are acknowledged and incorporated into the school experience. Addressing the needs of all children will require experimentation and risk taking, reflection, trust, and communication. Above all, the concept of unity of purpose is present in all of the values and practices of the school, a necessary approach to the inclusion of all students in a common school dream.

POWERFUL LEARNING

The three principles and nine values of Accelerated Schools are all used to create what are called powerful learning situations (Hopfenberg et al., 1993, chaps. 6-10). A powerful learning situation is one that incorporates changes in school organization, climate, curriculum, and instructional strategies to build on the strengths of students, staff, and community to create an optimal learning situation. What is unique about this approach is that changes are not piecemeal but integrated around all aspects of the learning situation. This contrasts sharply with the usual attempts to transform schools through idiosyncratic reforms involving the ad hoc adoption of different curriculum packages, instructional practices, and organizational changes to address each perceived problem that the school faces. Over time, some of these are pruned and others are added, without any attempt to integrate them into an overall philosophy and vision of the school. Powerful learning builds on the strengths of all community members and empowers them to be proactive learners by developing skills through intrinsically challenging activities that require both group work and individual endeavor.

Accelerated Schools also emphasize the connections between the big wheels of the school and the little ones. The *big wheels* refer to the overall school philosophy and change process that are shared collaboratively by all members of the school community. The *little wheels* refer to the informal innovations that grow out of participation by individuals or small groups in embracing the school's philosophy and change process. These little wheels result from the internalization of the school philosophy and change process into the belief system of school members, resulting in changes in their individual decisions and commitments in classrooms and in individual and group interactions.

Empowerment Processes

To have empowerment evaluation, there must be empowerment. Consonant with this premise, the Accelerated School is largely a self-governing community that has established a system of governance and decision making over its own destiny. The process that is used requires intense scrutiny of both practices and results, problem solving, and the sharing of information with the larger community. All staff members, parent representatives, and student representatives are expected to participate in decisions. The school initiates its empowerment by taking stock of its resources, activities, students, community, and other dimensions (Hopfenberg et al., 1993, pp. 60-73). All members of the school community are involved in taking-stock activities, from identifying which dimensions of the school to investigate to setting out the questions that must be answered and the methods for answering them. Data are gathered using these methods and analyzed by taking-stock committees that are devoted to particular school dimensions. As the "school is the center of expertise," it must immediately engage itself in developing expertise about itself through setting out research questions and answering them through systematic inquiry. Analysis of the data and reflection provide contributions to a schoolwide report to which everyone contributes. Especially central is the search for school and community strengths and resources as well as challenges. The school will ultimately build on these strengths to overcome its challenges.

When the taking-stock process is completed, the school has developed a baseline of information that has also brought the school community together around a common understanding and enhanced familiarity through working together. This is followed by establishing a vision for the future that the school will be dedicated to (Hopfenberg et al., 1993, pp. 74-81). All participants work in groups to design a dream school that would be worthy of their own children, a learning community, a high level of professional development for staff, and active engagement of parents. Students are asked to participate in the design of their dream school, incorporating the concepts in academic work through essays, interviews, artistic renderings, comparisons of different perspectives, and so on. Although the vision may be summarized by a vision statement, it is the system of goals, beliefs, and shared practices that arises from the months of working together that truly constitutes the vision. The words of the vision statement are only a reminder. This hard work and its resulting inspiration and excitement are celebrated with a community-wide vision celebration, which launches the next stage. The school community has become empowered to develop its own dream of what it will become rather than to aspire toward a stultified set of goals set out by higher levels of authority.

At this point, the school compares its baseline data from taking stock and its vision and compiles a list of differences between the two. To get from baseline to the vision requires substantial changes in the school. These necessary changes are compiled and clustered according to major areas. Through extensive discourse, the school chooses just a few priority areas that it will begin to work on initially (Hopfenberg et al., 1993, pp. 82-85). Given the popularity of these priorities, there is no difficulty in obtaining participation from all staff, parent, and student representatives to choose one of the areas to work on, and the power to address those issues that they feel are most crucial. These self-selected groups are called *cadres* and are responsible for carrying out the most intensive analysis in their particular priority areas. Typically, there are between three to five cadres per school depending upon the size of the school community. Cadres may focus on priorities that vary from school organization and resources to parental involvement to curriculum areas.

In addition to cadres, there are two other levels in the governance structure of the school (Hopfenberg et al., 1993, pp. 86-94). A *steering committee* comprises representatives from each cadre along with the principal and at-large representatives from among teachers, support staff, parents, teachers, and the larger school community. The steering committee is responsible for coordination of activities, distribution of information, monitoring the progress of cadres, and approving the recommendations of the cadre for submission to the *school-as-a-whole* (*SAW*). The SAW is the primary decision-making body and includes all persons who wish to participate in the school community. Cadres are expected to meet weekly, the steering committee meets biweekly, and the SAW on a quarterly or as-needed basis. Announcements of minutes and agendas are provided in advance for the entire school, and minutes are made available in a timely fashion to enhance discussion and communication.

The entire school is trained in both group dynamics and meeting management (Hopfenberg et al., 1993, chap. 5) as well as in group problem solving (Hopfenberg et al., 1993, chap. 4). Empowerment in schools requires not only a wide scope of discretionary authority by those who will be affected by decisions but also the skills and experience to make informed decisions. Problem solving follows an inquiry method in which cadres and committees take sufficient time to define and understand the challenges that they are addressing followed by generating hypotheses on the causes of the problem. Hypotheses are tested by gathering information through surveys, interviews, school records, and observations. That is, the cadres learn how to do research and continually hone their research skills. Those hypotheses that are confirmed are used as a basis for brainstorming for solutions with a search for solutions both within the school community from those with expertise and successful experience as well as outside the community to find what others have done. The cadre formulates an action plan that is pilot tested, revised if necessary, and submitted for approval to the steering committee and the SAW. The plan includes the details and logistical requirements for implementation as well as an assessment framework. That is, the plan must include answers to the question of how the school will know if it has succeeded in solving the challenge. Thus, an assessment plan must be set out in advance. Based upon the assessment results, the school may wish to modify its intervention or continue intact while undertaking a new challenge.

Evaluation for Empowerment

We have set out as one requirement for evaluation empowerment that empowerment must exist. The Accelerated School is organized to empower parents, students, and staff around the development of the dream school that they would want for themselves and their own children. The process of getting there requires taking stock of the school and school community, creating a collective dream that incorporates everyone's input, establishing priorities and a system of governance that maximize participation, and instituting a process of problem solving and informed decisions to reach that dream. Such an approach requires a system of evaluation from the very beginning.

TAKING STOCK AS EVALUATION

The very launch of an Accelerated School immerses the school community in evaluation through taking stock. Taking stock is largely a summative evaluation activity with its emphasis on working collectively to establish a baseline for the school. Much of taking stock is descriptive, summarizing the community, programs, facilities, and achievements and seeking to find both strengths and challenges. The empowering aspect of this activity consists of the inclusion of all staff, and parent and student participation, in this initial evaluation as well as its emphasis on the participants constructing the questions that they will ask and the methods by which they will answer the questions. Although guidance is provided by a coach through questioning strategies, the actual evaluative decisions are made by the participants and are summarized for the benefit of the entire school community. Even beyond this process there is empowerment in the information that is derived, a comprehensive compendium of knowledge about many dimensions of the school in which "knowledge is power."

BUILDING AND EVALUATING THE SCHOOL DREAM

Constructing a collective dream for the school is an evaluation activity that requires drawing upon and combining the perspectives of students, staff, and parents as well as setting out criteria for assessing whether the dream is becoming a reality. As a community endeavor that follows taking stock, the different constituencies are asked to join together in sketching their own dream schools and then combining individual, group, and constituency dreams into a school dream. Each participant must construct his or her own dream and share with others until a collective vision is forged for the entire school. This requires a dynamic and sensitive process of sharing, discussion, debating, and synthesizing. A major test of whether the vision is just a collection of words or has captured the ambitions and aspirations of the beholders is to set out criteria for ascertaining how the school will know that the dream is fulfilled.

For example, if part of the school vision is that the school will be a caring environment, it is necessary to set out the criteria that will enable the school to know if that part of the vision is met. This means that the school must state its vision in terms of observable outcomes including those events, processes, and feelings that will be present. What will happen in a caring school? Will each child feel that she belongs and that others will miss her if she is absent? Will older children automatically assist younger children without being asked by teachers? Will we observe second graders bending over to ask kindergartners what is wrong or to tie their shoelaces? Will all students feel that it is natural to assist others, to tutor others, to engage in teamwork? These are the kinds of evaluation questions that must be discussed constantly in setting out the vision and evaluating whether it is being achieved, and they must be done for each part of the vision.

EVALUATING THE PROCESS

Long before the school reaches its vision, it must participate in the Accelerated School transformation process. Periods for critical reflection are built into the Accelerated School process in which each school must evaluate whether it is following the process by asking crucial questions. For example, has the process of taking stock engaged all participants? Has there been a careful search for all of the important dimensions of the school? Are the taking-stock committees formulating their questions carefully? Are the methods of gathering information appropriate? Has the quest for school vision or constructing the school dream involved all of the pertinent constituencies? Has it taken a form in which all of the views are considered? Has the school selected its priorities carefully and properly constituted its cadres, steering committee, and school-as-a-whole? Are these governance and decision bodies meeting regularly and following the inquiry method and productive group processes? Are the action plans tested, and are decisions fully implemented, or do things stop once the decision is reached?

A major part of Accelerated Schools evaluation involves self-evaluation to ensure that the process, philosophy, and values are subscribed to and followed. If a school is not following this process and embracing the values and philosophy, it is unlikely to be transforming itself in the way that the Accelerated School model requires. Without such transformation, we would not expect results that are consistent with the model. Moreover, if the process is followed but decisions are not fully or properly implemented, the governance outcomes will have died stillborn and the results are unlikely to be forthcoming. Thus, an Accelerated School must monitor its own progress and ensure that it is fully meeting the various requirements of the process. An Accelerated School is a self-evaluating entity, examining not only the depths of its own commitment or dream but also the fidelity of the process that will get it to its dream.

EVALUATING OUTCOMES

An Accelerated School strives to achieve powerful learning and accelerated outcomes for all of its students by creating a school that makes and implements decisions toward that end. But, for Accelerated Schools, empowerment means not only the ability to make decisions but the ability to bear responsibility for the consequences of those decisions. The evaluation of outcomes is integral for taking responsibility for consequences. Such evaluation is undertaken by all three governance groups: the cadres, the school-as-a-whole, and the steering committee.

Built into the inquiry process is an assessment stage. As cadres or the steering committee identify and define challenges, set out and test hypotheses on why the problem exists, brainstorm for solutions, formulate action plans that are pilot tested and (we hope) approved, they must set out an assessment plan. That is, a plan to get more parental involvement or to improve school writing or to expand a program of hands-on science must set out the conditions under which the cadre will know that the program has succeeded. Indeed, that is the way that the question is formulated. At the problem consideration stage, it is clear what the existing problematic situation looks like. But it is equally important to stipulate what success would look like. Cadres need to set out their criteria for success and indicators of success that meet those criteria in advance of implementing the decisions coming out of the inquiry process. These criteria and indicators can be used to design the system of assessment to see if the promise of the intervention is being met. Of course, as the school gets more and more experience with the intervention, it may improve its system of assessment. But, even at an early stage, the assessment criteria must be used to evaluate the overall solution and the action plan to implement it. If there are still shortcomings in results or new horizons for improvement identified from this assessment, the initial stages of inquiry are undertaken once again to consider the necessary challenges and required changes. An effective application of this inquiry approach views assessment as central to decision making. Such evaluation is part and parcel of the quest for continuous improvement with inquiry.

Periodically, the school-as-a-whole examines the overall accomplishments of the school in moving toward its vision. This is done through revisiting the process. The original baseline data from the first taking-stock activity can be compared with those from a new taking-stock activity a few years later. The difference between the two periods provides a clear picture of the progress and improvement in educational outcomes that have taken place. But, at the same time, the Accelerated School process will have generated experience and sophistication that will enable the school to revisit its vision and enrich it. All of these activities require assessment and evaluation. In short, evaluation is embedded in virtually every part of the Accelerated Schools process. By combining systematically the theme of school empowerment with a reflective and continuous process in which assessment and evaluation are firmly embedded, the Accelerated School appears to embody rather fully the concept of empowerment evaluation.

References

Chasin, G., & Levin, H. M. (1995). Thomas Edison Accelerated Elementary School. In J. Oakes & K. H. Quartz (Eds.), *Creating new educational communities, schools, and classrooms where all children can be smart* (pp. 130-146). Chicago: University of Chicago Press.

English, R. A. (1992, October). *Accelerated Schools report*. Columbia: University of Missouri, Department of Education and Counseling Psychology.

Fetterman, D. M. (1994). Empowerment evaluation. *Evaluation Practice, 15*(1), 1-15.

Finnan, C. (1994). School culture and educational reform: An examination of the Accelerated Schools Project as cultural therapy. In G. Spindler & L. Spindler (Eds.), *Cultural therapy and culturally diverse classrooms* (pp. 93-129). Thousand Oaks, CA: Corwin.

Gardner, H. (1983). *Frames of mind*. New York: Basic Books.

Hopfenberg, W., Levin, H. M., Chase, C., Christensen, S. G., Moore, M., Soler, P., Brunner, I., Keller, B., & Rodriguez, G. (1993). *The Accelerated Schools resource guide*. San Francisco: Jossey-Bass.

Knight, S., & Stallings, J. (1995). The implementation of the Accelerated School model in an urban elementary school. In R. Allington & S. Walmsley (Eds.), *No quick fix: Rethinking literacy programs in American elementary schools* (pp. 236-252). New York: Teachers College Press.

Levin, H. M. (1986). *Educational reform for disadvantaged students*. West Haven, CT: NEA Professional Library.

Levin, H. M. (1987a). Accelerated Schools for disadvantaged students. *Educational Leadership, 44*(6), 19-21.

Levin, H. M. (1987b). New schools for the disadvantaged. *Teacher Education Quarterly, 14*(4), 60-83.

Levin, H. M. (1988). *Towards Accelerated Schools*. New Brunswick, NJ: Rutgers University, Center for Policy Research in Education.

McCarthy, J., & Still, S. (1993). Hollibrook Accelerated Elementary School. In J. Murphy & P. Hallinger (Eds.), *Restructuring schools: Learning from ongoing efforts* (pp. 63-83). Newbury Park, CA: Corwin.

Rappaport, J. (1981). In praise of paradox: A social policy of empowerment over prevention. *American Journal of Community Psychology, 9*, 1-25.

Rappaport, J. (1985). The power of empowerment language. *Social Policy, 16*, 15-21.

Wong, P. (1994). *Accomplishments from Accelerated Schools*. (Available from the National Center for the Accelerated Schools Project, Stanford University, Stanford, CA 94305)

Zimmerman, M. A. (in press). Empowerment theory: Psychological, organizational, and community levels of analysis. In J. Rappaport & E. Seidman (Eds.), *Handbook of community psychology*. New York: Plenum.

Empowerment Evaluation and the W. K. Kellogg Foundation

RICARDO A. MILLETT

The following chapter provides a brief historical overview of program evaluation at the W. K. Kellogg Foundation. It traces the foundation's ongoing attempts to improve its approach to program evaluation in the philosophy and values of W. K. Kellogg himself. This philosophy guided him not only to place his trust, but to invest his fortune, "to help people help themselves."

In the late 1980s, the foundation began a series of organization-wide activities to institutionalize program evaluation, based on Will Keith Kellogg's principles and values. The objective was to develop the means to use evaluation as a useful ally—not a painful process to be feared or mistrusted. This early planning included consultations with various practitioners in the field of program evaluation, such as Dr. David Fetterman and others, who helped to shape and inform our evaluation approaches. Although none of us used the term *empowerment evaluation* at the time, our attempt to approach program evaluation within the value framework of W. K. Kellogg easily led us toward these principles. Subsequently, the Kellogg Foundation has pioneered the evaluation of projects in "clusters." This approach helps build the capacity of the grantees to improve their projects by learning and by sharing lessons learned with grantees working on similar projects.

Our approach to project-level evaluation, which is a requirement for all grant recipients, is also guided by this objective. By increasing the grantees' capacity to control the evaluation process, we help them capture and use information that can enhance their projects' success. Accordingly, we negotiate with the grantee to establish evaluation methods that fit their nature and level of sophistication.

Importantly, the Kellogg Foundation is equally concerned with measuring project outcomes *and* with having evaluation implemented in a manner that increases the skills, knowledge, and lessons learned by individuals and organizations. This approach seeks to strengthen an organization's ability to use the tools of evaluation after the Kellogg Foundation's grant support ends.

As one of the world's largest foundations, whose expenditures exceeded $260 million in 1993-1994, including approximately $20 to $25 million to support evaluation activities, it is fair to say that program evaluation is a major investment at the W. K. Kellogg Foundation. It is an integral part of what the foundation does, and it is an activity that is taken very seriously.

The W. K. Kellogg Foundation has long supported evaluation as integral to the mission of "helping people to help themselves." Originally established in 1930 as the Child Welfare Foundation, the organization has for decades made grants in health, youth, agriculture and rural development, leadership, and education. Program evaluation has been a part of the foundation since the early days. A 1988 board report cites a 1930s evaluation of a Kellogg Foundation outdoor education program for youth and a $100,000 investment in an evaluation of the Michigan Community Health Project during the 1940s. In the 1970s, one staff member held the title of Vice-President for Dissemination, Evaluation, and Utilization, signifying the foundation's commitment to these ideas. Early evaluation efforts seem to have been driven by the curiosity, interest, and expertise of individual program directors.

In the 1980s, the foundation board and staff identified a growing need to learn about the foundation's grant-making investments. This marked the beginning of the foundation's renewed emphasis on evaluation.

Taking evaluation seriously is synonymous with taking programming seriously. And, at the Kellogg Foundation, programming is taken seriously because it is considered an investment in people, because it focuses on improving lives, and because it can potentially boost grantees' capacity to solve problems and sustain the gains made.

In 1990, as we began to plan our initiatives for this decade, we also reemphasized our long-standing support of evaluation.

Program Evaluation at the Kellogg Foundation

In great measure, the foundation's approach to programming and evaluation was shaped by W. K. Kellogg, who was a pioneer in the breakfast cereal industry. He was an industrial giant who, in the words of biographer Horace B. Powell (1956), "was one of the 20th century's great movers of ideas and shakers of tradition." He was modest about his wealth and considered his millions only a by-product of his will to succeed in business. It was Mr. Kellogg's hope "that the property kind Providence has brought me may be helpful to many others, and that I may be found a faithful steward" (p. 165).

Even before he started the foundation that bears his name, he donated to countless charities. His early personal philanthropies included aid for British war orphans, the blind, a number of hospitals, and medical programs. In June 1930, the W. K. Kellogg Child Welfare Foundation was established. Two months later, it was reorganized as the W. K. Kellogg Foundation to broaden its role to encompass "the promotion of health, education, and welfare of mankind" (p. 306). During his lifetime, W. K. Kellogg gave the foundation nearly $66 million in assets (largely stock in Kellogg Company). Today, the W. K. Kellogg Foundation carries on the tradition of Mr. Kellogg and his belief in "helping people to help themselves" (p. 61). These same values have shaped the foundation's approach to program evaluation.

EVALUATION VALUES: "WE KNOW BETTER THAN WE DO . . ."

Simply, one of those values is to apply what we know to the social problems that confront us. We do not mean to suggest that program and support staff are always engaged in action research and applied research. We try, rather, to evaluate and learn from our experience, on an ongoing basis, and to do even better in the future. We also believe that, in most areas of human endeavor, "we know better than we do."

EVALUATION CAPACITY BUILDING

A second value of the foundation is that of "helping people to help themselves." We seek to help people do what is important in a way most appropriate for them. To this end, the foundation has no a priori agenda and is not prescriptive in its approach to program evaluation. We believe that people in the communities and institutions we serve are in the best position to make decisions, to implement the programs that are the best suited for their circumstances at a given time, and to evaluate the lessons learned. That is why we believe in community problem solving.

EVALUATION AND COLLABORATION

A third value deals with the issue of collaboration. Communities are not made of one individual or one agency. They are made up of many people and many agencies trying to solve problems. To collect useful evaluation information, the various stakeholders must work together at different levels to develop a complete network of programs and services.

LEARNING BY DOING

Another value is "learning by doing." Within this context, the foundation philosophy has always reflected the objectives of empowerment evaluation. One of the more important elements of empowerment evaluation is involving participants in the process. In our case, this means the grantees. Our process engages grantees and other stakeholders at all levels of the evaluation. This process "asks" people—in a very respectful way—what kind of information they need and how they will use this information to achieve better project results. It is a process that ultimately empowers people, particularly at the so-called grassroots or neighborhood level, who may not have sophisticated research or evaluation skills.

This way of thinking suggests that, at the Kellogg Foundation, our job is to locate people with good ideas and to provide them with the support needed to find lasting solutions. The foundation's program staff are not in a position to know the solutions to the health, economic, youth, education, and leadership problems faced by Native Americans, Latinos, rural farmers, urban residents, South African women, or Mexican Americans along the Texas border. We strongly believe that solutions to social problems are likely to be more viable and long-lasting when they come from inside the community.

From these values, we have developed an evaluation approach that embodies the principles of empowerment evaluation. This is a function of both Mr. Kellogg's commitment to fostering self-determination and Dr. Fetterman's association with the foundation over the years. A discussion about our successes and challenges helps demonstrate how we use evaluation.

Key Features of Evaluation

Guided by the four values I've just discussed, evaluation at the foundation has now become an institutional function. In general, it is fair to say that evaluation at the foundation serves one prime function: "to improve, not prove." Specifically, the purposes of evaluation at the Kellogg Foundation are to improve projects, to improve programming, and to inform policy making. In pursuing these goals, the foundation seeks to

enhance the likelihood of individual project success;

assess the effectiveness of individual, or groups of, projects (clusters/strategies);

help determine whether a programming focus is having the desired effects; and

assess the extent to which foundation staff activities (e.g., networking conferences, project leadership development, evaluation assistance, grant development) contribute to project success.

Challenges for the Future

The challenges that will be faced by empowerment evaluation are the challenges that are already familiar to us at the foundation. These center in large part on the need to reconcile the rigor of methodology with the desired effect of maximizing program outcome—in other words, reconciling the potential tension between project implementation and the determination of effects. We are constantly refining an evaluation approach that brings into better alignment the need to "improve" with the need to "prove." To date, we have not been as successful with this as we would like to be. We are quite aware that our grant-making philosophy may be laden with what research methodologists term *biases*. There may be, in fact, some inherent skepticism in our philosophy about conventional evaluation approaches. From an empowerment point of view, that skepticism is founded in the legitimate concern that evaluation improperly conceived and conducted may inhibit the foundation's intention to be action oriented, responsive, and supportive of risk taking with our grant-making investments. The skepticism may also reflect the foundation's experience with professional evaluators for whom evaluation becomes an end in itself, for whom evaluation can become an enterprise that is concerned more with determining/implementing the appropriate methodology than with what effect their approach may have on building "people capacity."

Empowerment Evaluation

Through years of grant making, we have discovered there are certain tension points among grantees with respect to evaluation. Briefly stated, these tension points include the following:

- A need to reconcile evaluation research methodology with the need of the grantee to use outcome data to improve program implementation
- The evaluator's need to generate valid and reliable data documenting an evaluation approach that more often than not is sensitive to the grantee's capacity to be an informed participant or feel ownership of the data
- The inability of the above approach to enhance trust between the grantee and the grantmaker, which can cause fear and anxiety about evaluation as opposed to enhanced competence on how to use evaluation

In large measure, these tension points led us to characterize our approach to grantees and potential grantees as one that intends to "improve rather than prove." This characterization has been tremendously helpful in reducing fear, anxiety, and mistrust among our grantees and has increased their willingness to use evaluation as a management tool.

By sharing this experience, which helped to shape our approach, it should not be assumed that the Kellogg Foundation is more concerned with ideology than empiricism, or that we are dogmatic in our approach to evaluation. We do have a philosophy that is people focused. We do see evaluation as an integral part of programming, not as a separate function to determine program effects. In addition, we are also concerned with the "bottom line," that is, judging whether the program made a difference based on objective evidence. The balance we are trying to achieve deals with who is doing the judging and how this determination affects ownership and use of evaluation as a process and a product.

We constantly seek ways to strengthen program evaluation within the context of "balancing the need to improve [projects/programs] against the need to prove [project/program effects]." Over the years, we have learned at least one important lesson regarding this balancing act. This lesson is the necessity to calibrate how evaluation judgments are made. Useful judgments—that is, evaluation outcomes—are more often than not determined by who is making the judgments.

We are committed to strengthening and widening empowerment evaluation. We do not, however, believe that we have found the right way to do program evaluation, nor do we advocate it for all foundations. Our approach is consistent with a philosophy focused on people, "helping people to help themselves." For strict methodologists, the approach may not always yield evaluation data that meet rigid standards of validity and/or reliability. We are, however, open to learning how best to accommodate the desire for valid and reliable data as outcomes without sacrificing the desired outcome of supporting, enhancing, and sustaining the capacity of improvement at the grantee level.

As the foundation increases its grant making to urban neighborhoods and small rural communities, we have become increasingly aware of the anxiety that evaluation evokes among these grantees. They see evaluation as something that is being done to them. They view it as a tool to form judgments that will provide an excuse to kill their project. Unfortunately, traditional evaluation methodology has developed a bad reputation in this respect.

Another limitation of traditional evaluation is the use of the conventional research paradigm of applying randomized experiments and highly structured statistical controls. These methodologies are not always compatible with building capacity or active participation in the evaluation process (see Conrad, 1994; Fetterman, 1982). We respect the use of research methodology and its appropriate use in program evaluation. We believe, however, that methods should be applied with due consideration to social context, informational use/utility, or, more simply, the needs of the people—particularly people at the grantee level who will need or use the evaluation information. In short, for the foundation, evaluation must include, and give adequate weight to, the many different human, political, cultural, and contextual elements that affect what is being evaluated. Relatedly, the foundation program staff and evaluation consultants must interact with people in a manner that respects their dignity, integrity, and privacy. The methods used, however essential to reliability or validity of the evaluation process, are only one part of what must be carefully considered when designing an evaluation plan.

At the foundation, we continually try to minimize the anxiety and fear that evaluation might provide. We attempt to engage the grantees in a process whereby they will see evaluation, particularly at the project level, as another tool to assist in project management. Our challenge is to work with them so that they will embrace the evaluation process as helpful and collaborative rather than paternalistic and judgmental. The important thing is to have the foundation working together with the grantee as the project unfolds. We have learned that nothing has greater risk of undermining our desire to support the capacity of people to help themselves, particularly with nontraditional grantees, than a heavy-handed, prescriptive, inflexible approach to evaluation.

A key principle of foundation evaluation is that this function should not interfere with doing important things and that a proactive orientation should be combined with a careful assessment of lessons learned.

During the past 3 to 4 years, we have worked with the Independent Sector,[1] other foundations, and many other nonprofit organizations to examine the program evaluation function. I think it is fair to say that more and more nonprofits are beginning to see the role/function of program evaluation the way we do at the foundation. A workshop on the Independent Sector's "New Vision for Evaluation" at the American Evaluation Association's annual meeting in Dallas (1993) can attest that a new vision is being cultivated among nonprofits. The new vision is clearly aimed at collaborating with and empowering people as well as determining the impact of programs. As it happens, the foundation has had considerable experience with the objective of this new vision and, obviously, we agree with it in principle and practice.

New Vision for Evaluation

Some of the key elements of this vision include the following:

- The output of evaluation is organizational learning.
- Evaluation is a developmental process, not a report card.
- Evaluation is everybody's job. Everyone in the organization gathers information and asks the question: What can we do to get better?
- Evaluation is not an event but a process, not episodic but ongoing, not outside the organization but ingrained in its day-to-day operations.
- From the onset of an evaluation process, there is a collaborative relationship between grantmaker and grantseeker as both seek to learn how the organization can solve a problem or deal with an issue more effectively.

Our challenge at the foundation is to improve and refine the evaluation skills of our staff and grantees continually. This challenge is captured in our recently developed Evaluation Mission Statement: "To provide continuous, high quality support to enhance the mission of the Foundation by developing, capturing, and communicating useful and usable information for key stakeholders and other audiences."

In May 1994, the foundation made a strategic decision to relocate the Evaluation Unit under a newly created vice president for communication and strategic planning. This joining of the communication and strategic planning functions brought the foundation's evaluation, marketing/dissemination, and public policy education activities under one umbrella. By so doing, the foundation is in a better position to increase its strategic grant-making objectives, that is, to realize an increase in data use and the capacity of our grantees to leverage change.

With audiences' needs primary, we believe that this alignment of evaluation, marketing/dissemination, and policy program functions will enable the foundation to balance the formative information needs of grantees with the summative information needs of our board of trustees better.

Any experienced professional in the world of philanthropy knows the critical importance and challenge that evaluation faces to be "relevant" to several constituencies. For evaluation data to be useful, they must respond not only to the needs of the grantee but also to the needs of the foundation's board, its program staff, and various allied audiences external to the foundation. In short, effective empowerment evaluation must focus on the needs of not only the grantee but other critical audiences as well.

Our approach at the Kellogg Foundation is to view the audiences' needs for information as primary. Integrating the functions of evaluation, marketing/dissemination, and policy education within this context has achieved a clarity of focus that helps all participants design and deliver more responsive evaluation. This strategy empowers grantees, gives our board the accountability data that they often require, and provides program staff with information to assist their grant management functions.

As a major stakeholder, evaluation will be useful to program staff if

   evaluation information allows staff to make more informed selection of initiatives and grantees;

   evaluation information allows staff to assess their strategic progress, and informs and guides program management;

   evaluation guidelines allow staff to play an even more helpful role with potential and actual grantees as they continually search for good answers to good questions and apply them to further their own development.

As a major stakeholder, evaluation will be useful to grantees if

   evaluation products and processes help grantees solve problems, improve their performance, and develop their capacity to undertake even more productive work;

   evaluation expectations are designed at a scale appropriate to the grantee's level of development as well as the significance of the project (this allows grantees to increase their evaluation skills so that they can better understand and communicate their activities and impact to their various stakeholders);

evaluation allows grantees to learn from other grantees, past and current, and provides answers or insights useful to their own development.

As a major stakeholder, evaluation will be useful to the board of trustees if

evaluation products and processes give the foundation's board a better view of progress toward the foundation's objectives and/or opportunities to have greater impact;

evaluation processes provide an opportunity to discuss and examine lessons learned, whether successes or failures, in a safe and instructive way.

As a major stakeholder, evaluation will be useful for allied audiences if

evaluation allows grantees to identify audiences important to them and allows grantees to be responsive to their own stakeholders, networks, and advocates;

evaluation facilitates the use (not just the receipt) of findings and lessons by allied audiences, to increase their impact;

evaluation helps articulate lessons and illustrates successful programs to help inform policy communities (policymakers and those that influence policy);

evaluation helps generate lessons and examples useful to other practitioners and contributes to the development of "social products" useful to the general public and their advocates;

evaluation helps inform other stakeholders in the philanthropic community of the principles and benefits of "strategic grant making."

By thinking about various stakeholders (including those who can influence policy) and what they will want to know about a particular project from its outset, the evaluation can be tailored to meet the information needs of key audiences. Evaluation information can then be used strategically to communicate with different audiences to improve and advance efforts supported by the foundation.

Conclusion

The W. K. Kellogg Foundation's approach to evaluation is not touted as a model that everyone should adopt. It is still evolving and there are areas where we may still be considered "underdeveloped" against the Program Evaluation Standards recently published by the Joint Committee on Standards for Educational Evaluation. As an approach taken by a foundation, as opposed to an academy or a research or policy firm or institution, this approach offers insights, ideas, and encouragement to ways evaluation can revive its relevance to people and organizations working at the "ground level." It suggests ways that evaluation can be considered not as the "enemy" but simply as a process to problem solve *with* grantees, to gather information, to interpret data, and to analyze data to improve successful program implementation at the "ground level." At the same time, it suggests ways to enhance program impact through the strategic integration of marketing/dissemination and public policy education, to inform progress at both the grantmaker and the grantee levels, and to share and disseminate lessons learned.

What makes evaluation useful to the Kellogg Foundation and its grantees is that it isn't directed at absolute determination of "success" or "failure"—to fund or not to fund. Rather, the emphasis is on helping people and their projects learn and improve as their projects evolve. Implemented in this manner, evaluation has relevance to the real world of work and can better the human condition.

We will continue to refine our approach to evaluation along the principles and purposes of empowerment evaluation. We invite professionals in the field of evaluation to join us in this effort.

Note

1. Independent Sector is a nonprofit coalition of over 850 corporate, foundation, and voluntary organization members with national interest and impact in philanthropy and voluntary action.

References

Conrad, K. J. (1994). *Critically evaluating the role of experiments* (New Directions for Program Evaluation, No. 63). San Francisco: Jossey-Bass.

Fetterman, D. M. (1982). Ibsen's baths: Reactivity and insensitivity (A misapplication of the treatment-control design in a national evaluation). *Educational Evaluation and Policy Analysis, 4*(3), 261-279.

Powell, H. B. (1956). *The original has this signature.* Englewood Cliffs, NJ: Prentice Hall.

CONTEXT

Empowerment Evaluation
and State Government

Moving From Resistance to Adoption

JOYCE KELLER

This is not your "normal" case study.

True, it is a description and analysis of two empowerment evaluation contexts. But their unfolding paralleled and even precipitated a personal odyssey of empowerment. Not only do those connected with a specific program undergo a metamorphosis during an empowerment evaluation effort, the evaluator herself can experience illumination and liberation, becoming an advocate not only for program stakeholders but for the centrality and universality of continuous self-evaluation. Such was my experience.

THE AUDITOR

The context of my work differs from other work described in this volume. I have been an auditor with the Texas Office of the State Auditor. This means that our "client" agencies typically have not invited us to evaluate their work or to assist them in self-evaluation techniques. When we say, "We're here to help," agency expectation is less than that of being empowered. Our increasingly "headline-grabbing" reports, as agencies refer to them, have often engendered fear and defensiveness on the part of these clients. Consequently, implementation of recommendations comes begrudgingly and is perhaps more a reflection of compliance than of eagerness to improve, even when the agency recognizes the soundness of the recommendation.

TRUE EMPOWERMENT

This type of response is not what empowerment evaluation is about. Empowerment evaluation involves ownership and the excitement that ownership elicits—ownership because the training in the tools for continuous improvement, imparted to constituents by the evaluator, has liberated program participants and has given them their strongest positions as program advocates. Their data, qualitative and quantitative, tell them how effective their programs are, what in their programs does or does not work, for which groups, and why. Their data tell them how efficient they are in their service delivery and the level of quality of their service. They no longer walk around wearing blinders, hoping "everything is going to be OK."

They know.

And knowledge is power.

THE QUESTIONS

These case studies describe the passage of two agencies from a compliance-driven passivity and defensive resistance of our audit itself to an open embrace and possession of self-empowerment evaluation, informing and facilitating ever-greater effectiveness, quality, and efficiency. They illustrate the power of key figures in government leadership who listen, believe in themselves and in their people, and are willing to take risks—*pioneers* in the best sense of the word. Finally, these cases exemplify the snowballing synergy of all involved in a self-empowerment effort, creating a force definitively greater than the sum of the parts. No one remains the same—not even the evaluator.

The characteristics of management in these two studies are similar to those identified by Fetterman (1994a) and Morrissey and Wandersman (in press). The authors describe successful empowerment managers as risk taking, actively supportive, and willing to share responsibility. Fetterman also asserts, however, that a prerequisite condition to empowerment evaluation is that a group requests assistance in that area. In both of our studies, management initially resisted our very presence in their agency; they certainly were not interested in requesting our assistance. What mixture of events and characteristics precipitates and produces a passage from resistance to request and adoption? Our empowerment initiatives used several strategies to effect this passage. The strategies reflect the five facets of empowerment evaluation: training, facilitation, advocacy, illumination, and liberation (Fetterman, 1994a). They included

1. helping management, and eventually all to be associated with the empowerment process, to see the need for self-evaluation, or illumination (Brocka & Brocka, 1992);
2. becoming advocates for the less powerful so that their needs would be heard;
3. raising the comfort level of management vis-à-vis our insignia as state auditors so that they would feel free to request assistance and explore problems with us as necessary (Juran & Gryna, 1988);
4. moving management and employees to a sense of ownership of the endeavor, creating excitement about the possibilities that self-evaluation presents, and liberating participants to take charge of their own direction (Brocka & Brocka, 1992; Juran & Gryna, 1993);
5. facilitating the implementation of the empowerment effort by providing the evaluation methodology and training to ensure success (Scholtes, 1988);
6. doing all this by speaking two languages: the language of upper management (money) and the language of lower management (products), and by providing supporting data wherever possible (Juran & Gryna, 1993).

These case studies also illustrate the four stages of empowerment evaluation described by Fetterman (1994b):

- Taking stock
- Setting goals
- Developing strategies
- Documenting progress toward goals

What happens to the evaluator and where does an evaluator go when the evaluation is over? Can an evaluator remain independent in such a setting? These questions and illustrations of the strategies and stages are addressed in the descriptions that follow.

Background

THE STATE

During the 1991 biennial legislative session, the Texas State Legislature began its movement toward outcomes-based budgeting. During that session, agencies were required to submit targeted performance goals, including measures of inputs, process outputs, and outcomes. The 1993 session carried this shift in focus further with the advent of budgeting by goals, objectives, and strategies rather than by line items. The preemptive question changed from one of compliance with regulations to efficient and effective service delivery to ensure the fulfillment of program and agency missions and goals.

That's the good news. Unfortunately, the compliance-driven mentality so pervasive at all levels of government—federal, state, and local—does not die easily. The performance measures developed for the legislative budget requests more often reflect efforts to comply with the new requirements than to identify indicators that would assist the agency in monitoring progress toward its goals. The focus on measurement has clouded critical qualitative considerations and obscured the need for holistic evaluations of programs. Increasingly tight state budgets have threatened to drive funding through a performance-based budgeting mechanism, using these inadequate measures as the basis for decisions.

In the middle of all this are agencies with long-standing processes and systems that may have functioned well at one point but have become outdated, redundant, and cumbersome. For example, as in private industry, state agencies have also unconsciously developed long cycle times in their processes. Many steps in the processes reflect a self-protection mentality as fingers in a management-by-control environment pointed across the table, casting blame on anyone who did not have protective mechanisms in place. The systems also often illustrate the indifference exemplified in the familiar phrase, "It's close enough for government work."

The setting then is one with virtually no integrated evaluative studies to determine effectiveness; inefficiencies due to outdated and self-protecting process development that must compensate for attitudes of indifference; and, most recently, dangerously inadequate performance measurement requirements, promising use for budgeting decisions. These factors make the system ripe for some very costly errors in decision making.

Two Empowerment Evaluation Efforts

The two groups involved in these initiatives are the Support Services Division of the Texas Department of Human Services and the Community Justice Assistance Division of the Texas Department of Criminal Justice. Although the evaluation emphases are different, the two cases illustrate the efforts and strategies of the Texas state auditor's office to illuminate and motivate these agencies to self-evaluate.

EMPOWERMENT EVALUATION OF THE
TEXAS COMMUNITY JUSTICE ASSISTANCE DIVISION

The empowerment initiative with the Texas Community Justice Assistance Division actually has several layers. It involved an audit of the adult probation system statewide, a methodology handbook on how to evaluate probation programs, strategies to empower state-level personnel, and initial empowerment efforts with the community supervision and corrections departments. The task of handing over the methods and mentality of evaluation to those who work in the programs is far from complete, but the groundwork has been laid.

The Setting. The adult probation system in Texas is managed on the state level by the Community Justice Assistance Division (CJAD), a branch of the Texas Department of Criminal Justice (TDCJ). On the local level, probationers are serviced through the Community Supervision and Corrections Departments (CSCD). Although the criminal justice system in Texas overall has received large increases in funding over the past several years, most of the money has gone to build more prisons. Until 1993, the Texas prison system was under the control of the federal courts because of the inadequacy of prison facilities and ensuing court cases. The courts mandated that Texas upgrade its prisons and for 10 years maintained control of them to ensure compliance.

Because of the additional requirements of prison facilities, many convicted felons were diverted from the prison system and into the probation system, resulting in a significant strain on the latter's resources, not only because of the added numbers but also because of the greater severity of the offenses committed by those admitted to the probation system. Numerous special programs sprang up, such as intensive supervision, electronic monitoring, community restitution programs, and substance abuse treatment facilities. These programs were designed for various purposes, including the rehabilitation of probationers, community protection, restitution to society, and reduction in revocation of probation.

The Illumination and Advocacy Opportunity: Conveying the Need for Self-Evaluation and Raising the Comfort Level of the State Department (CJAD). The efforts of our office with the adult probation system in Texas began with an evaluation of the system statewide in 1992. Our audit found that virtually no evaluation of the effectiveness of probation programs was being performed either by the Community Justice Assistance Division or by the corrections departments. Only data on probationers who participated in some special programs were collected. As a part of our endeavor to assist the probation system in evaluating its programs, we set up a model to evaluate substance abuse treatment programs and evaluated nine such programs in Tarrant County. We also created and illustrated a model for evaluating the effectiveness of probation overall, assessing the attainment of four commonly accepted goals of probation: community protection, restitution, rehabilitation of the offender, and reduced revocation of probation privileges. The models and results were published in our audit report (Office of the State Auditor, 1993). Our audit report recommended the continuation of these evaluations and an extension to all probationers so that the needs of the several groups—probationers, taxpayers, and the local community—could be better served. This meant that CJAD would need to accumulate much more data than it was currently doing.

The first draft of that report had been written in a heavily negative tone, with findings stating that the Community Justice Assistance Division had failed to set benchmarks, that no effectiveness evaluation was performed, and similar wording. This inflammatory writing style forced management of the Community Justice Assistance Division into a defensive mode. The division's written responses questioned our credentials for evaluating probation programs in the first place. Further, the division fought our recommendation concerning the use and integration of information from multiple risk assessment instruments to determine a probationer's supervision level. Yet the director herself had been urging corrections departments to use multiple instruments. Although the division had been performing only compliance evaluations before our audit, the division's response implied that the division had always planned to move in the direction of effectiveness evaluation, thus questioning whether our message added any value to what the division already knew. Although a power war was clearly occurring, there was very little *empowerment* in the positive sense of the term.

To reduce the polarization between our offices, we subsequently met with the division and discussed the points of contention. During that meeting, we agreed to revise the report, emphasizing our recommendations for improvement rather than the findings of inadequacy in the probation system.

By the time of our exit conference, the director of CJAD had become less defensive and voiced an interest in my assisting the division in program evaluation. The audit manager offered my services to write a booklet on how to evaluate probation programs.

Because the report had been issued during a legislative session, the director used the opportunity to request funds for evaluation efforts and was granted a seed amount. She reorganized her employees and their workload so that more effort could be focused on effectiveness evaluation. The division's 2-year plan, designed by the employees, contained an evaluation design and strategy very similar to that which our audit had used, including an evaluation of substance abuse treatment facilities and the accumulation of baseline data on all probationers statewide so that progress toward the four goals could be tracked. The division has implemented both ideas, with the database soon to be yielding baseline information.

*Liberation and Training/Facilitation Opportunities: Imparting Owner-
ship and Assisting in the Implementation.* While CJAD was reorgan-
izing, I wrote a handbook for evaluating probation programs (Office
of the State Auditor, 1994). This handbook used Stufflebeam's context-
input-process-product (CIPP) model to frame the methodology for
evaluating programs, regardless of the focus of the evaluation—efficiency, effectiveness, or funding decisions. To ensure that CJAD
would take ownership of the handbook when completed, I called the
director and requested that she and her staff review the draft of the
handbook and provide me with comments. After the staff had sent me
written comments, I called the director to get her feedback. She said
that she had enjoyed reading the booklet and suggested that her group
meet with me for further discussion on the revised handbook.

At the follow-up meeting, the director's representative said that
CJAD would use the booklet as the basis for the development of
training courses in evaluating programs and would send copies to
corrections departments around the state. He also stated that CJAD
had welcomed our audit report and any follow-up report we might
want to write about the importance of their evaluating their programs.
Such reports armed the division with support for their requests for
more funding from the legislature. Meeting attendees expressed the
same sentiments; they were clearly eager to evaluate their programs.

After receiving the booklet, the director sent copies to her advisory
committee. The committee members were enthusiastic about the
booklet and endorsed the creation of a research and evaluation
department at the state level to train and assist corrections departments in self-evaluations of their programs and to perform statewide
evaluations. They also asked Sam Houston State University to include
a presentation of the booklet on the agenda of the 1994 Annual Chief
Probation Officers conference to be held in Huntsville in October.

Empowerment at the Local Level. I presented a 4-hour workshop
to the chiefs at the conference. During the conference, one of the
chiefs questioned why they should put together data on the efficiency
and effectiveness of their programs. After all, wasn't somebody going
to try to hit them with the data? I replied that they could count on
someone trying to do that. But, with state budgets as tight as they are,
a chief could not defend the continuation of funding for the department's programs without data to support effectiveness. Furthermore,

leaving it up to the state to mandate measures could result in an inappropriate or incomplete set of measures. Being proactive and performing holistic evaluations of their programs would arm the chiefs with the kind of support needed to safeguard funding for effective programs. This use of the language of management, money, is critical to gaining the attention of management.

A second discussion during the workshop centered on the development of benchmarks. I remarked to the participants that our audit had found the 7-year recidivism rate to be 70% and asked if that was acceptable. This started a discussion of benchmarking. The chiefs said that they had no benchmarks and that CJAD had not provided them with any. I suggested that 10 or so corrections departments with similar characteristics (size, geographics, etc.) compile recidivism and other data on the probationers in the 10 departments. They could benchmark on each other and did not need to wait for CJAD to do it for them. They appeared receptive to doing this.

Another discussion arose concerning some aspect of evaluation. One of the chiefs insisted that I tell him what my office would say is right or wrong about the issue. I told him that what was important was not what my office thought about it but whether or not it was important to him and what he would do about it. It was his program and his evaluation.

After the presentation, a number of chiefs expressed interest in further training and assistance in program evaluation and stated that CJAD should offer this training. I explained that the director very much wanted to do this but needed their support in convincing the funding agents of the importance of this training. They indicated that they would be willing to lend her their support in this endeavor.

CJAD plans to provide in-depth training in evaluation to the large corrections departments in the state and has requested the legislature to appropriate more funding for evaluation during the next biennium. For small corrections departments, CJAD will perform the more rigorous evaluations of programs but will also assist those departments in developing measures for ongoing self-evaluations.

EMPOWERMENT EVALUATION OF THE
TEXAS DEPARTMENT OF HUMAN SERVICES

The empowerment initiative at the Texas Department of Human Services (DHS) involves the convergence of several events: an Office of the State Auditor (SAO) audit of cost information systems in three agencies, an attempt by the Department of Human Services to develop a pricing system for charging the Department of Protective and Regulatory Services (PRS) for support services, an increasing number of complaints from that agency concerning the timeliness of the services, the state and national movement to curtail funding for human services, and, most recently, a legislative directive to the Texas Health and Human Services Commission to consolidate administrative services over the 14 health and human services state-level entities in Texas. It involves a switch from producing an audit report to developing a cost handbook, two surveys, total quality management methods, and thousands of travel vouchers. It has culminated in the author's exit from the state auditor's office to provide the Department of Human Services with the training and assistance needed to put a total cost/ quality management system in place.

The Setting. The Texas Department of Human Services has an annual budget of over \$3 billion. At the state level, it consists of several divisions including four large groups: support services, management information systems, client self-support services, and long-term care services. The current year's state-level budget, used primarily for higher-level administrative purposes, exceeds \$70 million. The Department of Human Services is generally conservative in implementing new management ideas, at least formally. To all appearances, total quality management (TQM) movements in the state have included a strong team-building component but have done little to address the nuts and bolts of process analysis for continuous improvement. Consequently, there is little to show for the dollars invested in TQM training. Seeing this, the Department of Human Services has been reluctant to jump on the TQM bandwagon.

At the same time, the agency's systems and processes for service delivery reflect their creation in prior years. In many instances, they are outdated and encumbered with unnecessary complexity. Cycle times in processing can be slow, and service delivery can be convoluted and circuitous.

During the 1993 legislative session, the legislature created a separate agency, the Texas Department of Protective and Regulatory Services, to provide protective and regulatory services for vulnerable adults and children in Texas. Such services had previously been administered by the Department of Human Services. The new agency has continued to use the services of the Support Services and Management Information Systems Divisions of DHS. It has voiced dissatisfaction with some of the services, however. The Department of Human Services, in addition to trying to improve its delivery, has also been developing a better system to price its service, using an activity-based costing approach. This state-of-the-art costing method, while eventuating in superior service cost information, is difficult to develop because so few examples exist in such areas.

Adding to these issues is the current state and national movement to curtail spending on human services, particularly on welfare. For the DHS Support Services Division in particular, a recent rider to the Appropriations Bill of the Texas Health and Human Services Commission has given legislative authority to that commission to combine the administrative services provided by the 14 health and human services state-level entities in Texas. This directive will significantly affect the DHS Support Services Division, making the consolidation of services a survival of the fittest. These events place the Department of Human Services in a highly vulnerable position, with budget cuts threatening this agency more than others. The Support Services Division faces the threat of extinction.

Illumination and Advocacy Opportunities: Conveying the Need for Self-Evaluation and Raising the Comfort Level. The Department of Human Services did not welcome my audit. I had recently been part of the audit team that had evaluated the JOBS program administered by DHS. Unfortunately, the unfavorable results, coupled with the inflammatory writing style of the Office of the State Auditor, had driven a wedge between our offices. Because I had been responsible for perhaps the most incriminating of the evidence, the numbers, I was in particular a persona non grata.

The audit I was currently managing involved a review of cost information systems as they informed five major decision areas: planning; budgeting; pricing and rate setting; consolidation and outsourcing services; and improving the efficiency, quality, and effectiveness of services. To get a preliminary assessment of this, we designed two surveys, one to nonmanagement workers, and the other to their

supervisors. Among other questions, the employee survey asked workers to estimate the percentages of their time they spend reworking
things to meet requirements, waiting on others to finish their work so
that they can do their own, repeating activities unnecessarily, and inspect-
ing the quality of other people's work. This last item is considered
back-end inspection in contrast to building quality into the processes
in the first place. The survey to the managers asked how they measured
the efficiency, quality, and effectiveness of their processes; how often
they measured these things; what they did with the results; and what
backlogs, reworks, and gaps they were experiencing. We analyzed the
results both for the agency as a whole and for each division.

Through indirect channels, I learned that the DHS deputy commissioner of support services was alarmed when she learned that we
would be surveying the employees. Human resource issues were in her
area and so her concern was somewhat understandable. In addition,
sometime before our audit began, this deputy had initiated an activity-
based costing project (ABC) to price the services of her division to
the Department of Protective and Regulatory Services and to the
Department of Health. Although ABC is a state-of-the-art approach
to costing and has been implemented by many of the leading private
companies in the country, it requires significant insight and dedication
on the part of many in the company to develop it. The project had
been difficult, partly because of the dearth of published applications
of ABC to the types of services her division offered. With a budget
exceeding $16 million, her division offered a wide array of services
and handled large volumes. The project, therefore, affected many
managers and employees; they did not always welcome the work
involved in this project. Our audit of cost information systems had the
potential of making her difficult position impossible. Further, a survey
tallying employees' opportunity for input into system changes and into
time spent in rework and redundancy did not promise to help her
cause.

In a frank conversation with the deputy, I asked what her concerns
were. She remarked that, in the past, auditors had at times misinter-
preted what she had told them; she was concerned that we might say
things based on the surveys that were misleading or distorted. I
explained to her generally accepted audit standards and my own
additional standards of auditing. I assured her that I am very careful.

I also told her that, as the project manager, I had significant control over what got reported and how it was written. I would make sure that it was balanced, fair, and sensitive. After extensive discussion about this issue, I was able to calm her fears. She expressed interest in seeing the results of the survey for her division.

My project had initially focused on the production of a handbook on the cost decision areas and systems. Along the way, it had shifted to being an audit. After discovering that the Department of Human Services had developed an ABC system, and for other reasons, I concluded that the state would be better served if we wrote the handbook than if we issued an audit report. This would also increase the comfort level of the agencies so that they would be more receptive to our message. SAO management consented to the change. I conveyed this change in plans to DHS management; they were obviously relieved.

Shortly after compiling the results of the surveys, I met with the support services deputy to discuss the results. Before doing so, however, I met with her TQM expert to get him on board and to appraise his competency level. Although he had strong expertise in team building, he acknowledged that he knew nothing about how teams were to analyze processes to determine how to fix them. In a telephone conversation, I informed the deputy about our office's TQM expert and suggested that he would be willing to give the deputy some direction based upon the results. The four of us attended the meeting. I told her that fixing the problems noted through the surveys would give her the chance to double her spending power. She then said that she had some projects she wanted to do, using the structured approach to process analysis that our TQM expert knew. We agreed to have our expert meet with theirs to pursue the projects.

The deputy commissioners of the three other large divisions had also expressed an interest in seeing division results. After I met with the deputy commissioner of the Management Information Systems Division, he took the results to the executive deputy commissioner, relating that I had said they needed cross-functional teams to solve some problems that were costing significant amounts of money. The executive deputy then called me and asked that I present all the divisions' results to him.

In my meeting with the deputy commissioner of the Client Self-Support Services Division, I explained that our office would not be issuing a report. She became much more receptive to the survey results and also expressed an interest in improvement projects. She asked if our TQM expert would be willing to assist her division in this initiative.

During my meeting with the executive deputy commissioner, he expressed great concern about some of the waste identified by our surveys. I told him, as I had told his deputies, that some people might take those percentages of estimated waste, multiply them by the divisions' budgets, and suggest that this much money could be saved if the processes were fixed. This statement got his attention also. DHS is currently considering forming a cross-functional team to address the leading problem we identified to the executive deputy.

Liberation and Training/Facilitation Opportunities: Imparting Owner-ship and Assisting in the Implementation. I knew that if DHS wanted to go forward with learning to self-evaluate for the continuous im-provement of their processes, the agency would need help. I requested that our office schedule our TQM expert on my audit. The office's scheduling team could not understand why I needed an expert in TQM on a cost audit; after all, he had no financial background. Sometimes the link between efficiency of processes and dollars saved is difficult to make. After much hassle and negotiation, he was placed on the project.

DHS support services has experienced slow cycle times in its travel voucher processing. Employees of both DHS and PRS had complained about the delay in obtaining reimbursement for these expenses. Our TQM expert has assisted DHS in putting together a cross-functional team to study the problem, has trained the team in the concepts and tools of continuous improvement and process analysis, and is currently acting as team facilitator to gather data on the process and suggest solutions. He is also working with two other groups, a training and an employment services group, to identify opportunities for improve-ment. The employment services group has already identified a key problem in the human resources information system, that is, that the system has no mechanism to ensure the continual updating of its information. The team has identified a solution and, with management approval, is implementing it. This implementation will greatly facili-tate the work of other DHS groups statewide.

Because of the heavy involvement of management in this initiative, the travel voucher team has been able to convey problems through the levels of management as the issues emerge, clearing the way to find solutions. The team has developed a detailed process flowchart that illustrates the significant need for a reengineering of the process. The team is gathering baseline data to determine the process's current capability and to identify the root causes of the long cycle time in processing travel vouchers. This information will be used in developing recommendations for reducing the cycle time. As DHS implements the changes, those involved in the various activities of the process will continue to gather data to determine the effectiveness of the changes in meeting the goal.

The DHS Support Services Division is committed to continuous improvement. The division directors are working as a team to ensure that continuous improvement, customer search, and benchmarking projects optimize the use of division resources. The directors are currently working together to identify projects for another round of continuous improvement initiatives. They are deciding upon products for customer-search teams. They are determining the prioritization of processes for the development of detailed process flowcharts; such flowcharts often identify opportunities for immediate improvement of processes.

Having left the state auditor's office to be of greater assistance to this division in their endeavor, I am facilitating the team charged with developing plans for making this an integrated system. The team has taken stock of what work has been done so far, that is, the preliminary identification of activities, some costing of the activities, fragments of process flowcharts, and the development of some performance measures. The team has also determined the goals of the total cost/quality management initiative:

- An accurately priced set of products with customer needs identified
- The processes that produce those products defined
- Trained teams of workers searching for ways to improve those services and increase the efficiency of the processes
- Benchmarks to ensure that the services and products are cost competitive with private industry

The team has, in addition, outlined seven steps for achieving those goals and is currently assisting support service departments in implementing the strategies. With the approval of the division's directors, the team has set the dates for completion of several of these steps.

Our role as empowerment coaches continues. We continue our efforts to help management and employees understand that continuous improvement is not a one-time quick fix, a permutation of "putting out fires." We continue our efforts to help them understand that it is not just something employees do or that some employees sometimes do or something that management does to employees, but is a way of thinking and doing that everyone owns. It is total ownership, the epitome of empowerment evaluation. These are not easy concepts to convey.

The Support Services Division plan goes beyond solving isolated process problems or developing a price mechanism. The deputy commissioner is dedicated to a total cost/quality management system, integrating the identification of customer needs with more efficient, high-quality processes. She plans to achieve this through a system that links the costs of services to those services and through ongoing process analysis and improvement, letting measures of process performance guide cross-functional teams of employees and managers in this endeavor. She has conveyed this message formally to her managers and workers in two lengthy meetings. She has assured them that their jobs are not in jeopardy because of this effort. The travel voucher team described their work at one of the meetings to further acquaint management with continuous improvement teams. These meetings served as the kickoff of the total cost/quality management initiative.

Identifying problems is a first step, but finding and implementing the solution is far more challenging. Usually, as evaluators, we can stay outside the setting of the evaluation and facilitate self-empowerment. But sometimes, if the magnitude of the project is great enough, we need to be more involved. At times, the teaching of the concepts must be subtle and delivered over time. It is often not feasible to do this as an outsider. I am no longer an outsider; I have jumped into the picture to help make this happen.

Discussion

Our earlier premise concerning the requisites to move clients from a position of resistance to enthusiastic adoption suggested that evaluators can use several strategies:

- Convince management of the need for ongoing self-evaluation (illumination).
- Serve as advocates of the needs of those who cannot be heard (advocacy).
- Assure management that you really want to help, not hit (facilitation).
- Impart ownership of the evaluation movement to the clients (liberation).
- Facilitate this ownership through training and ongoing assistance during the implementation phase (training).
- Use transformative language, that is, the language each group understands.

In both of these studies, our evaluation team worked to help management see the need for self-evaluation. With the adult probation management, we effected this through the audit report and discussions of the recommendations. The audit report also allowed us to be an advocate for probationers, who need effective rehabilitation programs. We additionally did this through a workshop with the chiefs of the corrections departments. With the Department of Human Services, we used two surveys as vehicles to deliver this message. These surveys were our mechanisms to serve as advocates for the needs expressed by front-line workers and line management.

To raise the comfort level of the state-level management of the adult probation system, we redirected the focus of our audit report to one that was recommendations based instead of findings based. With the Department of Human Services, we met with management to discuss their concerns and switched from the writing of an audit report to that of a handbook.

To facilitate the CJAD director's efforts, we suggested that the chiefs support her before the funding groups. To facilitate the DHS deputies' efforts, we met with the executive deputy to explain the findings and what was needed to address the problems uncovered. We have also facilitated employees in their concerns by encouraging management of DHS to put together cross-functional teams and to put a communications mechanism in place for improvement teams to convey their concerns to management.

To create a sense of ownership and the liberation that ownership provides, we incorporated the suggestions of CJAD personnel into the handbook. The CJAD director herself has empowered her staff to find the best way to organize themselves and their evaluation training services to corrections departments. With DHS, we used the language of money to create excitement and a sense of ownership in upper management. The creation of cross-functional teams to address problems in processing travel vouchers, training, and employment services has created a sense of ownership in the workers and other levels of management. There is an air of promise for improvement that excites and liberates employees of the Support Services Division.

Our training efforts with respect to the adult probation system have consisted of the development of the methodology handbook and the workshop in Huntsville. We are heavily involved in the training component at the Department of Human Services, guiding TQM teams, training team facilitators, and providing a framework for integrating the cost, process, and product components of the total cost/quality management system.

Much of the language we used to facilitate this transformation was dual. We spoke of cost savings and preserving funds to upper management. We spoke of identifying what works and improving processes to lower management and workers. We provided preliminary baseline data to DHS management through our surveys. We provided baseline data on probationers in special programs to CJAD management. Money, products, and data are important linguistic tools in communicating the need for self-empowered evaluations.

The Four-Stage Process of Implementation. Fetterman's four-stage process of implementation is unfolding in several layers in these two settings. The state-level CJAD had done no effectiveness evaluations but it has set a clear goal for future work, that is, to evaluate the effectiveness of programs offered by the probation system, beginning with substance abuse treatment programs and the probation system overall. It has developed strategies for meeting those goals: reorganization of resources, creation of a database system, continuous search for funding, and training classes for large corrections departments. It is monitoring progress; the baseline data on probationers are almost complete and the basic study designs are in place.

The Department of Human Services is also developing its layers of implementation. The travel voucher team is collecting baseline data; its initial goal for cycle time has been set. The employment services team has identified a problem in the information system, discovered the root cause, suggested a strategy to correct the problem, and is implementing that recommendation. The team charged with integrating the system has taken stock of where it is, set its goals, developed strategies for achieving those goals, and begun the implementation. Division directors are guiding and overseeing the project and have approved deadlines to monitor progress. These cases illustrate that the four steps are serviceable at any level of a project.

The Leadership. A reflection on the characteristics of the management in both these studies may provide some further insight into why our efforts were successful. In both instances, the directors were open, believed in themselves and in their people, and were willing to take risks. In state government, change means risk. Evaluating yourself exposes your weaknesses to others who, in a political arena, are all to ready to capitalize on those weaknesses. The two directors have been willing to do this because they see the opportunity to improve their services through the information provided by ongoing evaluations. The directors were open to our message, despite the fear that our position as auditors engendered. They were able to separate their fear from their realization of the value of the message. Finally, both of these directors believed in themselves and in their people. They have turned over much of the planning for change to their employees, valuing the input of those who are closest to the work. At the same time, the two leaders have defended their beliefs. The director of CJAD has actively pursued funding for this endeavor; the support services deputy commissioner has continued her efforts for an integrated system despite rumblings and criticism. It is doubtful that our efforts as empowerment evaluators could have succeeded without this type of leadership in place.

Conclusion

Both the Community Justice Assistance Division and the Support Services Division of the Department of Human Services have embraced the empowerment of self-evaluation and are working toward full implementation throughout their divisions. It is not just corrections departments and support services that need to do ongoing evaluations of their programs, however. Job training and education programs, transportation projects, insurance applications, and a host of other programs could benefit from the ownership of self-evaluation strategies. Every government entity should be evaluating the efficiency and quality of its services to ensure continuous improvement. But agencies and departments need training and assistance in evaluation if they are to be successful. The Office of the State Auditor, although providing such assistance on an ad hoc basis, has chosen not to offer it formally or systematically. I have submitted a proposal to the Office of the State Comptroller to locate training and ongoing consultation in empowerment evaluation in that office.

And what is the role of the professional evaluator or the auditor? Perhaps it is time to reassess the adequacy of the traditional position of the evaluator as essentially an outsider who reviews, comments, and departs. Evaluators have a wealth of experience to impart. Their contacts with many programs and players have given them insights often not found in those who work in individual programs (Patton, 1994). Their training and experience in evaluation can be of great assistance to program people if evaluators decide to impart some of that knowledge to participants.

The business world is well aware of the need for continual self-evaluation. No manager would think of waiting for outsiders to evaluate the company's performance, not bothering to accumulate and evaluate performance information on a continual basis. The company would not survive. Those entrusted with the public's funds must take the same view and find the mechanisms to ensure accountability, not only regarding compliance with regulations but also with respect to the efficiency, quality, and effectiveness of programs and service delivery systems.

The evaluation world is poised for a pivotal role in this movement. Attempts to protect its territory through refusal to assist or even acknowledge the validity of the movement may result in the reduction of the evaluation profession to trivial and nonsignificant work, ending "in a boring whimper as voices of objectivity drifting off into the chaos. The ultimate development" (Patton, 1994, p. 312).

Programs that wish to continue must become self-evaluating, and they will, with or without the assistance of the evaluation profession. Although there may always be a need for some external evaluation, the degree of external evaluation work may be minute compared with that of ongoing internal evaluation. There is relatively little customer interest in a better mousetrap, an improved phonograph player, or a one-time evaluation of a multiyear program. The evaluation world should not become preoccupied with the shape of its product, reducing evaluation to having a single goal (Fetterman, 1995), and ignore the need underlying customer demand for the product. The context of customer need has changed from one of assured long-term funding regardless of effectiveness to that of a continual fight for survival. The customer of evaluation information today requires continual information and so must learn to evaluate its success on a continual basis. The evaluation profession can choose to fill that need or to remain entrenched in the product that worked in some past era but may have relatively little customer demand today.

As for me, I won't pay a penny for a better phonograph player; my CD player works just fine.

References

Brocka, B., & Brocka, M. S. (1992). *Quality management: Implementing the best ideas of the masters*. Homewood, IL: Irwin.

Fetterman, D. M. (1994a). Empowerment evaluation. *Evaluation Practice, 15*(1), 1-15.

Fetterman, D. M. (1994b). Steps of empowerment evaluation: From California to Cape Town. *Evaluation and Program Planning, 17*(3), 305-313.

Fetterman, D. M. (1995). In response to Dr. Daniel Stufflebeam's: "Empowerment evaluation, objectivist evaluation, and evaluation standards: Where the future of evaluation should not go and where it needs to go." *Evaluation Practice, 16*(2), 321-338.

Juran, J. M., & Gryna, F. M. (1988). *Juran's quality control handbook*. New York: McGraw-Hill.

Juran, J. M., & Gryna, F. M. (1993). *Quality planning and analysis: From product development through use*. New York: McGraw-Hill.

Morrissey, E., & Wandersman, A. (in press). Total quality management in health care settings: A preliminary framework for successful implementation. In P. R. Keys & L. H. Ginsberg (Eds.), *New management in human services* (2nd ed.). Silver Spring, MD: National Association of Social Workers, Inc.

Office of the State Auditor. (1993). *The verdict on probation: Its effectiveness is unknown* (SAO Report No. 3-037). Austin, TX: Author.

Office of the State Auditor. (1994). *Evaluating probation programs: A methodology* (SAO Report No. 94-118). Austin, TX: Author.

Patton, M. Q. (1994). Developmental evaluation. *Evaluation Practice, 15*(3), 311-319.

Scholtes, P. (1988). *The team handbook*. Madison, WI: Joiner Associates.

5

The HIV Prevention Evaluation Initiative

A Model for Collaborative and Empowerment Evaluation

CYNTHIA A. GÓMEZ
ELLEN GOLDSTEIN

In 1995, acquired immune deficiency syndrome (AIDS) became the leading cause of death for adults aged 22 to 45 living in the United States (Centers for Disease Control and Prevention, 1995). The complexity of the human immunodeficiency virus (HIV) that causes AIDS has made the short-term prospect for a preventative vaccine unlikely. Therefore, the only strategy currently available to decrease the spread of HIV is to prevent its transmission.

With the knowledge that HIV is transmitted primarily through sexual and drug use behaviors, researchers have been working to

AUTHORS' NOTE: We would like to thank the Northern California Grantmakers AIDS Task Force, the Center for AIDS Prevention Studies (CAPS) scientists and staff involved in the project, and the 11 participating community agencies for their courage and commitment to HIV prevention. This project was also supported through funds from the National Institute of Mental Health Grant No. MH42459-09 and the Ford Foundation.

develop theoretical models of behavior change while community practitioners have been implementing intuitive strategies to promote HIV prevention. Lacking in these efforts has been the mechanism to link the two together. Researchers and practitioners have traveled on separate train tracks, if you will, while attempting to reach the same destination: providing effective strategies to stop the spread of HIV. Unfortunately, what each has learned as they have gone along their journey has not often been shared even though one could benefit from the knowledge of the other. It has become painfully clear that the tracks must converge to create the knowledge that is needed to move forward in the fight against AIDS.

This chapter will outline some of the current impetus for collaborative models as well as historical barriers to science and service collaborations, and will describe an empowering model that allows participants to overcome some of these barriers while creating important advances in the field of HIV prevention. The principles of empowerment evaluation—that is, training, facilitation, advocacy, liberation, and illumination (Fetterman, 1994a)—are embedded within this model. Our efforts exemplify how different fields are reaching similar conclusions regarding the importance of helping communities and their practitioners to learn the skills and benefits of evaluation.

Impetus for Collaboration: HIV Prevention Community Planning

The concept of community ownership as a necessity for HIV prevention programs to be most effective was incorporated at the national level by the Centers for Disease Control and Prevention (CDC). As of 1994, the CDC required all prevention funding distributed through state and local health departments to have a community planning process. Community representatives, researchers, people with AIDS, and health officials were to come together to the table to develop the priorities of their community regarding HIV prevention efforts. This new model created a demand across the country for community-based organizations and other service agencies to learn whether their programs were working. Likewise, researchers were expected to provide data that informed community planning bodies

The MORT Syndrome[1]

Money. Funding has long been a bone of contention between researchers and providers. Providers argue that money spent on funding research would be better spent on increasing capacity or enhancing existing interventions. Financially independent universities and research firms are frequently viewed with skepticism and animosity by impoverished community-based organizations (CBOs). CBOs are often naive about the reality of the immense costs of rigorous science, whereas researchers may not understand the financial constraints under which CBOs are providing service and evaluation. Funding agencies contribute to this division by providing opposing incentives: Service providers must meet units of service requirements but researchers must show proven effectiveness.

Ownership. The question of ownership occurs before, during, and after a research or evaluation project is implemented. Decisions regarding research questions, design, and interpretation of findings are often sources of controversy in community-based research. Researchers' desire to disseminate findings through professional journals often prohibits or delays the community agencies' ability to do the same through community publications. Service providers, on the other hand, at times lay claim to the research by virtue of implementing a protocol that can lead to misuse of data if community providers are less familiar with data analysis and interpretation.

(continued)

on which strategies and interventions were most effective and which theories of behavior change had been proven. Researchers and practitioners alike involved in HIV prevention were faced with the reality of how little was actually known. The need for the train tracks to converge has become even more clear, and the need to assist both practitioners and researchers on how to collaborate is essential to future success of HIV prevention.

The MORT Syndrome (Continued)

Rigor. The disparity between scientific questions and client (participant) needs in community-based studies has complicated social science research. Sometimes researchers have been forced to involve practically oriented, and often nonscientific, agency staff to provide interventions and insight into the study findings; practitioners have sometimes been forced into a relationship with researchers whom they view as theory oriented and nonrealistic. Issues of study assignments and modifications of research agendas, based on the needs of the participants, again come to bear. Researchers endeavor to maintain integrity and usefulness of the science but are judged (perhaps erroneously) by providers as impervious to the needs of the clients.

Time. Finally, the time expenditure for research is another area of discontent between community-based agencies and researchers. Practitioners have the immediate need to provide services to their clients, but researchers tend to concentrate on macrolevel findings that will benefit a large number of people at some time in the future. Providers, who rarely have enough time to provide the desperately needed services, oppose the inclusion of time-consuming activities such as data collection into their service delivery. The importance of formative research, instrument development, and pretesting are also viewed as luxuries rather than necessities by practitioners. Researchers become frustrated as they attempt to maintain the integrity of the science under such immediate time-oriented conditions.

Historical Barriers to Collaborative Relationships

In some ways, researchers and community providers come from different cultures. A lack of understanding of each other's language, context, and value system often impedes the development of a common ground from which to cultivate effective HIV prevention programs. The MORT Syndrome (an acronym symbolizing a "death" in communication) has been used to describe some of the key differences reported by researchers and providers that contribute to the historical mistrust (Gómez & Goldstein, 1993). The acronym MORT refers to *money, ownership, rigor,* and *time.*

In an effort to begin to address some of these perceived and real inequities, the Center for AIDS Prevention Studies (CAPS) at the University of California, San Francisco, developed a workshop to assist community-based organizations to learn about program evaluation. What ensued from this small project was an awareness of the need for "intercultural education," that is, helping researchers understand the realities of research application and helping service providers learn the benefits of research and program evaluation.

Mechanisms to create an environment that would lead to this intercultural education were solicited from both researchers and practitioners. Three important strategies emerged: (a) Funding agencies could provide funding to community agencies with separate money targeted specifically for program evaluation; (b) funders could require researchers to include the expertise of a community-based agency on projects that involved community-level research, and service funding could require scientific experts for the evaluation of any service provision; and (c) there could be a request for proposals that specifically encouraged the development of projects that would provide both direct service and rich data. These projects would be developed together from the beginning and would require a collaborative, empowering, and evaluative application.

The common thread in all of these recommendations was the reality that any efforts to collaborate would be driven by the requirements of the funding agencies; therefore, it would be necessary to find a funding agency willing to participate in what could become a model for collaborative partnerships. To this end, CAPS approached the Northern California Grantmakers (NCG) AIDS Task Force, a consortium of philanthropic organizations, and together they developed what would become known as the HIV Prevention Evaluation Initiative. NCG would provide all program funding and CAPS would contribute most funding and personnel for the evaluation component of each award.

What follows is a description of the goals of this initiative and the specific process that took place in its development. Each step of the process will include a list of lessons learned to assist others who may be interested in developing similar collaborative and empowering evaluation-based models.

The HIV Prevention Evaluation Initiative

PROJECT GOALS AND OBJECTIVES

The primary goal of the HIV Prevention Evaluation Initiative was to support activities within community-based agencies that would target those individuals at highest risk for HIV infection who had not been effectively reached through existing efforts. The funding agency, NCG, informed community agencies that they would make significant changes in their approach to funding. In their request for proposals, NCG stated: "With its relatively limited, but flexible dollars, the Task Force is seeking to support significant HIV/AIDS prevention efforts which can be evaluated and have a strong likelihood of changing high-risk behaviors and advancing the field of knowledge in HIV prevention" (Northern California Grantmakers Program Announcement, 1992).

The funders, who had given over a million dollars toward HIV prevention and service efforts over the previous 4 years, also recognized that they had not been able to assess whether their funded programs were having any impact on the AIDS epidemic. This initiative would provide important facts to the private contributors and increase the likelihood of their continued commitment to fund HIV- and AIDS-related projects.

The objectives of the initiative were to fund 6 to 10 HIV prevention programs for 18 months of operation, work with funded agencies over a 24-month period to evaluate the effectiveness of those programs in changing high-risk behaviors, and help disseminate the findings of those evaluations to providers and policymakers.

The HIV Prevention Evaluation Initiative would have a threefold purpose: to pilot a model of collaboration between service providers, academics, and private funders; to answer real-world questions about effective HIV prevention activities; and ultimately to disseminate the evaluation findings. In essence, this initiative would create research questions that were community generated, scientist modified, and private-sector funded. Moreover, the initiative would use evaluation as the tool to facilitate empowerment for members of the collaborative.

DESCRIPTION OF THE PROCESS

The project was initiated by a request for proposals (RFP) issued by the funders. As opposed to their usual open funding cycle in which they selected grantees from a variety of AIDS service, prevention, or care arenas, this HIV Prevention Evaluation Initiative (HPEI) solicited proposals that would incorporate innovative, replicable, and evaluable prevention programs with the evaluation component to be conducted in collaboration with the University of California, San Francisco, Center for AIDS Prevention Studies. The RFP established that the programmatic funding would be augmented by evaluation funds, and that collaboration with the researchers on program evaluation was a required component for the award.

A multistage proposal process was established including a community orientation session to respond to questions about participating in this novel collaboration; required letters of intent, which were reviewed by teams from both the funder and the research institution; a required 32-hour evaluation theory and proposal-writing workshop for those applicants invited to submit a full proposal; and, finally, the submission of full proposals, which were received and reviewed. Then, 2-year awards were announced. The particulars in each stage of this proposal process are discussed in more detail with specific recommendations for replication.

Community Orientation Session

Community-based organizations (CBOs) were invited to attend an optional orientation session to review the specifics of the RFP. The purpose of the session was to provide a more detailed description of the RFP process, eligibility requirements, and selection criteria.

Many CBO representatives attending the orientation session were clearly unfamiliar with the nature of evaluation, yet had been facing increasing pressure by funders to evaluate their programs for proof of effectiveness. It was this lack of experience, training, and financial backing that the collaboration sought to address. It recognized that prevention programs that had routinely been funded and accepted as "standard of care," yet not evaluated, deserved the supportive evaluation funding and scientific rigor that this collaboration allowed. Approximately 25 agencies attended this orientation session.

Lessons Learned. The community orientation session was the first opportunity to engage in a dialogue with community agencies concerning this new collaborative effort. It was an important aspect of the process as it allowed an opportunity to begin to spread the word around the community with plenty of time for CBOs to consider whether they wanted to participate. The norm in AIDS funding was that evaluation was required, but not funded. The novel requirements of this RFP, that program money was tied to evaluation money, necessitated increased advance work in the community. This emphasis on evaluation was a "paradigm shift" in funding. The orientation session occurred 2 months prior to the due date of letters of intent and nearly 6 months prior to the due date for the final proposals. This session also weeded out agencies that really were not interested or able to participate in such a collaboration.

Letters of Intent

Characteristic of this collaboration, the letters of intent (LOIs) were reviewed by both the research institution and the funder. Scientists reviewed the LOIs for novel research questions and approaches, potential benefits to prevention literature, and likelihood of achieving significant results. Funding agency reviewers, many of whom had experience funding the agencies in the past, were similarly given a selection of the LOIs to evaluate for agency capability, replicability of the intervention proposed, and prioritization of the population targeted. Three reviewers, one from NCG and two from CAPS, reviewed all LOIs to achieve a broad view of the potential grantees so as to assure a diversity of target populations, geographic distribution, and type of intervention proposed. The letters of intent were the seed of the CBOs' empowerment, for in these initial proposals, the CBOs defined questions that they sought to answer rather than agreeing to participate in a researcher-initiated project. Still, these LOIs revealed the need for additional training in developing a pointed research question and protocol. In all, 54 agencies submitted letters of intent to participate in the HIV Prevention Evaluation Initiative. Not all agencies who attended the community orientation session submitted LOIs. Differences in the quality of LOIs based on attendance at the community orientation session were not evident or expected given that agencies were allowed to call the funder for further explanation of the RFP. After final review, 18 agencies were invited to participate in the evaluation workshop.

Lessons Learned. Although the letters of intent differed significantly from the final proposals (in part due to the 32-hour workshop on writing evaluation proposals), they allowed the reviewers to select for diversity of projects, priority of population being served, and originality of the intervention being evaluated. Although we screened for agency's commitment to the evaluation process, changes in staffing from this stage to the award and implementation stages foiled our efforts. In the future, broader agency involvement—that is, staff commitment to evaluation at all levels—will be assessed at this step to increase the likelihood that agencies can complete the process.

Evaluation Workshop

Critical to the notion of empowerment evaluation is to teach individuals or agencies the skills necessary for them to conduct their own evaluations (Fetterman, 1994a). The HIV Prevention Evaluation Initiative, although collaborative with a research institution, ultimately sought to train community providers to begin to become capable of conducting evaluations independently or with minimal consultation.

The workshop, titled "Evaluating Behavior Change Efforts," was conducted by faculty from CAPS, the research institution. The goal of the workshop was threefold: to train community-based HIV/AIDS service providers in the rudiments of evaluation research, to coach these grantees toward the development of a solid evaluation proposal for the prevention initiative, and to evaluate each CBO for likelihood of success in collaborating on the evaluation of their prevention program.

The workshop was conducted over 4 consecutive weeks, for a total of 32 hours of instruction. The format was a combination of didactic presentations by CAPS faculty and small group discussion sessions in which two staff each from up to four CBOs met with two CAPS faculty to present, review, and modify their proposal. Each small group maintained stable membership over the 4 weeks to allow for group development and provide for increased familiarity with each other's proposals. Even though this was a group of CBOs in competition with each other for funds and under review from potential funders, the group discussions maintained an earnest grappling with the technical information and generous provision of feedback and assistance. The small group format provided the CBO staff the opportunity to develop peer support for the learning they were doing.

Workshop Content. Each participant was given extensive reading and informational materials. Specific reading and homework assignments were attached to particular segments of the workshop. The workbook provided to participants also included resources such as an inventory of funding sources for AIDS-related projects, an inventory of CAPS survey instruments, instructions on how to use the UCSF library, and a glossary of terms used in evaluation research. The outline of the 4-day workshop is presented in Table 5.1.

Following the workshop, applicants were given a month to complete the proposals. They were encouraged to request consultations with researchers and continue working with each other in the development of their evaluation plans. It is notable that several CBO staff maintained contact, even forming an ad hoc support group during this time, with plans to meet for a "celebration/consolation drink" upon award notification with both research staff and funders being invited. The evaluation workshop served as the first opportunity to experience the potential collaborative relationships among providers, researchers, and funders. This level of collaboration and the development of a *CBO research community* was key to the success of the empowering and evaluative model.

Lessons Learned. Regarding the content of the workshop, we recommend a thorough introduction to the theory and practice of evaluation. Although much of the information will bear repeating as the CBO participants reach each stage of protocol design (sample selection, instrument development, staff training, research implementation, and data cleaning and analysis), having an overview provides a necessary framework. Participant evaluations of the workshop noted the importance of simple, easy-to-understand definitions, handouts, and the use of examples to illustrate complicated concepts. The small group interactions (peer review) were noted as being a particular strength of the workshop.

Although the workshop and selection process were not strictly collaborative, with the workshop curriculum being set by the research

TABLE 5.1 Example of the Evaluation Workshop Schedule

EVALUATING BEHAVIOR CHANGE EFFORTS
Day One: 9:00 a.m. to 4:30 p.m.
Changing HIV Risk Behaviors: A Rationale for Program Development and Evaluation
  Didactic A
    —Rationale for behavior change interventions
    —Components of behavior change interventions
    —Evaluating behavior change
  Small Groups: Group members discuss the programs they wish to evaluate, what
    their programs are, what they expect the programs to do, and why they expect
    these outcomes.
  Didactic B
    —Elements of a good proposal
    —The FINER model: feasible, interesting, novel, ethical, and relevant
  Small Groups: Discussion and clarification of presentations; consideration of which
    expected outcomes are of greatest interest; and how to develop a research
    question to evaluate those outcomes.
  Wrap-up
    —Clarification of any key points
    —Preview and homework for Day Two
    —Session evaluation forms completed

Day Two: 9:00 a.m. to 4:30 p.m.
Selecting an Appropriate Study Design
  Didactic A
    —How to answer your evaluation questions
    —Measuring change: independent and dependent variables
  Small Groups: The research or evaluation questions under consideration by
    participants in each small group are discussed, reviewed, and revised.
  Didactic B
    —Types of study designs used in evaluation
    —Summary of major themes or problems that arose in small groups; discussion
      of selected questions of particular interest or that present special problems
  Small Groups: Discussion and clarification of didactic presentations; consideration
    of possible study designs for the research questions discussed in the morning
  Wrap-up
    —Clarification of any key points
    —Preview and homework for Day Three
    —Session evaluation forms completed

(continued)

institution and the grantees selected by the funder and researchers,
CBOs appreciated acknowledgment of their expertise and experience.
Workshop faculty should be familiar with the CBO environment and

TABLE 5.1 Continued

Day Three: 9:00 a.m. to 4:30 p.m.
Evaluating Behavior Change
    Didactic A
        —Types of program evaluation: formative, process, outcome
        —Selecting whom to study: Principles of sampling and sample size considerations
        —Data Collection: Instrument design, selection, and pretesting
    Small groups: Discussion and clarification of didactic presentations; discussion of
        participants' project designs and instruments (continued from previous day)
    Wrap-up
        —Clarification of any key points
        —Preview and homework for Day Four
        —Session evaluation forms completed

Day Four: 9:00 a.m. to 4:30 p.m.
Fine-Tuning and Submitting the Proposal
    Didactic A
        —Data management and analysis: what you need to consider when designing
            your project, that is, study variables, interpreting your data, presenting data
    Small Groups: Review first drafts of participants' proposals.
    Didactic B
        —Practical aspects of project implementation
        —Ethical issues and the Institutional Review Board approvals
        —What NCG and CAPS will look for in a successful proposal: content,
            rationale, answers to key questions, and budget guidelines
    Small Groups: Continue proposal reviews.
    Final wrap-up
        —Major themes, cautions, and questions
        —Session evaluation forms completed

---

the challenges inherent in conducting research in a service provider
setting.

The concern about competitiveness among CBOs during the workshop was noted in the workshop evaluations and in personal communications to the workshop faculty. One agency noted discomfort in
sharing their ideas for an innovative intervention because other agencies could "copy" their originality. Other participants noted the
"strange dynamic" in helping a competing agency develop a strong
proposal. Although the consequences of this dynamic were not clearly
evident during the workshop, the timing of the workshop, whether it
occurs pre- or post-award, should be taken into consideration.

Selecting Proposals for Award

The final phase of the proposal process was the selection of grantees. After the workshop, 14 agencies ultimately submitted full proposals for consideration. The 4 agencies that chose to excuse themselves from the application process after the workshop reported doing so because in the course of the workshop they realized the commitment of time, resources, and prioritization that an evaluation required and were not willing or able to dedicate the staff for such a venture. This drop of over 20% of potential applicants points to the burden of conducting evaluations in community-based service agencies, which are often understaffed, underfunded, and working to address the complicated task of promoting behavior change among clients who might not want to change. Although the funding in the initiative sought to ameliorate the financial burden of learning about and conducting a self-evaluation, it remains a daunting task.

The 14 proposals were distributed for review to both researchers and funders. The reviewers included both faculty from the initiative and researchers from outside the core who were unfamiliar with the CBO staff personally. This was to provide additional safeguards in the review process. Each review team rated the proposals and noted any questions regarding the proposal, the CBO's capability, or the rationale for the research question. Several applicants were invited to come before a joint funder/researcher review board to respond to these questions so as to give novice evaluation proposal writers the ability to reply to concerns and clarify their intentions. Again, the criteria for selection were the strength of the science, the novelty of the intervention, the likelihood of achieving results, prioritization of the target population to be reached by the intervention, and perceived agency capability and commitment to conducting an evaluation of their prevention program. Ultimately, 11 proposals were funded, representing 20% of the original letters of intent. These agencies represented five Bay Area counties and targeted in- and out-of-school youth, gay men, incarcerated men, immigrant women, gay men of color, and recovering substance abusers (see Table 5.2). Awards were granted for program funding of up to $50,000 per year for up to 2 years and evaluation up to $10,000 per year. The next step was to match each CBO to a scientist as a collaboration partner.

TABLE 5.2 Participating Community Agencies

Coalition of Immigrant & Refugee Rights & Services (CIRRS): Impact of Mujeres
    Unidas y Activas on HIV risk behaviors among Latina immigrant women
East Bay Community Recovery Project: Impact of improvisational theater as a
    method for HIV/AIDS prevention among substance abusers in recovery in
    Alameda County
Face to Face/Sonoma County AIDS Network: Peer-led street outreach to high-risk
    youth and their parents in two neighborhoods of Santa Rosa in Sonoma County
La Clínica de la Raza—Fruitvale Health Project, Inc.: Multilingual peer education
    program for school-based Latino youth and their parents in Oakland
Marin AIDS Project: Peer-led HIV/AIDS education to inmates at San Quentin prison
    at entrance and, for a subset, prior to release
Mid-Peninsula YWCA: Peer- and adult-led prevention interventions with youth in
    Santa Clara County
National Task Force on AIDS Prevention: Impact of intensive community outreach
    and prevention intervention with black gay, bisexual, and transgender men in San
    Francisco
The New Conservatory Theater Center: Educational theater to youth and their
    parents in four sites in Napa, Solano, and Sonoma Counties
STOP AIDS Project: Modification of STOP AIDS meetings for gay men in preexisting
    groups
Tri-City Health Center: Outreach and small group prevention intervention with
    high-risk youth in central and southern Alameda County
Youth Advocates, Inc.: Impact of a 10-session prevention intervention for high-risk
    youth in San Francisco

In the discussion of selecting the CBO participants in the collaboration, it is important to note that the researchers participating in this
collaboration were likewise carefully selected from those at CAPS as
people with experience and dedication to community-based research.
To heal the rift of the MORT Syndrome, both service providers and
scientists were selected on the basis of their demonstrated or potential
"biculturalism" and ability to appreciate the inherent value of the
other.

Lessons Learned. The changes in staffing between the workshop
and the implementation of the programs disrupted the trust and
common understanding built by the workshop. In the future, we plan
to deliver the workshop after awarding funds. In addition, we plan to
begin the postaward process with a full-day reorientation in the
process and intent of the evaluation funding. Furthermore, we plan a

half-day orientation in evaluation for each agency's full staff. The lesson here is about the importance of building a good foundation, both scientific and relational, before embarking on the collaborative research. Roles need to be specified, and a clear understanding of responsibility and task assignment is essential.

COLLABORATION STRUCTURE

The structure of the collaboration was designed to institute the philosophy of group participation and support for the new "community researchers." Following an empowerment evaluation model, the scientists were to become facilitators rather than implementors of the evaluation research (Fetterman, 1994a). Each CBO was matched with one primary scientist who was responsible for consulting on the agency's research. This included working together on all aspects of the study: refining the research question and protocol; designing the survey instruments; in some cases, training CBO staff in data collection techniques; and conducting data analysis.

The research institution provided an ethicist who shepherded each research protocol through the University Committee on Human Research application process and two statisticians who consulted on sample size requirements, reviewed all data collection instruments, and were to conduct data analyses in concert with the CBO-researcher dyads. The research institution additionally provided language translation services for data collection instruments and consent forms to agencies working with bilingual or monolingual client populations. Finally, two senior administrators and one support staff provided problem-solving services, troubleshooting, oversight, and backup for the CBO-researcher dyads.

On the funders' side, one staff person was designated the liaison between the funding collaboration and the initiative, providing oversight and financial assurance to the collaboration. The chair of the AIDS Task Force maintained involvement and support through attendance at the monthly meetings. This level of involvement and visibility of the funding agency was a unique and welcomed experience for both CBOs and researchers.

The staffing for the CBO arm of the collaboration was a critical component in ensuring its success. Grants provided for both program and evaluation staff at the HIV/AIDS service agencies. Program staff included outreach workers, peer educators, program managers, and case managers. Depending on the nature of the intervention, evaluation staff could include those listed above as well as clerical staff to manage data or provide administrative support.

The provision of experts for specific portions of the evaluation further reinforced the empowerment ideal of the collaboration. Having the ethicist and statisticians as consultants reinforced the model of the researchers and CBO as a team instead of the scientist either unilaterally directing the CBO's evaluation or conducting the evaluation as an "outsider." In this model, the researcher was one of several coaches who trained his or her CBO partner in evaluation skills. In turn, the CBO staff tailored the evaluations to be relevant and feasible in the service settings. The negotiation taught all members of the team about what is required to conduct community-based, *community-driven* research.

Lessons Learned. Both the selection and the orientation of the participants are critical to the success of CBO/research collaborative evaluation. Those with previous experience or familiarity with each other's realm are more likely to be able to conduct effective evaluations. With novice CBO self-evaluators, the group nature of this particular evaluation provided much appreciated peer support. The team of experts possible through this research center might not always be available, but the availability of administrative/neutral support as well as contact with other CBOs conducting evaluation is strongly recommended.

PROGRAM STRUCTURE

Although each dyad established its own style of working together, the initiative maintained an initiative-wide structure to support the collaboration and further the training of the participants in community research. Once a month, the entire group—10 scientists, three university administrators, two funder representatives, and one to four staff members from each of 11 CBOs—gathered for 2 hours for a lunch meeting. These monthly meetings had several explicit, as well as several implicit, goals. They were a place for additional technical training, for networking and brainstorming among the agencies, and for CBOs to share their knowledge and expertise with the group. The monthly meetings provided support to agency staff who were frustrated by the barriers to conducting research in their service settings. They allowed the project administrator a chance to check in and possibly prevent any brewing difficulties within dyads. And, finally, group meetings discouraged reinforcement of the power dynamic that says *science* equals *knowledge*. The resources, wisdom, and experience of the collective group were essential in problem solving some of the barriers each study faced.

The initial meetings began with a check-in by two CBOs regarding their studies' status. This gave each agency an opportunity to share their program's knowledge as well as ask for assistance and feedback *from both the group of researchers and their CBO counterparts*. The second half of the meeting was reserved for additional technical assistance, with technical training topics including how to complete the Committee on Human Research (CHR) application, information regarding questionnaire design, and how to track research participants. After 5 months of technical assistance agendas, selected primarily by the research institution, it became clear that the didactic format needed retooling. An open discussion at a monthly meeting revealed that, although the CBO participants appreciated the involvement of and training by the research institution, they craved an opportunity to problem solve with each other and have more time to hear about each other's programs and experiences with instituting evaluation. Subsequent meetings alternated between open discussions initiated by the presentation of a single agency's experience (often including representatives from the CBO's direct service staff) and training topics as requested by the group.

These monthly meetings provided a common knowledge base, vernacular, and resource pool for both the CBOs and the researchers in their efforts to conduct community-initiated research. The early development of group cohesion originating in the 4-day workshop led the group to talk candidly about the difficulties of conducting the evaluation, without fear of reprisal from the funder or derision from the researchers. Similarly, when a dyad was facing the effects of a political initiative that threatened its ability to isolate the impact of the intervention, the researcher posed the methodological dilemma to the group, allowing CBO participants to offer suggestions on an evaluation difficulty as peers. This openness allowed for constructive problem solving and empowerment for both the CBOs and the researchers.

Although the monthly meetings allowed the larger institutionalized support to hold the collaborative structure, and provided a larger context for each study (i.e., the development of a model for collaboration), each dyad developed its own character. Depending on the need of the CBO partner, any previous experience with conducting research and evaluation, and the communication styles of each dyad, the contact ranged from many phone calls each week to consultations on an as-needed basis. Researchers met with their CBO partner either at the research institution or at the service delivery site. Several researchers met with staff who would be conducting the program under evaluation to include them in the evaluation partnership.

Lessons Learned. Some researcher-CBO teams instituted regular weekly or biweekly meetings, and others functioned by setting deadlines and meeting for consultation at pivotal points in the evaluation process. Either way, it is important for the CBO to have access to support and information as it engages in this unfamiliar area—conducting research and evaluation.

The monthly meetings became a key place for sharing information, supporting each other, and exploring solutions to the methodological barriers in conducting research and evaluation in a service setting. Although time-consuming for busy service providers and researchers, this type of group meeting should be a part of any collaborative model, as it fosters a true sense of intercultural exchange.

PRELIMINARY EVALUATION OF THE COLLABORATIVE MODEL

A formal evaluation of this three-way collaborative model, funded by the Ford Foundation, is being conducted by an independent agency and is still under way. The goals of the meta-evaluation are to assess the impact of the collaboration on the researchers' work, on the greater evaluation efforts by the CBOs, on the funders' decision making, and on the field of prevention as a whole. The evaluation will determine whether or not this initiative resulted in changes in prevention programs, prevention research priorities, funding priorities, and

CBO self-evaluation capabilities. This effort is responsive to the Joint Committee's meta-evaluation standard.

Although the results of the formal meta-evaluation are not yet available, several salient lessons have emerged as keys to successful collaboration and should be considered in efforts to replicate this model of evaluation.

Resources

Although funders and policymakers decry the lack of rigorous evaluation of community programs, the resources often have not been available to conduct such evaluation. In this collaboration, program funds were specifically separate from evaluation funding, with the evaluation component funded at up to 20% of the program costs. Also significant was the 2-year time commitment from the funders and research institution. These provisions respond to the *money* and *time* requirements in the MORT model. In addition, scientific consultation, data entry and analysis, assistance with the CHR process, and programmatic oversight were all provided pro bono by the research institution. Although not all collaborations will have the luxury of such resources, their costs must be considered. It is not enough to develop the relationship, it must be supported with resources.

Peer Support

One of the strongest lessons from this collaboration is the importance of peer support as an empowerment evaluation tool. The monthly meetings allowed novice CBO evaluators to compare their processes and brainstorm solutions. Just as peer support and peer education prove a powerful educational tool in promoting health behaviors, peer support empowered the members of this collaboration to learn new skills. CBOs learned to question their own assumptions as well as the assumptions of the researchers. Researchers were challenged to rework assumptions about appropriate methodologies for field settings. When conducting field evaluations became frustrating, the support of the entire collaboration helped maintain enthusiasm and commitment to the process. In replicating this program, we strongly recommend peer support as key for empowerment evaluation.

Prioritization

Conducting an evaluation under the auspices of a service agency is a difficult task. Staff need to be retrained and reoriented. Unfamiliar skills, such as instrument development, must be incorporated. Client interactions are changed or modified to include the research protocol. Although the value of learning evaluation skills and conducting an evaluation on one's own services is great, so is the burden. To balance this, the agency itself must have a commitment to engage in evaluation, which must become a priority. Some of the agencies in the HPEI staffed the evaluation with only one person as opposed to others who involved several levels of staff in the research. One year later, we've learned that the more integrated the research *and the commitment to research* are in a CBO, the more successful the evaluation process. This prioritization of conducting evaluation demands that the CBO incorporate the importance of *rigor* from the MORT model. Although research protocols for service settings might not incorporate strict experimental designs, the evaluation must involve rigor to achieve useful results.

This agencywide commitment is especially necessary where high staff turnover is a problem. The interest and skills cannot reside in a single person or set of persons. Learning should be integrated into the agency. According to Fetterman (1994b), empowerment evaluation "demystifies evaluation and ideally helps organizations internalize evaluation principles and practices, making evaluation an integral part of program planning" (p. 305). Changes in staffing can paralyze the evaluation if it is not integrated into the agency. The agency must perceive *ownership* of the evaluation, and not merely be used as a source for subjects in someone else's research.

Related to the importance of CBO prioritization for conducting their own evaluation is the variable of perceived benefit. Agencies that sought funding through this mechanism for the most part were motivated by a sincere interest in learning about evaluation and about their program's efficacy. In cases where the funding was sought as a means to continue programs, with interest in evaluation a distant second priority, evaluations have suffered. The benefit of conducting evaluation and of learning evaluation skills must be prized by the CBO.

Flexibility

This model of collaboration has been a three-way learning experience. Although the first year taught CBOs about the process and skills of evaluation, scientists and funders also have learned about the difficulties of conducting field research in the context of a service agency. Several evaluation designs needed modification once CBO staff had the chance to field the intervention. In one notable situation, the peer educator observed that she was modifying her teen support groups to comply with the research protocol, and that what she felt worked about the intervention was lost. Rather than destroy the heart of the intervention, the researcher and community provider modified the protocol to better evaluate "what made the intervention work." In another case, the CBO proposed a modification to their existing intervention and wanted to see if the modification worked better. After a year of difficulty enrolling participants in the modified groups, the researchers suggested a switch of interventions midstream.

This flexibility comes from the close relationship developed among funders, researchers, and community providers as well as a sincere interest in supporting each other's efforts. This type of funding flexibility may be crucial to any model that seeks to evaluate innovative HIV-prevention strategies.

Finally, the funders have learned about the costs of evaluation. Although they have regularly requested evaluation of programs they fund, this collaboration has shown them the extent to which evaluation is only initiated if it is funded. Although not all programs need to be evaluated all the time, a judicious use of evaluation grants can balance the funders' interest in funding services versus funding the evaluation of programs. Empowering CBOs with the necessary skills and resources for self-evaluation can set the stage for institutionalizing evaluation, making it a part of the planning and management of the CBO (Fetterman, 1994b). Eventually, agencies where empowerment evaluation has been institutionalized will require minimal support for quality evaluations—leaving more resources for direct service monies. More complex, collaborative evaluation efforts such as HPEI must be well funded if they are to meet the intended goals.

Conclusion

Although the outcome of the overall collaboration is not yet apparent, the process of collaborating has proven successful. Service providers see evaluation as a necessary and possible part of program delivery. Already, several of the CBO collaborators have approached the research institution for assistance in evaluating other programs at their agencies. Evaluation has been turned from something punitive and external into something desirable and within reach. CAPS has also experienced the collaboration as positive and plans to continue participating in the collaboration with the funder after this group of CBOs has "graduated." NCG has also committed to continuing to fund evaluation via this collaborative model after this funding cycle.

The effects of this positive collaborative experience are just beginning to ripple. The model is already being replicated on a university-wide level across the state, and there has been interest by both state and federal funding agencies in establishing their own versions of the model in future funding cycles as they recognize the potential benefits of intercultural education for both science and application.

This model of evaluation, which is designed to empower rather than judge, to share skills and knowledge rather than to find fault, to improve services rather than to shut them down, provides for the kind of learning necessary in the second decade of the AIDS epidemic. The ability for community providers, researchers, and funders to collaborate successfully is evident from the experiences described in the HIV Prevention Evaluation Initiative.

An African proverb states that "one who does not cultivate one's field will die of hunger." Funders, researchers, and community providers must and—as this empowering and evaluative collaboration proves—can work together in assisting communities to cultivate effective HIV prevention interventions and save lives.

Note

1. This MORT description was adapted from Rodriguez, Villa-Barton, and Faruque (1994).

References

Centers for Disease Control and Prevention. (1995). Update: Acquired immunodeficiency syndrome—United States, 1994. *Morbidity and Mortality Weekly Report, 44,* 64-67.

Fetterman, D. M. (1994a). Empowerment evaluation [American Evaluation Association presidential address]. *Evaluation Practice, 15*(1), 1-15.

Fetterman, D. M. (1994b). Steps of empowerment evaluation: From California to Cape Town. *Evaluation and Program Planning, 17*(3), 305-313.

Gómez, C. A., & Goldstein, E. (1993, August). *The view from the community: Developing collaborations between AIDS mental health service researchers and providers in the trenches.* Invited address in Opportunities and Challenges in AIDS Mental Health Services Research: A Research Symposium to Stimulate the Field, held by the American Psychological Association's Office on AIDS and the Public Policy Office at the 101st Convention of the American Psychological Association, Toronto, Ontario, Canada.

Rodriguez, G. M., Villa-Barton, C., & Faruque, S. (1994). Strategies for community involvement in research. In *Policy report: AIDS and drug abuse research and technology transfer in Hispanic communities* (pp. 25-37). Rockville, MD: National Institute on Drug Abuse.

Empowerment Evaluation

Building Upon a Tradition of Activism in the African American Community

CHERYL N. GRILLS
KAREN BASS
DIDRA L. BROWN
ALETHA AKERS

According to Fetterman (1994a), empowerment evaluation is the use of evaluation concepts and methods to help others to help themselves. It is designed to foster self-determination rather than dependency. As Fetterman notes, Rappaport's principles of empowerment provide the guiding framework within which our form of empowerment evaluation is conducted. This applies to our efforts as well. This chapter will focus on the use of certain evaluation concepts and strategies to enhance the community organizing, public policy work, planning, and prevention strategy development of the Community Coalition for Substance Abuse Prevention and Treatment.

At the Community Coalition, evaluation techniques and tools are used in conjunction with community organizing strategies to achieve coalition goals, which are conceived and implemented by the community.

Community empowerment within the context of alcohol, tobacco, and other drug abuse (ATOD) prevention is the coalition's ultimate goal. Evaluation is viewed by this community-based organization as a necessary and valued activity. In addition, it is understood that evaluation is being conducted within a community that is already self-determining. In this example, the evaluation process does not create self-determination within the community but facilitates a preexisting sense of self-determination within a community that has its own philosophy, style, strategies, and history.

The Community Coalition: A Paradigm Shift

The coalition[1] was formed as a response to the manner in which the war on drugs was being played out in the city of Los Angeles. It was a response to actions such as those of a recent Los Angeles police chief's launching of Operation Hammer, which was characterized by weekend sweeps (mass arrests) of African American and Latino youth. He touted this effort as a cure for the gang and crack cocaine problems of Los Angeles. He enlisted a tank with a batonlike device on the end known as the battering ram to smash the doors of crack houses in South Central Los Angeles. This type of action received national and local headlines but did not touch the sources of the problems.

The founders of the Community Coalition were longtime civil rights and community activists who believed that a comprehensive and more culturally relevant response was needed if the real issues were to be addressed. A comprehensive response would include addressing the economic problems that contribute to cocaine trafficking, the educational disparity that leaves people without the skills to compete in the workforce, and the overconcentration of liquor stores in severely economically and socially strained communities. It would also involve advocating for more traditional services for addressing ATOD and other socioeconomic problems of the community. Because of the long-standing hostility between the community and the Los Angeles Police Department, the coalition enlisted the support of city departments and elected officials to clean up drug- and alcohol-related problems in individual neighborhoods. Soliciting direct assistance from the police department could easily have backfired if they decided to conduct mass arrests and sweeps of the area or publicize the collaboration with the coalition, which would have undermined the coalition's ability to develop a relationship built on trust with area residents.

A basic premise of grassroots community organizing is the power of strategic mobilizations of large numbers of people to pressure decision makers. Mainstream power comes from wealth, race, and position in society. An assumption was made by the coalition founders that campaigns would have to be developed involving hundreds of people to pressure those in power to share the city's resources. Based on this assumption, the traditions and experiences of the civil rights movement and other grassroots organizing movements were used. The life-and-death situations faced on a daily basis by the coalition's organizers required that activists and leaders constantly debrief, critique, evaluate, and refine their work. These situations and the sincere dedication that was at the core of the movement meant that they put their egos aside to conduct exhaustive debriefing reviews. Within academic circles, these analytic reviews would be called evaluation.

In typical grassroots organizing settings, there are few resources. Because you are challenging power, you cannot afford to make mistakes with the resources on hand. Consequently, the need to review, critique, and look for ways to improve is an essential component of the organizing effort. In light of this, evaluation can be seen as a welcomed luxury and was indeed viewed as such by the coalition founders. The opportunity to evaluate the work in accordance with standard quantitative scientific procedures was an asset these activists had never had before. Some coalition members who were social service providers were particularly skeptical at first. They had previously had negative experiences with evaluators who stood "above the folks" and critiqued with a misunderstanding of the target communities, and had *no* understanding of their cultural or sociopolitical context. In this instance, however, the members quickly moved past skepticism to share the viewpoint that the evaluator and evaluation component were an asset to their ongoing work and a means of empowerment.

The coalition founders also believed that to acquire the necessary resources to be able to address ATOD problems, improve the quality of life, and effect change, the disenfranchised community of South Los

Angeles would have to shift power relationships in the city. For some communities that have Center for Substance Abuse Prevention (CSAP)-funded partnerships, the paradigm shift is a matter of bringing all of the right players to the table and getting them to talk to one another. The paradigm shift is based on the assumption that all of the players need to be on the same side and simply need to deepen their understanding of alcohol, tobacco, and other drug (ATOD) problems. In disenfranchised communities where the powers that be, at best, have written off the neighborhoods and, at worst, are openly hostile to them, however, the paradigm shift is an ongoing fight with *all of the players* to recognize the fact that the neighborhoods should be allowed to come to the table at all. For the Community Coalition, an organization rooted in activism, the people involved are committed to building something as opposed to simply fulfilling a partnership grant written by an agency that then hires staff to implement the project. The staff of the Community Coalition bring an agenda and philosophy to the table and combine it with the scientific techniques brought by the evaluator, who is equally invested in the cause.

According to *Webster's* (1983), to *enfranchise* means "to admit to the privileges of a citizen; to admit to political privileges or rights." This merger of the community activism tradition and social science evaluation technology promotes the coalition's goal to enfranchise its disenfranchised communities. South Central Los Angeles and many other inner-city communities are disenfranchised. Citizens do not perceive themselves as having the privileges or access to basic city services that more affluent white communities have. For example, when citizens complain to city government offices about a pothole, illegal dumping in alleys, or a nonworking street light, the response time tends to be so long that people often give up. Even calling the police while a violent crime is being committed does not receive a quick response according to a *Los Angeles Times* study that analyzed police response times in different communities (Schwada, 1995). For a disenfranchised community to effect change and improve the quality of life, a struggle needs to occur to shift the balance of power relations in the city so the disenfranchised section will receive the same types of service as the more affluent, "enfranchised" sections of town.

When residents of affluent communities complain and demand city services, they receive a quick response and are viewed as responsible citizens or homeowners who are working to maintain their neighborhoods. When they go to city hall to attend public hearings regarding their concerns, they look just like the people making the decisions. However, when South Central homeowners make the same demands and present a show of community concern they are viewed not only as complainers but as the cause of the deterioration in their neighborhoods. City bureaucrats ignore their concerns because South Central residents are presumed to be a powerless lobbying group. Bureaucrats *assume* that there will be consequences if they ignore the demands of affluent homeowners, that those citizens will complain to their superiors.

The coalition's organizational and evaluation work is very methodical. Pressure campaigns are developed to force the people in power to respond and bring about the change that is needed. The first step in building a campaign is to research and document the issue or problem that needs to be changed. The second step is to identify the decision makers with the power to bring about the desired results. The third step is to identify the types of pressure the decision makers will respond to, then organize their constituents to deliver that type of pressure. Some examples of pressure tactics are meetings with the decision makers, inviting them to community meetings, holding press conferences in front of their offices, filing petitions, and talking to their bosses. At each step, strategies are evaluated, reviewed, critiqued, and summarized. The outcomes are likewise subjected to critical appraisals.

Using the traditions of the civil rights movement (Morris, 1984), the coalition sought the involvement of consultants skilled in training direct action community organizing techniques. This combined with the expertise of the program evaluator enabled the coalition staff and volunteers to begin the needs assessment process. The decision was made to use the CSAP-required needs assessment as a method for involving current coalition members, expanding the base of volunteers, and training staff in volunteer recruitment and community organizing techniques. The coalition institutionalized needs assessment surveys as the method of documenting problems around which to build campaigns. Needs assessment surveys also serve as a recruitment tool; they are a first step or "excuse" used to talk to neighbors about community problems and then to encourage their involvement by returning to the neighborhoods, holding small community meetings, and reporting on the results. When neighbors see that the coalition is serious enough to produce the survey results and report the information in a professional scientific manner, it builds their confidence in the coalition as a serious organization that can produce results. At the community meeting, their involvement in crafting the campaign is then elicited.

INTEGRATION OF EVALUATION INTO THE COALITION STRUCTURE

The coalition uses an external evaluator who works with an evaluation team also external to the coalition. The evaluator has been with the project since the inception of the coalition. Evaluation subcommittees are formed on an ad hoc basis to work with the evaluator. These subcommittees typically provide input at each stage of the process (i.e., from generating hypotheses to be tested, to scale construction, pilot testing, and implementation and interpretation of results).

Procedurally, evaluation plays a key role in four areas: (a) the incorporation of process evaluation data into ongoing community mobilizing and organizing activities, (b) the employment of survey research data in campaign strategies designed to have a policy impact, (c) the employment of survey research methods and data into community organizing and campaign strategy development, and (d) the incorporation of outcome evaluations into ongoing organizational assessments. Each of these means for using evaluation has substantively served the advancement of the coalition's aims. The staff and membership of the coalition learned quickly that survey research methods and postevent debriefings (a version of process evaluation) are useful means of identifying issues, clarifying problems, prioritizing agendas for action, developing strategies with policymakers, defining goals, selecting targets for organizing, and critically assessing actions taken.

As previously noted, within African American and Latino communities, evaluation is seen as a necessary process and an integral part of any advocacy or community development work. Although there may be skepticism for evaluations conducted by those outside of the community (particularly given the contemptible reputation of Western social science in historically oppressed communities) (Ani, 1994; Guthrie, 1976; Jones, 1991), and although it may not be labeled *research* or *evaluation,* elements of evaluation are nonetheless operative in the praxis of many community groups and organizations. There is a natural willingness to "cut it up to make it better." These groups recognize that not doing so can carry serious repercussions. In a very practical sense, evaluation is part of a survival issue. Given that they exist within a reality in which their institutions are under constant attack from both internal and external entities, people who have been historically oppressed recognize the need to be open to self-critique. For example, little known to mainstream society, within the Black Panther Party, evaluation was one of the 20 points of their platform and an integral part of how they operated as a community organization.

For the Community Coalition then, evaluation is viewed as an extension of a process started and historically used by community-based organizations, movements, and leaders of African American and Latino communities (i.e., Marcus Garvey and Malcolm X, the NAACP, SNCC, and other organizations in the civil rights movement; the National Hispanic Leadership Conference; the National Conference of La Raza; the Political Association for Spanish-Speaking Organizations; and the Mexican American Legal Defense and Education Fund).

The Framework of Empowerment Evaluation

Dunst, Trivette, and Lapointe's (1992) delineation of the elements of empowerment best captures the manner in which empowerment evaluation was approached at the Community Coalition. There are six dimensions to empowerment: It can be characterized as a philosophy, paradigm, process, partnership, performance, or perception.

PHILOSOPHY

The philosophy of empowerment employed in the evaluation work of the Community Coalition borrows from Rappaport's (1981) three guiding principles of an empowering philosophy. According to Rappaport (1981):

1. All people have existing strengths and capabilities as well as the capacity to become more competent.

2. The failure of a person to display competence is *not* due to deficits within the person but rather to the failure of the social systems to provide or create opportunities for competencies to be displayed or acquired.
3. In situations where existing capabilities need to be strengthened or new competencies need to be learned, they are best learned through experiences that lead people to make self-attributions about their capabilities to influence important life events.

As a result, the coalition's approach to evaluation explicitly rejects the paternalistic and patronizing quality characteristic of many well-intended traditional evaluation approaches. These approaches are replete with assumptions that are not accepted carte blanche by communities of color (Stufflebeam, 1994). For the coalition, evaluation is conducted *with* the community, not *on* the community. There is no assumption that the community or its institutions are incapable or devoid of an evaluative tradition of their own. There is no assumption that there is something inherently better about professionals and their value-laden methods, that somehow their methods grounded in the Western scientific tradition are more sound, of greater ethical integrity, or steeped in insight and wisdom. Finally, there is no assumption that "The Program Evaluation Standards" set forth by the Joint Committee (1994) are sufficient to meet the evaluation standards of historically oppressed communities of color with divergent cultural styles and methods for characterizing reality (see also Fetterman, 1995). Rather, the existing traditions and strategies become a foundation for current efforts in process, outcome, and impact evaluation (Watts, 1993).

PARADIGM

The paradigm of empowerment evaluation for the Community Coalition entails the employment of a promotion model. This model operates from a mastery and optimization orientation. Emphasis is placed on the development, enhancement, and expansion of the coalition and the community's competencies and capabilities. Although this proactive approach values the development of a sense of control, which is a prominent feature of the coalition's philosophy, what is emphasized is a sociohistorically grounded, realistic sense of control. The ever-present political, economic, and racial climate within which the community constantly seeks to assert itself is never underestimated or forgotten. This helps to avoid misplacing attributions of failure on the community for not changing forces that are formidable and require concerted, oftentimes long-term efforts and systematic strategies to bring about change. Furthermore, promotion efforts are strengths based rather than deficits based, which is not commonly the case in strategies engaged in by those who seek to work within disenfranchised communities. To achieve empowerment both within community mobilization efforts and in empowerment evaluation, we operate from the assumption that "by building on strengths rather than rectifying weaknesses, people become more adaptive in not only dealing with difficult life events, but in achieving growth oriented goals and personal aspirations" (Dunst et al., 1992, p. 116).

In South Central Los Angeles, the coalition staff, membership, and resident constituency increased their competency in the use of the Western scientific methods of investigation, evaluation, and validation of reality. These methods are considered by those empowered sectors of society (i.e., policymakers, corporations, foundations, the academic community, the social sciences) to be legitimate ways and means of accessing knowledge. They have now become a part of the coalition's storehouse of tools to conduct and evaluate prevention work in their communities.

PROCESS

Evaluation is used as an enabling tool and process by the coalition. The coalition's astute capacity for "reading the temperature" of the community has been complemented by training offered by the evaluation team in the use of survey research methods. Survey research methods have become commonplace to the coalition's ongoing work in mobilization and policy change. Between 1991 and 1994, the coalition conducted eight substantive, empirically based investigations that have furthered their work in prevention (see Table 6.1).

Using both qualitative and quantitative process data to review their organizing efforts and campaign activities, the coalition integrated their philosophy of self-determination and self-assessment (whether it was in the form of evaluation, program planning, or community organizing) with program evaluation techniques. Within this context, the evaluator serves as a consultant helping the coalition to evaluate the work it deems necessary and important to do. This merger again

TABLE 6.1 Surveys Conducted by the Community Coalition for Substance Abuse Prevention and Treatment 1990-1994

Survey Name	Topic	Purpose	Total Number of Surveys Completed
Resident Survey	Needs assessment of South LA residents	To assess the extent to which residents of South LA perceive ATOD issues to be a problem in their community	1,125
Alcohol Availability Survey	Retail outlet nuisance problems	To document extent of nuisance problems associated with liquor stores	385
East/West Neighborhood Survey	East and West South LA neighborhoods' perceptions of their community	To obtain community perceptions of critical issues affecting their daily lives	202
Assessment of Tobacco Sales to Minors	Illegal tobacco sales to minors	To determine the extent of tobacco sales to minors and related concerns	30
Billboard Advertising Survey	Tobacco and alcohol billboard advertisements	To compare billboard advertising in two LA council-manic districts	237
Earthquake Survey	Northridge earthquake	To assess the emotional, social, and structural impact of the earthquake on South LA residents	1,751
Agency Director's Survey	Agency director's perception of the ATOD problem in South LA	To compare agency directors' perceptions of the ATOD problem with the perceptions of the residents of South LA and the agency line staff	15
Line Staff Survey	Agency line staff's perceptions of the ATOD problem in South LA	To compare agency line staff's perceptions of the ATOD problem with the perceptions of the residents of South Los Angeles and the agency directors	38

reflects the inherently empowered nature of community-based organizations (CBOs) to engage in efforts that demand accountability, critical analysis, strategy, and careful planning in all that is done.

In the early stages of the coalition's formation, the evaluator played a more significant role in helping to identify issues requiring further investigation (i.e., in the form of needs assessment). Often, the evaluator would identify initial variables and standard methodologies for investigation; however, coalition organizers and members quickly saw the utility of evaluative activities and began to identify their own areas of investigation, drafting their own measures, creating their own methodologies, and identifying their own targets for investigation using the evaluator as a sounding board for review of the methodological integrity of their plans.

The first investigation was a door-to-door survey of 1,125 South Los Angeles residents conducted as part of a multimethod needs assessment and community mobilization effort. Since then, the coalition has routinely engaged in random neighborhood surveys and empirically based studies to assess critical community problems such as alcohol and tobacco billboard advertising, illegal sales of tobacco to minors, public nuisance problems associated with retail liquor outlets, and the emotional impact of the Northridge earthquake on South Central Los Angeles neighborhoods ignored by the media. Because the surveys often yield critical data about community concerns, problems, and attitudes, they have become a primary component of the coalition's organizing and mobilization strategies.

Typically, the community is first approached for their opinions. Soon thereafter, follow-up visits are conducted to show them what their collective community concerns are and to ask what they want their next steps to be. *This* is empowerment. It is affirmation and validation of the community and acts as an impetus for action with direction. For example, because we were able to verify a pulse in the resident survey, we were well positioned so that, when the 1992 uprising occurred, it was natural for the coalition to take on a leadership role. The alcohol campaign was already in progress, and the 2 years prior to the uprising had been spent setting the stage for action. The coalition was ready to act.

Another example is the Northridge earthquake. In the aftermath of the January 1994 earthquake, the world's attention was riveted on Northridge while the forgotten residents of South Los Angeles quietly struggled with what is now unquestionable anguish, loss, suffering, and frustration. To identify immediate community needs and to determine the emotional and sociostructural impact of the earthquake on residents in South Central Los Angeles, a door-to-door survey was conducted by trained coalition members in collaboration with the evaluator. The survey was also administered to area schools and agencies. Information was gathered on a number of issues related to the earthquake such as neighborhood demographic characteristics, types of personal damage caused by the earthquake, its emotional impact, coping strategies employed to deal with the earthquake, disruption of vital area services, types of assistance sought and received, and desire to receive earthquake relief services and coalition resources. The most surprising information regarding the impact of the earthquake on South Central Los Angeles residents was the extent to which they experienced suffering from an inordinate number of emotional, stress-related symptoms including worry, nervousness, fear, and insomnia. In addition, two thirds of our residents did not know where to go in the community for assistance with substance abuse or stress-related problems.

Some of the lessons learned by the coalition regarding the use of survey research include the knowledge that (a) it is a doable technique that can generate useful information; (b) it is not difficult to design; (c) the community is receptive to this approach for gathering information; and (d) it can be used as an organizing tool to publicize the coalition, to generate enthusiasm for coalition initiatives, to establish community contacts for future follow-ups, and to provide important documentation needed to argue the coalition's positions on various policy matters, to enhance the understanding of the media about the community, and to negotiate with community institutions and policy-makers.

Survey research has become second nature within the work of the coalition. In fact, the evaluation team can barely keep pace with the volume of survey work initiated by the organization. Surveys are seen as a means of keeping in touch with community needs and determining what issues people are willing to get involved with. They help to verify the coalition's perceptions of the issues. They provide an excuse to talk with and recruit community residents and organizations. Entire constituencies have been built upon the foundation laid by the use of community surveys. Surveys are also used as a reentry point and a feedback/information dissemination mechanism. The coalition often goes back to the communities interviewed and involves members of the community in future actions. It is not a one-sided process in which the coalition simply takes from the community.

Beyond survey-based evaluation and community needs assessment, the coalition and its membership routinely engage in internal critical self-analysis and evaluation at their weekly staff meetings, team meetings, coalitionwide meetings, and board meetings to ensure that community goals and needs are clearly defined. They systematically develop strategic action based on evaluation findings, critically debrief and conduct analyses of coalition actions and campaigns, and review plans for needed follow-up activities. A typical strategy session at the coalition consists of a number of steps derived from the coalition's training in community organizing, history of community mobilization efforts, early exchanges with the evaluation team in process evaluation, and participation in annual and semiannual evaluation retreats facilitated by the evaluation team but that involved self-analysis along with external evaluation analyses.

PARTNERSHIP

The coalition's empowerment evaluation process is very much a partnership among the coalition staff, coalition members, the board, and the evaluation team. As such, this partnership is characterized by mutual respect and trust, reciprocity, open communication, shared responsibility, a shared appreciation for the cultural and sociohistorical context within which the community and its organizations operate, and a common cultural style of communication and interaction.

The balance of power intentionally rests with the coalition and the community. This, however, does not compromise the validity of the evaluation process. What it suggests is that the evaluator does not single-handedly operationalize critical variables or define the outcomes, process, or issues warranting investigation. In addition, the evaluator has never assumed an authoritative position with respect to the interpretation of findings. Interpretations are solicited from the coalition and community and synthesized with those of the evaluator. In empowerment evaluation, each perspective informs and enhances the other. The implications of these varying interpretations—the community's perspective versus objectivist social science perspectives—

are not minuscule, given that the way a problem is defined or interpreted has a direct bearing on the solutions devised and the accepted measures of success.

From the outset, it was clear that the target communities had their own theories of human behavior and the human condition. These culturally informed theories are often not given their due respect and recognition within academic, social science methods. In the face of documented problems of alcohol and other drug addictions in South Central Los Angeles, the coalition does not view the presence of substance abuse behavior by residents in the narrow, academic social science sense and does not label it merely as a skills deficit, lack of community standards, low self-esteem, or community disorganization. Regardless of the social sciences' inability to substantiate empirically the relationship between the availability of substances and consumption patterns, the community and the coalition do not dismiss the role of economic conditions, racism, housing deficits, unemployment, and cultural disalignment as critical contributors to the use and abuse of substances. Substance abuse, from the community's perspective, is merely symptomatic of a much larger, more pervasive set of issues. In their eyes, to simply remove the symptoms without substantively altering their causal agents would be an inadequate measure of success and lead to nothing more than symptom substitution. By allowing the community to evaluate and articulate its problems and propose interventions, the coalition has been able to understand the community's worldview and use it as a foundation for the creation of viable solutions.

The astonishingly high number of liquor stores in South Central Los Angeles is an example of a community-articulated problem that the coalition has helped to articulate. A mayoral task force reported that in a 40-square mile area, there are 682 liquor licenses. This equates to "17 licenses per square mile, compared to 1.6 per square mile in the remainder of LA County" (City of Los Angeles, 1992, p. 1). During the summer of 1991, a series of meetings were held with coalition members to discuss drug and alcohol issues in South Central Los Angeles. In August 1991, coalition members and staff decided to organize a community response to the overconcentration of liquor stores in the area. A leadership team was established consisting of neighborhood residents, social service providers, and recovering persons. Using survey data and systematic documentation of public nuisance conditions associated with liquor stores, the team developed strategies and made decisions regarding problem outlets in South Central. One important result of the coalition's efforts was the establishment of a Mayor's Task Force to examine the problems associated with area liquor stores and a commitment from the mayor to appoint a zoning administrator to focus on license revocation.

PERFORMANCE

The performance feature of empowerment evaluation of the Community Coalition can be seen in the blending of process, outcome, and survey evaluation principles with the mobilization and organizing techniques of community prevention. The role of evaluator relative to the community is critical to the effective performance of the evaluation work incorporated into the coalition's activities. There is a shared understanding of the community, its history and people, the issues and challenges confronting it, and the natural resources within South Los Angeles. The evaluator does not approach the community as a guru with skills and expertise to be bestowed on its residents. Rather, the evaluator is at the same level as the community and possesses no greater level of power. The evaluator is not giving anything to anyone but is engaged in a collaboration in which all parties bring equal, though perhaps different, forms of expertise to the table. There is one exception. The elders of the community are not seen as members on an equal footing with any of the players at the table: the community members, community organizations, or the evaluator. They are not on the same plane but are seen as operating on a higher level and are accorded their due respect in that regard. This includes having influential input into the process, tools, interpretations, and application of evaluation procedures.

PERCEPTION

The coalition's perception of empowerment entails a realistic view of the context in which its work occurs. This includes recognizing the historical partitioning of power outside of the community. It sees a distinction, however, in the community sense of solidarity mixed with a desire to bring about changes and to gain control of certain conditions in the community and in its social, political, and economic realities, which must happen on a much broader scale to permit an equitable distribution of the power and resources within this country. It is unfair to a community with very real contextual and historical realities that militate against self-determination to embrace a Pollyannaish posture toward empowerment. This does not mean that the community does not recognize its power or that it assumes a helpless role. Rather, the community accurately appraises the reality of the mitigating conditions levied against it and the multitude of levels at which change must occur. It recognizes that there are many small increments in the process of change and does not make unrealistic or unfair attributions of success or failure in the process of community empowerment. This perspective is crucial to empowerment evaluation because it informs how problems are defined and identified, the paradigm that drives community efforts, and the measures of both outcome and impact success (see Fetterman, 1994b).

The evaluation team consistently weaves these contextual factors into evaluation activities that contribute positively to community perceptions of evaluation work. Coalition members who were neighborhood residents and social service providers accepted the evaluator because she was from the same background as they were. It is critical that evaluators treat everyone with respect and not use their education and skills in an elitist fashion. In communities of color, evaluators traditionally have not come from the communities they evaluate and have acted like welfare workers who check up on and discredit the work people do. Consequently, it is not a lack of sophistication that has led to skepticism toward evaluation work but the result of real experiences. The coalition's evaluator has been compared with evaluators who are actually intimidated by the people they work with and who sometimes use language and scientific skill as a form of power that they hold over the community's or organization's head.

Finally, the evaluator and other coalition staff were viewed by neighborhood residents (who were primarily senior citizens) as the kids that benefited from the civil rights movement. They were seen as those who went away to school and returned with skills that could benefit the community. We in turn were honored and viewed the community elders as experienced veterans of their neighborhoods who had as much to teach us as we had to offer them. The work is a mutual form of empowerment; we teach each other.

Conclusion

Empowerment evaluation within the context of the Community Coalition involves a multidimensional application of empowerment that has operated at the level of the individual coalition member, the structure of the coalition, and the communities within South Central Los Angeles. We begin with the premise that our communities are already self-determining, and that we are "collaboratively" working toward the achievement of common goals employing a process whose "ingredients and terms" may look new or foreign but whose end result is a familiar, critical self-appraisal. Empowerment evaluation has been defined as the use of evaluation concepts and techniques to foster self-determination (i.e., the ability to chart one's own course in life; Fetterman, 1994a, 1994b; Rappaport, 1987). Evaluation with historically oppressed groups such as those found in South Central Los Angeles has required an appreciation of the reality that to be self-determining is not a luxury but a necessity for survival. These members of society must be self- determining. They do not have any choice but to do so given their historical experience of repeated marginalization. In addition, real self-determination implies resources and power. Until they really have the appropriate resources and power, their self-determination will be tempered by the constraints of social realities. This reality cannot be denied. This is not seen as a deterrent to action, but it must factor into the practical application of skills typically associated with the manifestation of empowerment such as the ability to identify and express needs, to establish goals or expectations, and to plan the actions necessary to achieve them.

Note

1. The Community Coalition is funded by the Center for Substance Abuse Prevention. It serves the South Los Angeles community, which is composed of African American and Latino residents. South Central Los Angeles is one section of South Los Angeles.

References

Ani, M. (1994). *Yurugu: An African centered critique of European cultural thought and behavior.* Trenton, NJ: Africa World Press.

City of Los Angeles. (1992, November 18). *Mayor Bradley's South Central Community/ Merchant Liquor Task Force—report.* Los Angeles: Author.

Dunst, C. J., Trivette, C. M., & Lapointe, N. (1992). Toward clarification of the meaning and key elements of empowerment. *Family Science Review, 5*(1-2), 111-130.

Fetterman, D. M. (1994a). Empowerment evaluation. *Evaluation Practice, 15*(1), 1-15.

Fetterman, D. M. (1994b). Steps of empowerment evaluation: From California to Cape Town. *Evaluation and Program Planning, 17*(3), 305-313.

Fetterman, D. M. (1995). In response to Dr. Daniel Stufflebeam's: "Empowerment evaluation, objectivist evaluation, and evaluation standards: Where the future of evaluation should not go and where it needs to go." *Evaluation Practice, 16*(2), 321-338.

Guthrie, R. V. (Ed.). (1976). *Even the rat was white.* San Diego: Dunbar.

Joint Committee on Standards for Educational Evaluation. (1994). *The program evaluation standards.* Thousand Oaks, CA: Sage.

Jones, R. L. (Ed.). (1991). *Black psychology.* Berkeley, CA: Cobb & Henry.

Morris, A. D. (1984). *The origins of the civil rights movement.* New York: Free Press.

Rappaport, J. (1981). In praise of paradox: A social policy of empowerment over prevention. *American Journal of Community Psychology, 9*, 1-25.

Rappaport, J. (1987). Terms of empowerment/exemplars of prevention: Towards a theory for community psychology. *American Journal of Community Psychology, 15*(2), 121-148.

Schwada, J. (1995, January 25). Alatorre urges news ways to deploy police. *Los Angeles Times,* p. B1.

Stufflebeam, D. L. (1994). Empowerment evaluation, objectivist evaluation, and evaluation standards: Where the future of evaluation should not go and where it needs to go. *Evaluation Practice, 15*(3), 321-338.

Watts, R. J. (1993). "Resident research" and community psychology. *American Journal of Community Psychology, 21*(4), 483-486.

Webster's new twentieth century dictionary of the English language. (1983). (J. McKechnie, Ed.). (2nd ed.). New York: Prentice Hall.

Realizing Participant Empowerment in the Evaluation of Nonprofit Women's Services Organizations

Notes From the Front Line

ARLENE BOWERS ANDREWS

For the past three decades, local nonprofit service organizations throughout the United States have evolved programs to support the poorest of the poor, those who have been chronically homeless and in perpetual crisis. Initially, most programs primarily offered comfort and such emergency subsistence aid as food, shelter, clothing, and medication. In recent years, transitional programs have emerged for those people who are ready to shift from transient living to relative stability. In one southern U.S. city, a group of seven women-run

AUTHOR'S NOTE: Beryl Miller has been the primary research associate for this project, which is based on the work of a study group that includes the directors of seven organizations: Anita Floyd, the Women's Shelter Transitional Program; Beebe James and Margaret McFaddin, St. Lawrence Place; Mary Nichols, the Nurturing Center; Pat Mikelson, the Cooperative Ministry; Lou Anne Pierce, Connections; Nancy Barton, Sistercare; and Mary Jane Scott, Habitat for Humanity. It has been supported in part by a small research grant from the Research and Productive Scholarship Fund at the University of South Carolina.

programs that provide extended transitional support to women and families with children has gathered to focus on a question of mutual concern: "What works, and why?" The essential goal is to answer this question and promote coordinated effective action that is informed by systematic research. In the course of exploring alternative approaches to addressing the question, the study group has relied upon principles and practices of empowerment evaluation. This chapter examines the relevance of these principles and practices to programs in small organizations that have traditions of empowerment but have been unable to realize adequate evaluation processes as their organizations have evolved. It also illustrates the feasibility of using evaluation to foster self-determination. First, the interagency study group is described.

The Ways and Means Study Group

The purpose of the study group is to study the ways and means whereby women lead themselves and their families out of transient living into relative stability. The group is particularly interested in how their organizations and the community resource system facilitate or hinder the transitional process.

NEEDS AMONG THE TARGET POPULATION

People who have extended histories of homelessness, running away, family separation, or chronic transience are often searching for a place and social group they can call home. The population searching for stability includes adults, some with families, who have lived on the streets or in emergency shelters, repeatedly moved or lived as guests in others' households, been repeatedly at risk of eviction, had children removed from their custody because of neglect, and been institutionalized for psychiatric hospitalization, addictions treatment, or incarceration and have no place to go upon release. This also includes transient young adults (ages 16 to 25) who have been chronic adolescent runaways or lived in residential care such as foster homes, group homes, or children's institutions. To overcome transience and family disruptions requires stable residence, adequate economic resources, reliable social relationships (e.g., family and friends), child care, health and mental health care, and personal skills for coping with life's demands.

In the study group of programs, the predominant focus is on serving women who are alone, youth who have been in foster care, and women with their families, including male partners (when present) and children. Single men are excluded because in this community, as in many others, the programs that serve them are separate from those that have the capacity to serve children and youth; women's programs tend to be integrated with these systems. The vast majority of services recipients are African American, an overrepresentation given that the percentage of African Americans in the general population of the community is about 38%. They are in the lowest income brackets, having no income at all, sporadic income through pickup jobs, or Aid to Families With Dependent Children (AFDC), an amount well below federal poverty levels in the particular state where this study is under way. Their participation in the programs is entirely voluntary. Almost all participants have been involved with a wide range of community organizations including emergency, social, mental health, and public health services; schools; criminal justice agencies; housing authorities; legal aid; and so on. Although they are participating in the organizations under study, they need full life assistance, not just particular, focused "services" (see Notkin, Rosenthal, & Hopper, 1990, for an excellent description of the type of families served in this network and Hopper, 1993, for an example of how ethnographic findings were translated into policy action). A sense of urgency prevails as the work of the organization becomes substantially infused into their daily lives. They have been called multiproblem and high-risk families (Kaplan, 1986; Kaplan & Girard, 1994) and, derogatorily, the "underclass" (Wilson, 1987).

The goals of all the programs include helping these participants achieve self-determination and a sense of empowerment—major challenges. Almost all participants have been extensively exploited in their personal lives and bear wounds of childhood and adult sexual abuse, physical abuse, and/or neglect. A majority are struggling with addictions, in various stages of recovery. All have extended stories about unfair treatment by and disappointment in organizations in their lives. In general, they are wary, slow to trust, and often appear to be alienated. In traditional research and evaluation studies, representative subjects from this population are known to have high attrition rates.

CHARACTERISTICS OF PROGRAMS IN THE STUDY GROUP

The search for ways to attain stability can be challenging (Berry & Weitzman, 1993; Birdsall, Clifton, & Wood, 1992). Obstacles are plentiful throughout the environment, rising from factors associated with the person/family, services programs, community systems, and public policy (Billingsley, 1992; Couture, 1991; Lerner, 1991; Robertson & Greenblatt, 1992; Shinn, 1992). The effect for some is the "revolving door" phenomenon, the tendency for people to enter the services system, leave it, and reenter it due to repeated crisis (Kagan & Schlossberg, 1989). Others are locked in chronic dependence on the services system (Korbin, 1992; Vosburgh, 1988). Many nonprofit organizations have developed transitional programs, like those in this study, to help people shift from transient living to relative stability. They typically provide case management, counseling, life skills education, and social support facilitation along with supervised residences (6 months to 2 years) or home-based services. They serve as brokers, facilitating the elicitation of client competencies to use resources in the community while advocating enhanced availability of essential community resources. Thus the programs devote staff time to client-focused and community-focused activities.

The programs in the study group include

1. a transitional program based in an independent nonprofit woman's shelter organization, where single women can live for up to 18 months in scattered apartments and participate in support services;
2. a transitional program based in a nonprofit organization affiliated with a religious community, where up to 30 families can live in a supervised apartment complex for up to 18 months and participate in support services;
3. a program based in an interfaith nonprofit coalition, which provides support partners such as small groups from within churches to partner with families in transitional housing;
4. an emergency shelter and follow-up community support services program of a United Way-affiliated battered women's organization;
5. the follow-up program of a therapeutic home-based and center-based program for families of preschool children who have been abused or neglected, based in an independent nonprofit organization;

6. the family support program of the area Habitat for Humanity organization that helps families build their own homes; and

7. a newly formed independent nonprofit that provides support counseling to, and plans to provide residences for, youth living on their own.

This particular study group includes managers from the seven organizations and a university-based study facilitator, all women. The facilitator had previous experience as a consultant or volunteer with all the organizations and had been founding director of the battered women's agency. All but one of the represented organizations predominantly comprise women board members, staff (paid and volunteer), and services recipients, although men play significant partnership roles in board and staff positions and are represented in small numbers among the service recipients. One of the organizations is predominantly male but the small transitional program is managed by a woman and the majority of services recipients are women. The agencies range in size from no paid staff to 10. All rely heavily on volunteers. All agency managers are involved in direct client service. Workers tend to be personally committed, working extended and flexible hours and describing the work as a way of life rather than just a job. Their budgets are tight relative to need; large numbers of people must be turned away at each agency. Only two have resources allocated for evaluation—less than 2% of their budgets in both cases, with the rest devoted to agency survival. All seven are resource and asset poor with regard to the capacity for information processing.

The group coalesced to share their ideas and generate mutual resources to increase the evaluability of their programs and the community resource system of which they are a part. Each organization has unique needs as well as common interests with the partners in the group. The study process is designed to address individual and communal needs.

Applying Principles of Empowerment Evaluation

All organizational partners in the study group were founded on principles of empowerment and had been working throughout their development to facilitate realization of empowerment for service recipients and workers in the organization. Realizing a capacity for evaluation had been more frustrating because of time and resource constraints. As the study process has begun, applying empowerment evaluation principles has been easily accommodated by the partners (Fetterman, 1994a, 1994b; Gutierrez, GlenMaye, & DeLois, 1995).

(1) Self-Determination. The notion of self-determination applies to the group that may be called "workers" (paid and volunteer board members and staff) as well as service recipients. The organizations have been self-initiated; all still have founders involved with them. Much of the recent published literature about evaluation has emphasized its utility in organizations that have sponsors such as municipal or state governments or large traditional bureaucracies. These larger, sponsored organizations have been involved in reform as principles of democracy, inclusivity, and accountability are infused into modern management throughout the world. For such organizations, moving toward self-determination is often a departure from old ways, like the imposition of standards and regulations from higher sources of authority and reliance on external evaluations. The small organizations in this study have always had first accountability to their program participants and responsibility for designing and directing the course of organizational practice. The study partners tended to be wary of the term *evaluation* because of its association with traditional external reviews and perceived irrelevance to their self-determining ways. Workers in these organizations have been pioneers in the type of care they provide, discovering effective practices and policies as they go. Even the decision to pursue more formal evaluation methods is a self-initiated effort.

Similarly, self-determination is a significant principle applied to the service recipients. All services are voluntary and designed to facilitate increased participant control over their own lives. Empowerment evaluation calls upon service recipients to extend this determination by contributing to communal responsibility for the program of which they are a part. Many service recipients embrace this opportunity enthusiastically; others are less immediately trusting of the process and more challenged by the need to address their own needs. Promoting a balance in the evaluation process through which individuals can attend to their own interests while advancing the interests of the program community is a challenge.

In these organizations, change has been perpetual, based on informal evaluative processes that are integrated into business as usual. The small size of these organizations and intensive nature of their services facilitate close communication between workers and service recipients, allowing for continual feedback. The goal in developing this study has been to extend this informal evaluation process while adding more formal methods so that individual, program, and system factors can be more intensively and routinely examined. The partners have been quite protective of their capacity for self-determination as the process has evolved.

(2) Primacy of the Program Participants. All participating organizations are grounded in ideology that recognizes the historical oppression of the target populations. Their operations reflect the will to protect, nurture, and support the program participants, whether they are classified as workers or "clients" or both. Research ethics are closely monitored, with particular sensitivity to the situation that, historically, low-income populations have been denied personal power and treated unfairly (Heller, 1989; Sieber & Stanley, 1988). Members of the study team communicated caution about research with their service recipients, particularly raising questions about data analysis and interpretation, citing published works that have a negative bias. For example, several studies have described homeless people in shelters as intensely pathological (Burt & Cohen, 1989; Shlay & Rossi, 1992), when the group itself would have interpreted many of the same symptoms as normal reactions to extraordinarily abnormal living circumstances (Jencks, 1994). Program managers seem to be particularly wary of any study efforts that are designed primarily to enhance the researchers' career or produce information that will primarily be disseminated among professional "guilds" rather than community practitioners.

Informed consent that attends to rights of privacy, confidentiality, and freedom from coercion is critical. Empowerment evaluation creates incentives for program participants to participate in organizational management, but participants should be free to refuse participation. The term *informed* must be operationalized in a way that helps participants understand benefits to themselves while encouraging a sense of social responsibility toward the organization and other individuals with needs like their own. Services recipients must be protected from anxiety, frustration, or concern that choosing to abstain from participation will bring negative consequences.

Ideally, empowerment evaluation will enhance the therapeutic, developmental influence of the program on the participant. It goes beyond giving service recipients a voice; it means supporting them while they influence the operations and direction of the organization. It can be integrated into the skill-building services, encouraging self-assessment as a part of problem-solving and planning processes as well as consumer assertiveness in holding services organizations accountable for their actions.

This study process was also designed to be responsive to the needs of the service providers. All agencies are available, if necessary, 24 hours a day, seven days a week to their participants; some provide around-the-clock residential services. Their workloads are full; evaluation is a relative luxury. All procedures must be designed to interfere as minimally as possible with routine operations and to require as little extra time as possible from staff.

(3) Reciprocity and Feminist Principles. Women-run programs tend to be highly participative, intentionally designed to avoid interactions that might be perceived as authoritative or controlling. Traditional evaluation methods that rely solely on quantified measures or observations by "objective" outsiders are regarded by feminist scholars as inappropriate and unlikely to detect program outcomes and impacts (Shapiro, 1988). Furthermore, feminist perspectives of helping relationships emphasize mutual exchange, or reciprocity, rather than the charitable model wherein one person is perceived as more powerful or endowed than another. Empowerment evaluation helps organizations realize this ideal. Workers can be perceived as brokers, people who can help clients realize their own competence and access resources. Service recipients can, in turn, help workers by contributing to the development of the organization.

(4) Perpetual Innovation. The organizations in this study are continually assessing their activities, operating according to general plans and policies and revising as necessary. Each organization has a somewhat different process for doing this. One of the better-funded

agencies has frequent staff meetings with advisory consultants. Agency staff routinely set agendas for program areas to study and divide responsibilities for developing proposals for innovations. They test reforms and adopt, revise, or reject them as necessary. Another has open, unplanned discussions over the lunch table with staff, service recipients, and visitors, encouraging the sharing of informed and external ideas. All have an atmosphere of openness to innovation. One of the goals of the collaborative study that is under design is to develop resources to help each organization supplement their informal processes with more formal methods of evaluation, including the capacity to collect and report program and community resource information.

(5) Significance of the Ecological Context. Client and organizational outcomes are powerfully influenced by the broader community resource system, public policy, and evolving market demand for the type of assistance offered by these organizations. As this study was being developed, the populations in need expanded and the political environment became increasingly threatening. Traditional evaluation emphasizes program factors. The partner organizations in this study are keenly interested in monitoring the quality and effectiveness of their own activities. They are additionally interested in enriching their power by collaboratively examining their ecosystem, extending beyond organizational boundaries. Communities support and hinder families in transition through a variety of public, nonprofit, and privately sponsored resources that have developed without coordinated planning and no system of collective accountability. By creating a cooperative evaluation network, this group of agencies will empower themselves to better use and influence the broader ecosystem.

(6) Advocacy. Perhaps one of the most significant foundations of the approach to evaluation developed by this study group is the principle that evaluation is meant to inform advocacy. People who have experienced chaos or homelessness and the workers who stand by them are at risk of feeling powerless in the face of overwhelming challenges. Every day, participants in transitional programs are likely to confront inept or insensitive bureaucratic workers, rejection from potential employers or landlords, barriers to entitlements and opportunities, and discrimination on the basis of gender, race, disability,

age, or parental responsibility. Almost universal consensus exists about the need to reform the support systems for poor people. Yet little empirical information is available to inform policy and program redesign (Blasi, 1990; Brabeck, 1989; Brudney & Kluesner, 1992; Molnar, Rath, & Klein, 1990; Stolarski, 1988). The beauty of empowerment evaluation, particularly when it is applied to community systems as they influence programs, is that it can produce information of direct relevance for local policy development and resource allocation as well as for broader policy at the state and federal levels. The communicators of the information are the program participants themselves, with support from evaluation facilitators.

(7) Summary of Principles. The principles of self-determination, primacy of program participants, ecological context, women's ways of helping, perpetual innovation, and advocacy are inherent in the transitional services programs. Collectively, they provide a foundation for assuring that when evaluation science is applied in these organizations, the resulting information is likely to be relevant and used.

Empowerment Evaluation in Practice: An Interorganizational Case Example

The interorganizational study group has emphasized development of an evaluation process and is working to institutionalize this process as a part of their individual and collective operations. Thus far, the design of the process and pilot phases of the study are complete. The bulk of the work, instituting a multiyear plan and design for routine evaluation and special studies, is in the planning stages. These are notes from the front line, where work is currently under way. The experience thus far yields productive lessons about the application of steps in empowerment evaluation and tips for putting principles into practice.

EVALUATION PROCESS DESIGN

The study group has focused on establishing an infrastructure for support of evaluation processes. This includes stakeholder investment, team building, enriching the evaluability of the organizations through information systems development, and developing evaluation resources.

(1) Stakeholder Investment. The stakeholders in this effort include the service recipients, workers and managers in the agencies, board members, and donors, including various sponsors such as churches, the United Way, and county, state, and federal government sources. Typically, the donors have stated expectations for program accountability with varying standards regarding how the agencies are to report and verify their processes and outcomes. Decisions about design and management of evaluations are the responsibility of the agencies; the donors generally anticipate being an audience for evaluation reports. Only one participating organization has a major federal grant with specific expectations regarding evaluation, and even there the organization has considerable discretion regarding development of criteria and plans. The other stakeholders have less definitive expectations regarding accountability. As they develop their roles and responsibilities, they have affirmed interests as users of and participants in producing evaluative information.

In this project, the evaluation process has been facilitated by a university-based coach who has "insider" credibility because of previous experience in several of the agencies. In traditional evaluations, an insider-outsider dichotomy often exists that can generate mistrust and a risk that information on either side will be invalid. A degree of objectivity is essential for evaluation, but acknowledgment of the biases inherent in the various roles that stakeholders play with regard to an organization is also critical. Vested perspectives yield potentially in-depth information about the workings of an organization. Recognizing and valuing these differences is helpful to the evaluation process. The coach's role is to facilitate interaction among the various stakeholders with differing perspectives, to help them attain a unified, truthful impression of the organization.

A significant initial step is to promote stakeholder participation in the evaluation process (Greene, 1987). Roles and responsibilities of the participants have to be identified and accepted. In this study, agency personnel were eager to participate in defining the evaluation questions, adapting management procedures to facilitate evaluation processes, and preparing to disseminate findings so long as the costs were bearable and worth the gain in useful information. For them, time was the primary investment. They have been quite protective of time that is taken from other essential responsibilities for the sake of evaluation, insisting on efficient use of time. They also have responsibility for helping to gather resources, noted below, to support the effort. Theirs was the first buy-in promoted.

Ways to involve the service recipients were explored through a pilot study. A primary concern has been developing ways to compensate service recipients for their willingness to contribute to program evaluation. The study group was hesitant to ask them to donate their time to the organization, although some were willing to do so. Many service recipients are so pressed for time and money that volunteerism is beyond their capacity, particularly because they are already involved in considerable mutual support efforts with one another. The risk of exploitation, particularly of making service recipients feel coerced into assuming research responsibilities, was also considered. A small grant was obtained to compensate financially service recipients on an hourly basis for participation in a pilot study to determine the evaluation questions and assess applicability of potential methods. A plan was then developed to compensate selected service recipients through part-time employment as research aides and to continue more inclusive involvement of larger numbers of participants through hourly compensation. The aides will help gather information, interpret it, and report it in a way that is useful to services recipients (Whitmore, 1990). The roles of services recipients are thus similar to those of managers and workers: to contribute to defining research questions, conducting technical evaluation work, and disseminating information.

The plan for involving board members from each organization has been to provide oversight, including final selection of the evaluation questions based on program participant recommendations, and responsibility for approving action plans subsequent to evaluation reports. The board members will also be involved in disseminating evaluation findings to external groups, including donors and policymaking bodies. The coach will facilitate ongoing communication among the various stakeholders as the evaluation process proceeds.

(2) Team Building. This project has two levels of evaluation teams. Teams will operate within each organization and an interagency team

will operate among the organizations that are participating in the evaluation cooperative venture. The evaluation process has been designed to be adaptable to the needs of individual organizations and the study group as a whole. Process facilitation has emphasized meaningful participation at minimal cost to participants. Routine methods of communication are in the process of being established. These will rely heavily on the role of the central facilitator (coach) and the proposed evaluation facilitation team, which includes a professional specialist working across agencies and research aides working within the agencies. After initial team-building and decision-making processes are established, face-to-face meetings will be kept to a minimum because they are so time-consuming. They will be used for specific purposes throughout the evaluation process.

Procedures to protect confidentiality of information about participating organizations are being established. Beyond assurances that case information will be protected, findings of a sensitive nature that may be shared within the interorganizational group will also be protected. For example, in a network as small as this one, the evaluation may detect problems with performance for a particular worker in one agency. The group will develop guidelines for determining when, if at all, and how to discuss such information. Similarly, group norms for promoting consensus and handling disputes are in development.

(3) Evaluability Enrichment. All but two of the organizations in this study group have extremely limited capacities for collecting, managing, and reporting information for evaluation purposes. The technological resources for computer-assisted information processing are inadequate. Even the two organizations with sufficient technology need assistance regarding how to handle the technical aspects of evaluation. In addition, the human factors in processing information are a challenge for this group. Workers are oriented to efficient use of information and resistant to recording or producing anything that is perceived as excessive information or information for a purpose that fails to have obvious benefits for client services. Accuracy and completeness of records are hard to monitor because of workers' time demands and, in some cases, reliance on numbers of volunteer workers. The nature of program participation by service recipients creates hindrances to evaluation, particularly because attrition is

high, so gathering postinformation in pre-post designs may be a challenge.

The specification of clear, valid criteria for evaluation is an essential part of promoting the evaluability of these programs. The evaluation participants are uniquely qualified to understand the target populations' needs, the services, and the environmental context and thus are in positions to establish standards for program and systems assessment and procedures for the evaluation processes. Because no known standards exist pertaining to transitional services of the type provided by this network, the study group will rely on practice wisdom as well as standards for comparable programs like the Child Welfare League of America's (CWLA) standards for family services (CWLA, 1989).

The development of information systems and evaluation criteria requires technical support and considerable time and expertise in preparation for infusing more formal evaluation processes into the organizations. Recognizing this, the study group decided that an essential step was to generate resources to strengthen the infrastructure of each organization and the interorganizational network to promote evaluability and reduce demands on staff.

(4) Evaluation Resource Development. As noted earlier, the organizations are generally too poor to allocate specific resources for evaluation. Some have government grants that allow a percentage to be devoted to evaluating the specific project that is funded by the grant. The study group decided to collectively pursue resources that will support an evaluation facilitation team, which will include coaching regarding overall process and plan development and management, facilitation (including management information system development), and technical support to be provided by trained program participants.

The team will work with organizations individually and with the study group for purposes of examining questions of mutual concern.

EVALUATION PLAN

The evaluation plan for the study group has yet to be finalized. What follows are issues that have been identified for consideration as the questions, design, methods, and products are chosen.

(1) Questions. The purpose of the evaluation must be clear to all participants and potential users of the results. In this project, each organization must define its own purpose in participating in the cooperative, stipulating anticipated benefits and limitations, and the purpose for the group must be clarified as well.

The process of clarifying the purpose and selecting study questions is the model for all subsequent processes. By facilitating the inclusion of various stakeholder groups in the choice of questions, their power is recognized. In the type of organizations represented in this network, this is a complex process. The human needs addressed by these organizations have complex etiologies and manifestations; there are no simple solutions.

For example, in the pilot stages of this project, the study group produced a list of questions of mutual concern to which they would like to know the answers. The following is a sample of the questions:

- What is "success" for people who have lived in transience and/or chaos? How is "stability" defined and what does it mean?
- Given that such notions as "self-sufficiency" and "independence" are contrary to human needs for interdependence, what are appropriate goals for people who have been chronically transient or dependent on services agencies?
- Who comes to transitional programs? Who doesn't? Why?
- What are the characteristics of an effective worker-client relationship in transitional programs? How are issues of power and control handled?
- How do policies (e.g., agency policies, administrative rules of community agencies, laws) affect services recipients?

This preliminary list demonstrates that the group is interested in evaluating the basic missions and theoretical bases of the organizations as well as goals, strategies, and processes. The questions will be refined and operationalized as the evaluation plan is developed.

(2) Design. The questions must, of course, be framed in such a way that they can be answered through feasible research designs and methods. The overall approach may include multiple substudies with different designs to address questions that focus on such targets as participant progress, program outcomes, and services system barriers.

Simply describing the processes through which services recipients and workers go and the expected outcomes may be an essential first step in promoting the capacity for evaluation. Several of the participating organizations have been unable to produce routinely such descriptive information. The monitoring of longitudinal change, ideally with comparative or control data, is desperately needed to shed light on how people in transitional programs achieve stability.

(3) Methods. Again, program participants must be closely involved in developing study methods, particularly the selection of measures, data collection, and data analysis processes. The consensus in this study group is that certain quantitative measures may be credible and appropriate, but that qualitative information is essential for communicating the interpersonal nature of these programs and the context in which the care occurs (Gilgun, Daly, & Handel, 1992; Greene, 1988; Strauss & Corbin, 1990; Whyte, 1990). Interpretation of quantitative and qualitative data is regarded as a critical process, one that requires discussion and reflection by program participants. Implications for practice and policy will be developed.

(4) Products. Oral information is deemed by this study group to be as important as written information. A climate of exchange and interaction pervades the group. The essential intent of the evaluation is to produce *useful* information. It is anticipated that oral and written reports will be produced in various formats to meet the needs of specific stakeholders. For example, discussions will be facilitated among groups of services recipients about relevant findings, with handouts to support the oral presentations and suggestions for how to change what they are doing to help themselves, others, and the organization based on the findings. The organization's board will participate in reports that have the same overall data with discussion focused on other areas that pertain to their governance responsibilities.

Conclusion

The Ways and Means Study Group is in the midst of an evolving evaluation process that builds on the principles of empowerment evaluation. This approach fits the character of the participating organizations: independent, self-directed, nonprofit, women-run agencies that offer intensive support and care to relatively small numbers of service recipients. The agency personnel tend to be ideological, driven by principles and beliefs regarding self-determination, primacy of program participants, ecological context, women's ways of helping, perpetual innovation, and advocacy. The process of evaluation is thus considered to be critically significant, as well as the technical aspects. To formalize and intensify evaluation efforts is an option for these organizations, one that will be chosen only if the process is naturalistic and supportive of the growth that these organizations try to promote in their participants. The study group is developing an evaluation plan that incorporates sound science, credible standards, and the interpretive wisdom of program participants. The anticipated results will guide the development of realistic services recipient and worker goals and activities and improvements in agency management practices, community services systems, and public policy.

References

Berry, C. A., & Weitzman, B. C. (1993). Factors affecting housing comfort among formerly homeless families: Housing quality or individual vulnerabilities? *Community Psychologist, 26*(2), 27-29.

Billingsley, A. (1992). *Climbing Jacob's ladder: The enduring legacy of African-American families.* New York: Simon & Schuster.

Birdsall, C. T., Clifton, A. K., & Wood, M. F. (1992). Action research in planning housing for women. *Journal of Architectural and Planning Research, 9*(2), 149-157.

Blasi, G. L. (1990). Social policy and social science research on homelessness. *Journal of Social Issues, 46*(4), 207-219.

Brabeck, M. M. (Ed.). (1989). *Who cares? Theory, research, and educational implications of the ethic of care.* New York: Praeger.

Brudney, J. L., & Kluesner, T. M. (1992). Researchers and practitioners in nonprofit organization and voluntary action: Applying research to practice? *Nonprofit and Voluntary Sector Quarterly, 21*(3), 293-308.

Burt, M., & Cohen, B. (1989). *America's homeless: Numbers, characteristics, and the programs that service them.* Washington, DC: Urban Institute Press.

Child Welfare League of America (CWLA). (1989). *Standards for services to strengthen and preserve families with children.* Washington, DC: Author.

Couture, P. D. (1991). *Blessed are the poor? Women's poverty, family policy, and practical theology.* Nashville, TN: Abingdon.

Fetterman, D. M. (1994a). Empowerment evaluation. *Evaluation Practice, 15*(1), 1-15.

Fetterman, D. M. (1994b). Steps of empowerment evaluation: From California to Capetown. *Evaluation and Program Planning, 17*(3), 305-313.

Gilgun, J. F., Daly, K., & Handel, G. (Eds.). (1992). *Qualitative methods in family research*. Newbury Park, CA: Sage.

Greene, J. G. (1987). Stakeholder participation in evaluation design: Is it worth the effort? *Evaluation and Program Planning, 10,* 379-394.

Greene, J. G. (1988). Stakeholder participation and utilization in program evaluation. *Evaluation Review, 12*(2), 91-116.

Gutierrez, L., GlenMaye, L., & DeLois, K. (1995). The organizational context of empowerment practice: Implications for socialwork administration. *Social Work, 40*(2), 249-258.

Heller, K. (1989). Ethical dilemmas in community intervention. *American Journal of Community Psychology, 17*(3), 367-378.

Hopper, K. (1993). On keeping the edge: Translating ethnographic findings and putting them to use—NYC's homeless policy. In D. M. Fetterman (Ed.), *Speaking the language of power: Communication, collaboration, and advocacy: Translating ethnography into action* (pp. 19-37). London: Palmer.

Jencks, C. (1994). *The homeless.* Cambridge, MA: Harvard University Press.

Kagan, R., & Schlossberg, S. (1989). *Families in perpetual crisis.* New York: Norton.

Kaplan, L. (1986). *Working with multiproblem families.* Lexington, MA: Lexington Books.

Kaplan, L., & Girard, J. L. (1994). *Strengthening high-risk families: A handbook for practitioners.* Lexington, MA: Lexington Books.

Korbin, J. E. (Ed.). (1992). The impact of poverty on children [Special issue]. *American Behavioral Scientist, 35*(3).

Lerner, M. (1991). *Surplus powerlessness.* New York: Humanities Press International.

Molnar, J. M., Rath, W. R., & Klein, T. P. (1990). Constantly compromised: The impact of homelessness on children. *Journal of Social Issues, 46*(4), 109-124.

Notkin, S., Rosenthal, B., & Hopper, K. (1990). *Families on the move: Breaking the cycle of homelessness.* New York: Edna McConnell Clark Foundation.

Robertson, M. J., & Greenblatt, M. (Eds.). (1992). *Homelessness: A national perspective.* New York: Plenum.

Shapiro, J. P. (1988). Participatory evaluation: Towards a transformation of assessment for women's studies programs and projects. *Educational Evaluation and Policy Analysis, 10*(3), 191-199.

Shinn, M. (1992). Homelessness: What is a psychologist to do? *American Journal of Community Psychology, 20*(1), 1-24.

Shlay, A., & Rossi, P. (1992). Social science research and contemporary studies of homelessness. *Annual Review of Sociology, 18,* 129-160.

Sieber, J. E., & Stanley, B. (1988). Ethical and professional dimensions of socially sensitive research. *American Psychologist, 43*(1), 49-55.

Stolarski, L. (1988). Right to shelter: History of the mobilization of the homeless as a model of voluntary action. *Journal of Voluntary Action Research, 17*(1), 36-45.

Strauss, A., & Corbin, J. (1990). *Basics of qualitative research: Grounded theory procedures and techniques.* Newbury Park, CA: Sage.

Vosburgh, W. W. (1988). Voluntary associations, the homeless and hard-to-serve populations: Perspectives from organizational theory. *Journal of Voluntary Action Research, 17*(1), 10-23.

Whitmore, E. (1990). Empowerment in program evaluation: A case example. *Canadian Social Work Review, 7*(2), 215-229.

Whyte, W. F. (Ed.). (1990). *Participatory action research.* Newbury Park, CA: Sage.

Wilson, W. J. (1987). *The truly disadvantaged: The inner city, the underclass, and public policy.* Chicago: University of Chicago Press.

THEORETICAL AND
PHILOSOPHICAL FRAMEWORKS

Empowering Community Health Initiatives Through Evaluation

STEPHEN B. FAWCETT
ADRIENNE PAINE-ANDREWS
VINCENT T. FRANCISCO
JERRY A. SCHULTZ
KIMBER P. RICHTER
RHONDA K. LEWIS
KARI JO HARRIS
ELLA L. WILLIAMS
JANNETTE Y. BERKLEY
CHRISTINE M. LOPEZ
JACQUELINE L. FISHER

Empowerment evaluation is an evolving connection between apparently conflicting ideas. Traditional evaluation methods contribute to

AUTHORS' NOTE: Preparation of this chapter was supported by a grant from the Kansas Health Foundation (No. 9409082) to the Work Group on Health Promotion and Community Development at the University of Kansas. Thanks to Tom Wolff for his ongoing insights about empowerment in our collaborative work with community coalitions. We thank our former colleagues at the Decade of Hope Coalition, including Lee Martinez, Robert Martinez, Deanna Chechile, and Miriam Cachucha. We also thank our colleague on the adolescent health initiatives from the Kansas Health Foundation, Mary K. Campuzano, for her constructively critical feedback on this methodology. Special thanks to those citizens from collaborating communities who continue to teach us about promising approaches to evaluation in service of empowerment. We also thank Michele Scheppel and Jenette Nagy for their assistance in the preparation of this manuscript. Inquiries should be addressed to the first author, 4001 Dole Center, University of Kansas, Lawrence, KS 66045.

understanding (and perhaps improvement), with less emphasis on whether the capacities of those studied are enhanced by the inquiry. By contrast, empowerment evaluation offers the promise of using evaluation concepts and methods to promote self-determination (Fetterman, 1994a).

Empowerment is a core concept in the fields of community psychology (e.g., Rappaport, 1981, 1987; Rappaport, Swift, & Hess, 1984) and action anthropology (e.g., Tax, 1952, 1958). Its roots and application are also apparent in education (Freire, 1970), community organization (Mondros & Wilson, 1994), and public health (Eng, Salmon, & Mullan, 1992). *Empowerment* refers to the process of gaining influence over events and outcomes of importance to an individual, group, or community (Fawcett, White, et al., 1994). This construct highlights the value of individual strengths and competencies, natural helping systems, and social change (Zimmerman, in press; Zimmerman, Israel, Schulz, & Checkoway, 1992). Empowering processes and outcomes have been examined in a variety of contexts, including with children and youth (e.g., Mithaug, 1991), people with disabilities (e.g., Fawcett, White, et al., 1994), health care of minority populations (Braithwaite & Lynthcott, 1989), low-income elders (Minkler, 1992), and low-income and disadvantaged communities (Fawcett, Seekins, Whang, Muiu, & Suarez de Balcazar, 1984).

Empowerment evaluation as a capacity-building process is grounded in the tradition of participatory inquiry, research, and evaluation (Choudhary & Tandon, 1988; Whyte, 1991). This tradition reflects the core values of critical theory (Giroux, 1983), feminism (Lichtenstein, 1988), qualitative evaluation (Fetterman, 1988, 1989; Guba & Lincoln, 1989; Patton, 1980), and action anthropology (Fetterman, 1993; Stull & Schensul, 1987). Its aims are to legitimize community members' experiential knowledge, acknowledge the role of values in research, empower community members, democratize research inquiry, and enhance the relevance of evaluation data for communities. In participatory evaluation and empowerment evaluation, and some forms of intervention research (Fawcett, Suarez-Balcazar, et al., 1994), those studied help set the agenda for research, participate in collecting and analyzing data, and determine the use of the results.

Evaluation may enhance (or reduce) capacity to influence the environment, and to varying degrees. On one end of a continuum of evaluation, nonparticipatory evaluation can be completely coercive and proscribed, without input from those whose efforts are being appraised. At the other extreme, fully participatory (or participant-controlled) evaluation can be completely initiated, designed, and administered by the community initiative. Empowerment evaluation seeks to balance the legitimate interests of the field in promoting understanding with those of communities in fostering improvement and self-determination.

Community health initiatives provide a rich context for understanding and improving the practice of empowerment evaluation. These initiatives have attempted to build community capacity to address a variety of citizen concerns, including violence (Wilson-Brewer, Cohen, O'Donnell, & Goodman, 1991), substance abuse (Falco, 1992), injuries (Davidson et al., 1994), mental disorders (Fawcett, Paine, Francisco, Richter, & Lewis, 1994), and adolescent pregnancy (Nezlek & Galano, 1993). Community health initiatives often adopt a public health framework (Fawcett, Paine, Francisco, & Vliet, 1993; Green & Kreuter, 1991), using technical assistance and evaluation to help build local capacities to address identified community concerns.

This chapter explores the concept of empowerment evaluation in the context of several community health initiatives. First, we describe a conceptual framework that our Work Group uses to guide efforts to facilitate empowerment through evaluation. Second, we identify the contexts for case studies with community health initiatives with which we have collaborated: community coalitions for prevention of adolescent pregnancy and substance abuse in Kansas and a tribal partnership for prevention of substance abuse in New Mexico. Third, we outline and illustrate the evaluation process used to empower these coalitions: (a) assessing community concerns and resources, (b) setting a mission and objectives, (c) developing strategies and action plans, (d) monitoring process and outcome, (e) communicating information to relevant audiences, and (f) promoting adaptation, renewal, and institutionalization. Finally, we conclude with a discussion of the challenges and opportunities of empowerment evaluation.

Framework and Context for Empowerment Evaluation

This section outlines a conceptual framework used to guide our efforts to empower community initiatives through evaluation. We also describe the community health initiatives that serve as the context for using this methodology for empowerment evaluation. These serve as the basis for the several case studies of the process of empowerment evaluation that follow.

FRAMEWORK FOR EMPOWERMENT EVALUATION

Guided by models of community health and development (Fawcett et al., 1993) and enabling systems for community empowerment (Fawcett, Paine-Andrews, et al., 1995), we outline a framework for empowerment evaluation. Table 8.1 displays this framework and its four distinct elements: (a) agenda setting (assessing community concerns and resources), (b) planning (establishing or setting the mission, objectives, strategies, and action plans), (c) implementation (facilitating and monitoring processes and outcomes), and (d) outcome (documenting community competence and community outcomes). Enabling activities of the support team permeate every element of this conceptual framework—agenda setting, strategic planning, implementation, and outcome. Beginning with agenda setting, an outline of each follows.

Agenda Setting. Controlling the agenda—determining the problems and solutions for consideration and possible action—is the most potent form of citizen participation (Cobb & Elder, 1972). Community initiatives set the agenda by assessing community concerns and needs, and available resources for addressing them. The community's chosen agenda will create the context for its more specific planning efforts. To maximize community control of the initiative, the support team may prompt and provide technical assistance with assessing community concerns and resources, and gathering epidemiological data related to the identified concerns. Assessments of concerns and resources may include informal listening sessions as well as community surveys and community forums (Paine, Francisco, & Fawcett, 1994). Such agenda-setting efforts are often community-wide, with emphasis on involving underrepresented and marginalized parts of the community (Fawcett, Seekins, Whang, Muiu, & Suarez-Balcazar, 1982).

Planning. Planning is essential for an initiative to reach its goals, providing the grounding for the evaluation and related enabling

TABLE 8.1 Framework for Empowerment Evaluation and Related Enabling
Activities

Component	Related Enabling Activities
Agenda Setting	1. Assessing community concerns and resources
	2. Collecting epidemiological data (i.e., incidence and prevalence of identified problems)
Planning	3. Facilitating development of a vision, mission(s), objectives, and strategies
	4. Helping develop an action plan that specifies changes in programs, policies, and practices to be sought in relevant sectors of the community
Implementation	5. Monitoring and providing feedback on process and outcome (e.g., rate of community change)
	6. Helping communicate information to relevant audiences (e.g., grantmakers)
Outcome	7. Assessing community competence (e.g., in implementing interventions; evaluation)
	8. Detecting community outcome (e.g., community-level indicators of impact)
	9. Collecting qualitative information (e.g., about critical events and their meaning)
	10. Assessing adaptation, renewal, and institutionalization of the initiative and successful components

activities. Enabling activities may focus on creating a vision, setting
(or adapting) the mission, establishing objectives, and developing
strategies and action plans. With extensive community input, action
plans specify the changes in programs, policies, and practices to be
sought in relevant sectors of the community such as schools or
religious organizations (see, e.g., Fawcett, Francisco, Paine-Andrews,
Fisher et al., 1994).

Implementation. Empowerment evaluation should also facilitate
implementation of the initiative's plan for achieving intended outcomes. Monitoring processes and outcomes and arranging ongoing
feedback are critical enabling activities of the support team that can
empower the initiatives by allowing participants to make more informed decisions and identify where to focus efforts for maximum impact
(Francisco, Paine, & Fawcett, 1993). The support team also assists in
communicating information about the initiative's accomplishments (and

challenges) to relevant audiences, including community members, trustees, and current and prospective grantmakers.

Outcome. The community initiatives with which we collaborate strive to (a) enhance community competence to address health issues, (b) have a positive impact on health outcomes, and (c) create a structure that allows for adaptation, renewal, and institutionalization of the initiative. For example, collaboration with the original Project Freedom, a nationally recognized substance abuse coalition in Wichita, Kansas, resulted in enhanced competence in conducting interventions such as citizen-led "stings" on merchants who sold alcohol to minors (Lewis et al., 1994). The coalition facilitated community changes and improved community-level indicators related to the mission, adapting its strategic plan to reflect new opportunities and a change in leadership (Fawcett, Lewis, et al., 1994).

Enabling Activities of the Support Team. To maximize resources for capacity building, we join the often separated roles of technical assistance and evaluation in an integrated support system. The support team engages in a number of technical assistance and evaluation activities to enhance the capacity of community initiatives to influence valued outcomes throughout the development of the initiative. To enhance experience and competence of members of the community initiative, such as in collecting evaluation data, the support team may use informational materials, on-site consultation, or other modes of assistance. To enhance group structure and capacity, the support team may assist in developing a strategic plan, for example, using a planning retreat, regular on-site assistance, or other forms of facilitation. The support team may also enhance coalition capacity by conducting intervention research examining the effectiveness of specific interventions, such as a peer support program, using the results to strengthen the initiative (Fawcett, Suarez-Balcazar, et al., 1994). To remove social and environmental barriers to achieving intended outcomes, such as to building evaluation capacity, the support team or granting agency may provide computer systems to enable better and more efficient communication between sites and evaluators. Finally, to enhance environmental support and resources, the evaluation and support team may assist with obtaining grants and other potential resources and linking the initiative with local, state, and national informational resources.

The ongoing support provided to leadership and staff is often dynamic and individualized. In addition to providing information and skills building on topics that have been critical to previously successful endeavors, such as strategic planning, the support team responds to the needs of each leader and group. For example, a leader may schedule working phone calls with support team members to complete tasks she finds daunting, such as writing action plans. The support team listens to and learns from leaders, groups, and community members. Reciprocal relationships are formed. This type of support empowers leaders and initiatives to maintain progress toward achieving objectives consistent with their mission.

These elements—agenda setting, planning, implementation, outcome, and related enabling activities—outline a framework for empowerment evaluation. The process is *interactive:* The initiative's objectives affect the evaluation system, and data from a system for monitoring coalition activities and outcomes may suggest the need to modify the action plan. The process is also *iterative:* Success (or challenges) in attaining outcomes should influence the enabling activities of the support team, and enabling activities should contribute to agenda setting, planning, implementation, and outcome. Collaboration among community leadership, the support and evaluation team, and grantmakers should contribute to the mission and to the adaptation, renewal, and institutionalization of the initiative.

COMMUNITY HEALTH INITIATIVES AS CONTEXT

We have applied our methodology for empowerment evaluation with 17 different community initiatives for health and development (Fawcett et al., 1993; Fawcett, Lewis, et al., 1994; Francisco et al., 1993). Two case studies provide the context for this report: community coalitions for prevention of adolescent pregnancy and substance abuse in Kansas and a tribal partnership for prevention of substance abuse in New Mexico.

Community Coalitions for Prevention of Adolescent Pregnancy and Substance Abuse. Three Kansas communities—one rural (Franklin

County), one urban (northeast Wichita), and one military (Geary County)—received a grant from the Kansas Health Foundation to replicate a community-wide approach to prevent adolescent pregnancy first conducted in South Carolina (Koo, Dunteman, George, Green, & Vincent, 1994; Vincent, Clearie, & Schluchter, 1987). The three sites shared the mission of involving a variety of sectors of the community such as schools, health organizations, and religious organizations in modifying programs, policies, and practices related to reducing risk for adolescent pregnancy (Paine-Andrews et al., 1994). Their main objectives were to reduce unwanted pregnancies among adolescents, postpone the age of first intercourse, increase abstinence, and, for those who chose to be sexually active, increase use of contraceptives.

Three Kansas communities—one military/prison (Leavenworth), one university (Lawrence), and one military (Geary County)—also received a grant from the Kansas Health Foundation. This was to develop and implement community coalitions for prevention of adolescent substance abuse modeled after Project Freedom of Wichita, Kansas. The three sites shared the primary mission of reducing substance use among adolescents aged 12 to 17 years. Each coalition involved key influentials and grassroots groups from different community sectors including law enforcement, schools, health, and businesses in facilitating community change related to the prevention of substance use among adolescents.

Tribal Partnership for Prevention of Substance Abuse. The Decade of Hope Coalition was developed by several members of the Jicarilla Apache tribe on their reservation in Dulce, north-central New Mexico. The Jicarilla Apache number about 3,500 and live on approximately 1 million acres of land in the San Juan Mountain range. Although the community is traditionally agrarian (raising sheep, horses, and cattle), there is currently a very high level of unemployment, with the tribe being virtually the only employer. After oil and natural gas were found on the reservation in the early 1960s, funds from their sale were distributed to tribal membership, with a resulting decline in local agriculture. Many of the tribal elders have died from alcoholism, often from exposure during the long winters.

The coalition began its work in 1989, following a rash of suicides among young tribal members. Mothers of these teens and young adults decided that something needed to be done about the problem beyond the limited efforts of the Tribal Council. We began our involvement with this coalition near the end of 1991 when we were contacted by them to help with evaluation for a successful grant proposal to the Community Partnership Program of the U.S. Center for Substance Abuse Prevention (CSAP).

Process of Empowerment Evaluation

This section describes the process used to empower community initiatives through evaluation. Following Fetterman (1994b) and the framework described above, we outline six elements of the process of empowerment evaluation: (a) assessing community concerns and resources; (b) setting a mission and objectives; (c) developing strategies and action plans; (d) monitoring process and outcome; (e) communicating information to relevant audiences; and (f) promoting adaptation, renewal, and institutionalization. This is an interactive and iterative process by which the community, in collaboration with the support team, identifies its own health issues, decides how to address them, monitors progress toward its goals, and uses the information to adapt and sustain the initiative. Each aspect is described and illustrated with case examples.

ASSESSING COMMUNITY CONCERNS AND RESOURCES

In empowerment evaluation, the support team assists local initiatives in the initial and ongoing task of assessing local concerns and resources for change. Listening to community concerns should precede taking action; securing community input, ownership, and involvement is critical to sustaining initiatives. Support teams provide workshop sessions and on-site consultation to build the capacity of staff, leadership, and volunteers to hold listening sessions. Support teams also provide assistance with more formal needs assessments and inventories of community strengths and resources.

Case Study With Kansas Coalitions. Each local coalition for preventing adolescent substance abuse and pregnancy conducted listening

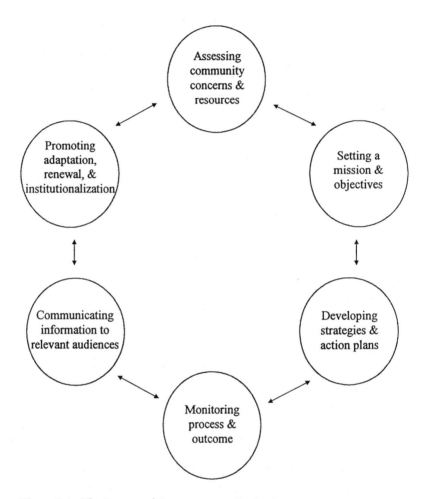

Figure 8.1. The Process of Empowerment Evaluation

sessions to assess community concerns and resources for addressing their missions. The listening sessions consisted of informal public meetings in which participants identified (a) problems or issues, (b) barriers or resistance to addressing the problem, (c) resources for change, and (d) potential solutions. The listening sessions were designed to involve key leaders, people affected by the problem, and people who could contribute to addressing the problem throughout all sectors of the community. For example, the listening sessions for the inner-city pregnancy prevention initiative in northeast Wichita involved grassroots leaders and people of color from this predominantly African American community. This included religious leaders, youth, parents, teachers, health officials, representatives from informal neighborhood groups, and other community organizations. Listening sessions continued to be held throughout the initiative to meet empowerment aims of (a) maintaining community involvement in setting (and adapting) the goals and objectives and (b) attracting volunteers to help implement the action plan.

Case Study With a Tribal Coalition. The coalition began its work with a social reconnaissance (or information-gathering process) conducted by local residents in conjunction with representatives from the U.S. Indian Health Service. This process involved conducting meetings with all tribal offices, as well as general meetings with community members. It resulted in identifying a number of goals and objectives for change. Members for the emerging Decade of Hope Coalition were solicited from the more vocal persons attending the town meetings, and they began to set priorities for action. With a grant from the Community Partnership Program of CSAP, a community mobilizer was hired from outside the community who took responsibility for implementing the objectives identified in the grant and mobilizing support for those activities in the community.

SETTING A MISSION AND OBJECTIVES

Although the mission of a community initiative may be largely defined by the granting agency, its form should be tailored to the community's own unique vision and circumstances. The support team assists community initiatives in identifying or adapting the mission and objectives. For example, a substance abuse initiative might modify its mission and objectives to include prevention of gang-related activity; an adolescent pregnancy initiative might do the same to include reducing risk for HIV/AIDS or sexual violence. Support teams facilitate workshop sessions with staff and membership of initiatives and provide on-site consultation to review and, if necessary, adapt these aspects of the strategic plan.

Case Study With Kansas Coalitions. For the community coalitions to prevent adolescent substance abuse and adolescent pregnancy, the mission and objectives were predefined by the funding agent, the Kansas Health Foundation. The initiatives were replications of previous community initiatives that showed some success in reducing adolescent pregnancy (Vincent et al., 1987) and community-level indicators related to substance abuse (Fawcett, Lewis, et al., 1994). Each grantee agreed to adopt the mission and objectives of the initiatives upon acceptance of the grant award. As the coalitions mature, the mission statements may be modified to embrace emerging community issues. For example, following a shooting in one community, the coalition modified its objectives and action plan to reflect community concerns about youth violence and associated substance use.

Case Study With a Tribal Coalition. The Decade of Hope Coalition emerged in response to community concern about suicides among youth of the Jicarilla Apache tribe. With support from CSAP and community concern about alcoholism and related issues, the coalition expanded its embryonic mission to include prevention of substance abuse. Its specific objectives, such as to establish a reward system for tips leading to arrests for drug sales and bootlegging, flowed from this mission. Information from community-wide surveys and listening sessions helped initiatives to modify established missions and objectives and adopt new ones.

DEVELOPING STRATEGIES AND ACTION PLANS

An important task for a community initiative is to develop strategies—the general approaches, such as coalition building or advocacy, by which it achieves its mission. Action planning, that is, identifying specific community changes to be sought in each relevant sector of the community, may be particularly critical to success. We developed action planning guides to support this process with community health initiatives, including those for prevention of substance abuse (Fawcett, Paine-Andrews, Francisco, Richter, Lewis, Williams, et al., 1994), adolescent pregnancy (Fawcett, Paine-Andrews, Francisco, Richter, Lewis, Harris, et al., 1994), youth violence (Fawcett, Francisco, Paine-Andrews, Fisher et al., 1994), and chronic disease (Fawcett,

Harris, et al., 1995). These planning guides enable community initiatives to identify changes in programs, policies, and practices to be sought in schools, businesses, and other relevant sectors that are consistent with known or suspected risk factors for the concern.

Case Study With Kansas Coalitions. The strategies for both initiatives were, in part, predefined; the initiatives replicated a model that emphasized coalition building as the dominant strategy. Both initiatives adopted a community coalition approach and involved key influentials and grassroots leadership from different community sectors such as schools, religious organizations, and law enforcement (for the substance abuse initiative). The adolescent pregnancy prevention initiatives, however, also were required to put in place several program components that were part of the replication effort (Paine-Andrews et al., 1994). These included a K-12 comprehensive and age-appropriate sexuality education curriculum, graduate training for teachers in sexuality education, parent-child communication about sexual issues, and enhanced access to contraceptives. Strategic planning emphasized how to get the core components in place, including dealing with opposition from those who do not support sexuality education or enhanced access to contraceptives.

With support from the Work Group, each site developed action plans specific to the community, the mission, and the core program components (if appropriate). The action plans listed proposed changes in programs, policies, and practices to be sought in each sector. For example, for an adolescent pregnancy prevention initiative, the following change was proposed for the school sector: "By September 1994, adopt and implement a comprehensive K-12 age-appropriate sexuality education curriculum." We used surveys inviting community input on the importance and feasibility of proposed changes to build consensus and set priorities for the action plan. The action plan also noted the specific steps needed to create the change, including who would do what by when. The action plans were used by coalition staff, leadership, and volunteers to keep them focused on working toward their mission. The action plans also contributed to the evaluation, identifying potential changes in the environment that served as early markers for the success of these prevention initiatives. Outside of the proscribed core components, coalitions were free to determine their own strategies and community changes for reducing pregnancy and substance abuse among adolescents.

Case Study With a Tribal Coalition. We collaborated with coalition staff to involve the Dulce community in more formal strategic planning, including identifying community changes to be sought and action steps for each objective. The staff, largely members of the tribal community, led a process that was sensitive to their predominantly Jicarilla Apache culture. Coalition staff first identified appropriate community sectors, such as the Tribal Courts, schools, and health services, through which the coalition could address its mission. They then used their experiential knowledge and the planning guide for substance abuse prevention to identify potential changes to be sought in programs (e.g., develop a wilderness experience program for at-risk youth), policies (e.g., establish a preferential hiring policy for tribal members for an oil and natural gas production facility), and practices (e.g., enforce mandatory age checks for purchasing tobacco and alcohol). After an initial draft of the action plan, staff worked with the community in identifying additional changes to be sought in all relevant sectors. Through this process, the coalition gained further acceptance as a catalyst for change in the community. Evidence of its enhanced capacity was an invitation to assist the Tribal Council in developing their strategic plan for dealing with alcohol and suicide problems in tribal government and the community.

MONITORING PROCESS AND OUTCOME

Our Work Group's evaluation system is used to help understand and improve how community health initiatives engage the environment and facilitate intermediate and ultimate outcomes related to the mission (Fawcett, Francisco, Paine-Andrews, Lewis, et al., 1994). Core components of the measurement system include (a) a monitoring system to assess process and intermediate outcomes (Francisco et al., 1993), (b) constituent surveys of process and outcome, (c) behavioral surveys, (d) measures of community-level indicators such as estimated pregnancy rate (for adolescent pregnancy initiatives), and (e) interviews with key participants to obtain qualitative information about critical events. All measures are refined, collected, and interpreted in collaboration with staff and leadership of participating community initiatives.

Case Study With Kansas Coalitions. Both initiatives used logs to keep track of how well the initiatives were being implemented and their accomplishments. Process measures included units of service provision and actions taken to create or modify programs, policies, and practices related to the mission. The primary measure of intermediate outcome was community change: new or modified programs (e.g., peer support groups), policies (e.g., a no-smoking ordinance), or practices (e.g., expanded clinic hours) related to the mission. In addition, surveys were used to obtain data on reported behaviors related to the mission such as use of tobacco and alcohol (for substance abuse coalitions) and age of first intercourse and use of contraceptives (for adolescent pregnancy coalitions). Archival records are used to assess community-level indicators, such as estimated pregnancy rate (for adolescent pregnancy) and nighttime single-vehicle crashes (for substance abuse).

The monitoring information was collected by coalition staff and sent to the evaluators. The evaluators coded and summarized the data and fed them back to staff and leadership on a regular basis. The monitoring system allowed initiatives to track activities and outcomes related to their mission, and provided a record of key actions taken to implement a particular outcome or community change. For example, one of the substance abuse coalitions used the monitoring data on community change to inform a new community mobilizer about previous coalition activities and outcomes. The data also enabled the coalitions to be accountable to their funding agents and the community. Because the data provide evidence of accomplishment, they can be used by the initiative staff to secure additional resources. Finally, as part of the participatory or empowerment evaluation, some tailoring of the monitoring system took place to meet the needs of individual coalitions better.

The behavioral surveys were collected by coalition staff or school officials and summarized by the evaluators. Data about the level of early sexual activity, for example, were used to help increase awareness about the problem and show improvement toward the coalition's broad health objectives. In response to community concerns, the proposed behavioral survey for the pregnancy prevention initiative was replaced with a more palatable survey that minimized explicit references to anatomy and put sexual risk in the context of other risky behavior of adolescents. This substitution of a more acceptable survey illustrates the interactive nature of empowerment evaluation.

Case Study With a Tribal Coalition. Early in our work with the Decade of Hope Coalition, we agreed to provide training and support for coalition staff and leadership in methods of community development and evaluation. Some of the evaluation budget was returned to the coalition to hire a tribal member to serve as on-site research coordinator. This local evaluator collected community-level impact data, such as emergency medical transports, and was the main contact person for completing logs for tracking process and outcome measures, such as community actions to change local policies, and community changes, such as change in the policy of a local bar to ban customers involved in fights. This monitoring system, and interviews with key members of the coalition and community, helped document the local nature and meaning of coalition process and outcomes.

COMMUNICATING INFORMATION TO RELEVANT AUDIENCES

Regularly sharing accomplishments and keeping constituents informed of progress are important to maintaining community support, obtaining additional resources, and ensuring accountability. Support teams provide data reports and training to enable coalition leadership and staff to communicate their data to coalition membership, boards of directors, current and prospective funding agents, and other important constituents.

Case Study With Kansas Coalitions. Data were shared with coalition leadership, grantmakers, and the community at large. The Kansas coalitions regularly provided data in reports to their primary funding agent, the Kansas Health Foundation. It was especially important to communicate data to the foundation given that annual renewal of grant funds was contingent on evidence of progress. The data on community change served to demonstrate progress, providing early indication of coalition success.

Coalition leadership, with support from the evaluation team, also reported the data to steering committees and coalition members. Keeping the coalition membership informed of progress helped to recognize contributions of members and other volunteers as well as maintain focus on and momentum for community change. Sharing data with committees and the community contributed to accountability to local constituents. Several coalition leaders presented their work at regional conferences and national conventions. This provided them with the opportunity to share their progress and to network with other initiatives with similar missions. Establishing networks and sharing information about the accomplishments of the initiative at local, state, and national levels are important for securing resources and commitments to sustain the initiative.

Case Study With Tribal Coalition. Feedback was provided to coalition leadership and membership in the form of cumulative graphs documenting the number of actions taken and community changes facilitated each month. Accompanying these graphs were printed reports itemizing those actions and accomplishments. In addition to monthly reports, quarterly reports were generated for the funding agent (CSAP). We discussed the evaluation data with staff, and trained them in how to use the data themselves during working sessions in Dulce. Data were used in subsequent grant applications to the Indian Health Service, Kellogg Foundation, and Robert Wood Johnson Foundation. Data in the hands of staff and coalition members proved to be instrumental in obtaining an award for excellence from the Indian Health Service and additional financial and community support.

PROMOTING ADAPTATION,
RENEWAL, AND INSTITUTIONALIZATION

In the life span of community initiatives, adaptation and renewal may be necessary to address a variety of predictable changes, including those in leadership, goals and objectives, and community conditions and concerns. Institutionalization of valued components of the initiative, including evaluation, may also be important to community initiatives. Support teams facilitate training and provide regular consultation to this end, but the ultimate success may be unknown for years after the evaluation.

Case Study With Kansas Coalitions. The monitoring data helped the coalition recognize accomplishments and redirect energies when necessary. For example, for one coalition for prevention of substance abuse, a high level of service provision and low levels of community action and change indicated to leadership, staff, and the funding agent that the coalition was becoming a service agency rather than a catalyst for community change. These data helped redirect the energies of coalition staff, leadership, and members away from service provision and toward creating community change. This adaptation was important and empowering because prevention initiatives may be more effective as catalysts for change than as new service agencies.

Changing leadership in another coalition for prevention of substance abuse required some reenergizing and refocusing of coalition members. The evaluator for this site worked closely with new leadership and the steering committee to share with them the intended purpose of the coalition, that is, to serve as a catalyst for change. This renewal of the coalition was necessary to reestablish and maintain its focus on facilitating community change related to the mission.

Similarly, the evaluation team provided training and consultation in strategies for promoting institutionalization of the initiative and its successful programs, such as securing personnel positions in city or county budgets or developing leadership in collaborating organizations. Initiatives gave consideration to sustainability as they designed project components. For example, one pregnancy prevention coalition used local data about the level of the problem to help initiate a new group at a local college that would assist in preventing unwanted pregnancies among college students and adolescents.

Case Study With a Tribal Coalition. The monitoring system, with regular reporting about key events, allowed the funding agents, staff, and local influential persons to keep track of and adjust activities of the coalition. Staff and coalition members were able to detect the immediate effects of their actions, such as a new program or policy change, and could suggest adjustments that would improve implementation. This process engaged all parties in the evolving process of evaluation and community development.

Major and repeated turnover in virtually all leadership and evaluation positions made renewal and institutionalization particularly critical issues with the Decade of Hope Coalition. Perhaps technical assistance with new staff, particularly with action planning and monitoring, would be helpful in coalition renewal and institutionalization of valued community changes.

Challenges for Empowerment Evaluation

There are a number of significant challenges to the practice of evaluation in service of empowerment. First, the ambiguousness of the construct of "empowerment evaluation" may make it difficult to detect good practice. Empowerment remains a vague concept, referring to both a process and a goal (Swift & Levin, 1987; Zimmerman, in press). Outcomes of empowerment evaluation, such as community competence to conduct evaluations, and community outcomes such as securing grants with evaluation data, may be difficult to attribute solely to the process of empowerment evaluation rather than empowerment evaluation and other intervention processes. Conceptual analyses of the process of empowerment evaluation, such as those offered in this book, may help explicate this important construct.

Second, it may be difficult to optimize the traditional goals of evaluation and empowerment in the same endeavor. As Stufflebeam (1994) noted in a critique of empowerment evaluation, the traditional goal of evaluation (Joint Committee, 1994) is "the systematic investigation of the worth or merit of an object" (p. 323). Empowerment evaluation offers a seemingly orthogonal goal: to promote self-determination. Although these complementary goals may not be easily maximized, creative applications of this process may suggest how the goals of assessing and contributing to worth can be optimized. The relationship has greater clarity when evaluation is viewed as the tool by which self-determination is fostered at every stage (Fetterman, 1995).

Third, empowerment evaluation, like other approaches to evaluation, must protect itself against charges of misuse (Stufflebeam, 1994). When involving community members and other stakeholders in the evaluation, we must acknowledge potential for bias and conflict of interest. To promote such collaborations is not to abdicate responsibility for assessing merit and worth; it is merely to share that duty with those most affected by the outcome. Evaluators can (and should) collaborate with key stakeholders, including community members and grantmakers, in setting the agenda for research, collecting and interpreting data, and communicating the findings to interested audiences. Evaluations must also meet the field's standards for propriety and accuracy when communicating findings to professional and other audiences (Joint Committee, 1994). Consistent with the aims of action science (Argyris, Putnam, & Smith, 1985), appropriate collaborations among evaluators and stakeholders can optimize the potential contributions of evaluation to both understanding and improving community initiatives.

Fourth, empowerment evaluation must also guard against potential confusion resulting from conflicting interpretations from various sources (Stufflebeam, 1994). As with qualitative evaluation (Fetterman, 1988, 1989; Guba & Lincoln, 1989; Patton, 1980) and other forms of relativistic evaluation, stakeholders contribute to the process of assessing merit. Although expanding participation always increases risk for conflict, it does not ensure it. If consensus on assessments of merit is valued, cases of differing interpretations of results could enable us to better understand the values that undergird judgments of merit. Such case studies may enable us to understand how evaluation methodologies can foster common assessment (and attainment) of worth.

Fifth, as with other approaches to evaluation, those that would promote self-determination must meet standards for feasibility (Joint Committee, 1994). Increased requirements for effort by either evaluators or communities could limit the practicality of implementation. For example, feeding back information about progress and providing training in communicating the data require time from evaluators. Similarly, if collaborating with evaluators diverts attention from the initiative's efforts to facilitate change, it may properly be viewed as another "unfunded mandate." When empowerment evaluations balance the interests of both evaluators and community members, they may be more likely to be judged as having met standards for feasibility and utility.

Benefits of Empowerment Evaluation

There are also a number of opportunities and benefits from evaluation in service of empowerment. First, empowerment evaluation encourages the creative coupling of technical assistance and evaluation. Because empowerment evaluation is designed to foster self-determination, it is essential that community stakeholders understand and can apply the methods. Enabling activities, such as technical assistance and training, are expected features of an integrated set of support activities. For example, in a more traditional paradigm, workshops in strategic planning or communicating evaluation results might be delegated to an outside contractor providing technical assistance. Because building capacity is among the primary ends of empowerment evaluation, an integrated support system, combining the functions of assessment of merit *and* technical assistance, is a more efficient and appropriate design.

Second, empowerment evaluation may enhance integration of qualitative and quantitative methods. One way in which evaluation processes extend community influence over the initiative is by inviting community members and other stakeholders to assess the value and accomplishments of the initiative. By integrating such qualitative information with quantitative data on accomplishments, we have begun to identify factors, such as action planning or monitoring and feedback, that may affect the functioning of community health initiatives. Such integration of qualitative and quantitative data may enhance the capacity of community initiatives to affect valued outcomes.

Third, capacity-building approaches may help demystify the process of evaluation. Collaboration increases ownership of the evaluation process, making evaluation practices an integral part of leadership activities. For example, community leaders help gather data on community change for the monitoring system and assess its significance for the mission using constituent surveys. Once initiative leadership become familiar with evaluation and recognize its empowerment capacity, they are more likely to understand the data and communicate the findings to relevant audiences.

Fourth, empowerment evaluation supports reinvention of evaluation methods and instruments. In collaboration with evaluators, initiative leadership rejected some proposed assessments as not feasible or useful, specified additional measures to be collected, modified measurement instruments to be used, and adjusted time lines for implementation of evaluation instruments. For example, coalition leadership associated with the Jicarilla Apache tribe rejected traditional survey methods as being too invasive, and staff and leadership of one of the pregnancy prevention initiatives modified and extended the time line for a consumer satisfaction survey to suit their needs better. When other evaluation standards are upheld, reinvention may enhance the value of self-determination while strengthening prospects for continued use of the evaluation methods after the evaluation team is gone.

Finally, a degree of self-determination in evaluation may promote institutionalization of the evaluation methods. Strategies for promoting institutionalization of evaluation include (a) promoting awareness of the value or need for evaluation data (Eng & Parker, 1994; Lefebvre, 1990; Mittelmark, 1990), (b) encouraging participation in developing the research goals and methodology (Eng & Parker, 1994; Fawcett, 1991; Fetterman, 1994a), (c) building community competence in designing and conducting the evaluation (Eng & Parker, 1994; Fawcett, 1991; Fetterman, 1994a), (d) incorporating evaluation into the structure of the initiative (Eng & Parker, 1994; Fetterman, 1994a, 1994b; Lefebvre, 1990; Price & Lorion, 1989), (e) providing needed resources such as people and materials, and (f) securing champions or change agents within the organization who will take responsibility for the evaluation (Lefebvre, 1990; Rogers, 1983; Seekins & Fawcett, 1984). Such empowerment strategies may facilitate long-term use of functional evaluation methods.

Conclusion

This chapter described the use of a framework and process for empowerment evaluation. The aim of this evaluation is to build community competence; optimize community outcomes; and promote adaptation, renewal, and institutionalization of community health initiatives. We illustrated this approach with case studies involving community initiatives to prevent substance abuse and adolescent pregnancy in Kansas and a tribal partnership to prevent substance abuse in New Mexico.

Although we have considered how to enhance the practice of empowerment evaluation, questions of *who* should be empowered and *for what ends* are largely questions of philosophy and ethics (Fawcett et al., 1982; Fetterman, 1994a). For example, evaluations of community initiatives to prevent youth violence may maximize benefits for at-risk youth and their families, or those who fear or resent them. Several questions may help clarify ethical issues in empowerment evaluation: (a) What are the vision, mission, and/or goals of the initiative? (b) Who experiences the health or social concern to be addressed? (c) Are those experiencing the problem among the primary beneficiaries of the initiative? (d) Are those affected by the problem involved in implementing the initiative? (e) Does the evaluation contribute to the community's capacity to address these and other concerns? (f) Does the evaluation contribute to a reduction in problems and other outcomes of importance to the community? Attention to these ethical issues may help maximize benefits for those whom evaluators would enable to help themselves.

Empowerment evaluation with community initiatives poses an apparent paradox: How do we simultaneously maximize the competing ends of community control *and* understanding of the processes and outcomes of community initiatives? For example, community input and control increase the number of stakeholders and potential competing interests that affect the evaluation. This can create confusion in the implementation of the evaluation and in the interpretation of evaluation results.

Maximizing community control and understanding may be precisely the sort of divergent problem that calls for apparently contradictory solutions (Fawcett, Mathews, & Fletcher, 1980; Fawcett et al., 1984; Rappaport, 1981; Schumacher, 1977). Empowerment evaluation optimizes both self-determination and understanding—even if it may not maximize each of these valued ends. Moreover, the varied forms of action science spawned by these efforts are challenging scientists and practitioners to reconsider how best to serve methodological rigor and relevance (Argyris & Schön, 1991; Fawcett, 1991). Such innovation should be good for action scientists, community practitioners, and the marginalized people and communities who are the intended beneficiaries.

References

Argyris, C., Putnam, R., & Smith, D. M. (1985). *Action science*. San Francisco: Jossey-Bass.

Argyris, C., & Schön, D. A. (1991). Participatory action research and action science compared. In W. F. Whyte (Ed.), *Participatory action research* (pp. 85-96). Newbury Park, CA: Sage.

Braithwaite, R., & Lynthcott, N. (1989). Community empowerment as a strategy for health promotion for black and other minority populations. *Journal of the American Medical Association, 261*(2), 282-283.

Choudhary, A., & Tandon, R. (1988). *Participatory evaluation.* New Delhi, India: Society for Participatory Research in Asia.

Cobb, R. B., & Elder, C. D. (1972). *Participation in American politics: The dynamics of agenda-building.* Baltimore, MD: John Hopkins University Press.

Davidson, L. L., Durkin, M. S., Kuhn, L., O'Connor, P., Barlow, B., & Heagarty, M. C. (1994). The impact of the Safe Kids/Healthy Neighborhoods Injury Prevention Program in Harlem. *American Journal of Public Health, 84,* 580-586.

Eng, E., & Parker, E. (1994). Measuring community competence in the Mississippi delta: The interface between program evaluation and empowerment. *Health Education Quarterly, 21*(2), 199-220.

Eng, E., Salmon, M. E., & Mullan, F. (1992). Community empowerment: The critical base for primary health care. *Family and Community Health, 15,* 1-12.

Falco, M. (1992). *The making of a drug-free America.* New York: Time Books.

Fawcett, S. B. (1991). Some values guiding community research and action. *Journal of Applied Behavioral Analysis, 24,* 621-636.

Fawcett, S. B., Francisco, V. T., Paine-Andrews, A., Fisher, J. L., Lewis, R. K., Williams, E. L., Richter, K. P., Harris, K. J., Berkley, J. Y., Oxley, L., Graham, A., & Amawi, L. (1994). *Preventing youth violence: An action planning guide for community-based initiatives.* Lawrence: University of Kansas, Work Group on Health Promotion & Community Development.

Fawcett, S. B., Francisco, V. T., Paine-Andrews, A., Lewis, R. K., Richter, K. P., Harris, K. J., Williams, E. L., Berkley, J. Y., Schultz, J. A., Fisher, J. L., & Lopez, C. M. (1994). *Work group evaluation handbook: Evaluating and supporting community initiatives for health and development.* Lawrence: University of Kansas, Work Group on Health Promotion & Community Development.

Fawcett, S. B., Harris, K. J., Paine-Andrews, A. L., Richter, K., Lewis, R., Francisco, V., Arbaje, A., Davis, A., Cheng, H., & Johnston, J. (1995). *Reducing risk for chronic disease: An action planning guide for community-based initiatives.* Lawrence: University of Kansas, Work Group on Health Promotion & Community Development.

Fawcett, S. B., Lewis, R. L., Paine-Andrews, A., Francisco, V., Richter, K., Williams, E., & Copple, E. (1994). *Evaluating community coalitions for the prevention of substance abuse: The case of Project Freedom.* Manuscript submitted for publication.

Fawcett, S. B., Mathews, R. M., & Fletcher, R. K. (1980). Some promising dimensions for behavioral community technology. *Journal of Applied Behavior Analysis, 13*(3), 505-518.

Fawcett, S. B., Paine, A. L., Francisco, V. T., Richter, K. P., & Lewis, R. K. (1994). *Conducting preventive interventions for community mental health* (Commissioned paper for the Committee on Prevention and Mental Disorders, Institute of Medicine). Washington, DC: National Academy of Sciences.

Fawcett, S. B., Paine-Andrews, A. L., Francisco, V., Richter, K., Lewis, R., Harris, K., & Williams, E. (1994). *Preventing adolescent pregnancy: An action planning guide for community-based initiatives*. Lawrence: University of Kansas, Work Group on Health Promotion & Community Development.

Fawcett, S. B., Paine-Andrews, A., Francisco, V., Richter, K., Lewis, R., Williams, E., Harris, K., & Winter-Green, K. (1994). *Preventing adolescent substance abuse: An action planning guide for community-based initiatives*. Lawrence: University of Kansas, Work Group on Health Promotion & Community Development.

Fawcett, S. B., Paine-Andrews, A., Francisco, V., Schultz, J. A., Richter, K., Lewis, R., Williams, E., Harris, K., Berkley, J., Fisher, J., & Lopez, C. (1995). *Using empowerment theory to support initiatives for community health and development*. Manuscript submitted for publication.

Fawcett, S. B., Paine, A. L., Francisco, V. T., & Vliet, M. (1993). Promoting health through community development. In D. Glenwick & L. A. Jason (Eds.), *Promoting health and mental health: Behavioral approaches to prevention* (pp. 233-255). New York: Haworth.

Fawcett, S. B., Seekins, T., Whang, P. L., Muiu, C., & Suarez-Balcazar, Y. (1982). Involving consumers in decision-making. *Social Policy, 13,* 36-41.

Fawcett, S. B., Seekins, T., Whang, P. L., Muiu, C., & Suarez de Balcazar, Y. (1984). Creating and using social technologies for community empowerment. *Prevention in Human Services, 3,* 145-171.

Fawcett, S. B., Suarez-Balcazar, Y., Balcazar, F. E., White, G. W., Paine, A. L., Blanchard, K. A., & Embree, M. G. (1994). Conducting intervention research: The design and development process. In J. Rothman & E. J. Thomas (Eds.), *Intervention research: Design and development for human service* (pp. 25-54). New York: Haworth.

Fawcett, S. B., White, G. W., Balcazar, F., Suarez-Balcazar, Y., Mathews, R., Paine, A., Seekins, T., & Smith, J. (1994). A contextual-behavioral model of empowerment: Case studies involving people with disabilities. *American Journal of Community Psychology, 22*(4), 471-496.

Fetterman, D. M. (1988). *Qualitative approaches to evaluation in education: The silent scientific revolution*. New York: Praeger.

Fetterman, D. M. (1989). *Ethnography: Step by step*. Newbury Park, CA: Sage.

Fetterman, D. M. (1993). *Speaking the language of power: Communication, collaboration, and advocacy: Translating ethnography into action*. London: Falmer.

Fetterman, D. M. (1994a). Empowerment evaluation. *Evaluation Practice, 15*(1), 1-15.

Fetterman, D. M. (1994b). Steps of empowerment evaluation: From California to Cape Town. *Evaluation and Program Planning, 17*(3), 305-313.

Fetterman, D. M. (1995). In response to Dr. Daniel Stufflebeam's: "Empowerment evaluation, objectivist evaluation, and evaluation standards: Where the future of evaluation should not go and where it needs to go." *Evaluation Practice, 16*(2), 321-338.

Francisco, V. T., Paine, A. L., & Fawcett, S. B. (1993). A methodology for monitoring and evaluating community health coalitions. *Health Education Research: Theory and Practice, 8,* 403-416.

Freire, P. (1970). *Pedagogy of the oppressed*. New York: Herder & Herder.

Giroux, H. (1983). *Theory and resistance in education: A pedagogy for the opposition*. South Hadley, MA: Bergin & Garvey.

Green, L. W., & Kreuter, M. W. (1991). *Health promotion planning: An educational and environmental approach* (2nd ed.). Mountain View, CA: Mayfield.

Guba, E., & Lincoln, Y. (1989). *Fourth generation evaluation*. Newbury Park, CA: Sage.

Joint Committee on Standards for Educational Evaluation. (1994). *The program evaluation standards.* Thousand Oaks, CA: Sage.

Koo, H., Dunteman, G., George, C., Green, Y., & Vincent, M. (1994). Reducing adolescent pregnancy through a school and community-based intervention: Denmark, South Carolina, revisited. *Family Planning Perspectives, 26,* 206-211, 217.

Lefebvre, R. C. (1990). Strategies to maintain and institutionalize successful programs: A marketing framework. In N. Bracht (Ed.), *Health promotion at the community level* (pp. 209-228). Newbury Park, CA: Sage.

Lewis, R. K., Fawcett, S. B., Coen, S., MacDonald, P., Jecha, L., Bell, D., Pippert, K., & Foote, C. (1994, November). *Using research information to influence public policy regarding sales of tobacco products to minors.* Paper presented at American Public Health Association Convention, Washington, DC.

Lichtenstein, B. (1988). Feminist epistemology: A thematic review. *Thesis Eleven, 21,* 140-151.

Minkler, M. (1992). Community organizing among the elderly poor in the United States: A case study. *International Journal of Health Services, 22,* 303-316.

Mithaug, D. (1991). *Self-determined kids: Raising satisfied and successful children.* New York: Macmillan.

Mittelmark, M. B. (1990). Balancing the requirements of research and the needs of communities. In N. Bracht (Ed.), *Health promotion at the community level* (pp. 125-139). Newbury Park, CA: Sage.

Mondros, J. B., & Wilson, S. M. (1994). *Organizing for power and empowerment.* New York: Columbia University Press.

Nezlek, J. B., & Galano, J. (1993). Developing and maintaining state-wide adolescent pregnancy prevention coalitions: A preliminary investigation. *Health Education Research: Theory and Action, 8,* 433-447.

Paine, A. L., Francisco, V. T., & Fawcett, S. B. (1994). Assessing community health concerns and implementing a microgrant program for self-help initiatives. *American Journal of Public Health, 84*(2), 316-318.

Paine-Andrews, A., Vincent, M. L., Fawcett, S. B., Campuzano, M. K., Harris, K. J., Lewis, R. K., Williams, E. L., & Fisher, J. L. (1994). *Replicating a community initiative for preventing adolescent pregnancy: From South Carolina to Kansas.* Manuscript submitted for publication.

Patton, M. Q. (1980). *Qualitative evaluation methods.* Beverly Hills, CA: Sage.

Price, R. H., & Lorion, R. P. (1989). Prevention programming as organizational reinvention: From research to implementation. In D. Shaffer, I. Philips, & N. B. Enzer (Eds.) and M. M. Silverman & V. Anthony (Assoc. Eds.), *Prevention of mental disorders, alcohol and other drug use in children and adolescents* (Prevention Monograph No. 2, DHHS Publication No. ADM 89-1646, pp. 97-123). Rockville, MD: Office of Substance Abuse Prevention and American Academy of Child and Adolescent Psychiatry.

Rappaport, J. (1981). In praise of paradox: A social policy of empowerment over prevention. *American Journal of Community Psychology, 9,* 1-25.

Rappaport, J. (1987). Terms of empowerment/exemplars of prevention: Toward a theory for community psychology. *American Journal of Community Psychology, 15*(2), 121-148.

Rappaport, J., Swift, C., & Hess, R. (Eds.). (1984). *Studies in empowerment: Steps toward understanding and action.* New York: Haworth.

Rogers, E. M. (1983). *Diffusion of innovations.* New York: Free Press.

Schumacher, E. F. (1977). *A guide for the perplexed.* New York: Harper & Row.

Seekins, T., & Fawcett, S. B. (1984). Planned diffusion of social technologies for community groups. In S. C. Paine, G. T. Bellamy, & B. Wilcox (Eds.), *Human services that work: From innovation to standard practice* (pp. 247-293). Baltimore, MD: Paul H. Brookes.

Stufflebeam, D. L. (1994). Empowerment evaluation, objectivist evaluation, and evaluation standards: Where the future of evaluation should not go and where it needs to go. *Evaluation Practice, 15*(3), 321-338.

Stull, D., & Schensul, J. (1987). *Collaborative research and social change: Applied anthropology in action.* Boulder, CO: Westview.

Swift, C., & Levin, G. (1987). Empowerment: An emerging mental health technology. *Journal of Primary Prevention, 8,* 71-94.

Tax, S. (1952). Action anthropology. *American Indigena, 12,* 103-106.

Tax, S. (1958). The Fox Project. *Human Organization, 17,* 17-19.

Vincent, M. L., Clearie, A. F., & Schluchter, M. D. (1987). Reducing adolescent pregnancy through school and community-based education. *Journal of the American Medical Association, 257*(24), 3382-3386.

Whyte, W. F. (Ed.). (1991). *Participatory action research.* Newbury Park, CA: Sage.

Wilson-Brewer, R., Cohen, S., O'Donnell, L., & Goodman, I. F. (1991). *Violence prevention for young adolescents: A survey of the state of the art.* Washington, DC: Carnegie Council on Adolescent Development.

Zimmerman, M. (in press). Empowerment theory: Psychological, organization, and community levels of analysis. In J. Rappaport & E. Seidman (Eds.), *The handbook of community psychology.* New York: Plenum.

Zimmerman, M., Israel, B., Schulz, A., & Checkoway, B. (1992). Further explorations in empowerment theory: An empirical analysis of psychological empowerment. *American Journal of Community Psychology, 20*(6), 707-726.

Empowerment Evaluation at Federal and Local Levels

Dealing With Quality

ROBERT K. YIN
SHAKEH JACKIE KAFTARIAN
NANCY F. JACOBS

This chapter presents a recent and ongoing case of empowerment evaluation. The case highlights empowerment as a group process as well as one in which the quality of the process and evaluation ideas are given explicit attention. In particular, we focused on three steps involving quality: the formal development of the evaluation ideas as an integral part of the empowerment process, which could then be shared and reviewed broadly; an assessment of the implementation process according to specific evaluation standards; and enumeration of specific outcomes from the process.

AUTHORS' NOTE: Under federal law, a work of the United States government is to be placed in the public domain and, neither the government, the author, nor anyone else may secure copyright in such a work or otherwise restrict its dissemination.

Our case derives from an activity related to the evaluation of community partnerships in implementing substance abuse prevention strategies. The background to this activity is described next.

Background

In November 1988, the Anti-Drug Abuse Act (Public Law 100-690) was signed into law. Among many provisions this law authorized the Office for Substance Abuse Prevention (OSAP), later to be renamed the Center for Substance Abuse Prevention (CSAP), to initiate the Community Partnership Demonstration Grant Program—to assist communities in developing partnerships of local organizations and to plan and implement community-based strategies for preventing the use of alcohol, tobacco, and other drugs. Two rounds of (mostly) 5-year grants were made, totaling 251 local partnerships, in 1990 and 1991.

The evaluation of this multicomponent program was mandated by Congress. Grantees were required to plan and conduct process and outcome evaluations of their projects to assess the attainment of their local goals and objectives. In addition, CSAP was required to conduct a cross-site evaluation, to identify and assess successful and innovative community-based partnership and prevention models and to assess common inhibitors to the design and implementation of effective substance abuse prevention programs.

The program design gave the 251 grantees the sole authority to select and contract with local evaluators. The range of evaluation skills and experience varied greatly. Moreover, evaluators experienced difficulty in community-based prevention projects, because their approaches to evaluation design were often inappropriate to the purposes or nature of the demonstration (Springer & Phillips, 1994). With specific respect to the community partnership program, grantees as well as many local evaluators were initially frustrated by existing evaluation methods, which they found ill-suited to the dynamics of evolving local partnerships.

Over the past year, this early frustration has begun to shift. The shift appears to be a direct result of an effort that might be characterized as practicing the principles of empowerment evaluation. The parties to the empowerment process, however, were more diverse than the evaluators and program participants commonly referenced. Further, as noted earlier,

A. Generic Model

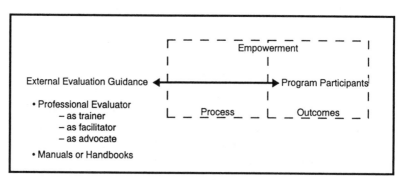

B. Model Followed in Present Study

Figure 9.1. Empowerment Evaluation: Participation and Process

the evaluation effort involved explicit attention to the quality of the empowerment process. The community partnership program, therefore, provided an unusual opportunity to the grantees and their federal counterparts to participate in an empowerment evaluation and also to attend to the issue of the quality of the process and its outcomes. Our experience, like other examples in this collection, provides more information on the process of carrying out empowerment evaluation and is intended to benefit future applications of empowerment evaluation.

The Participants in the Empowerment Process

Figure 9.1 portrays the basic configuration of participants, both in the generic empowerment evaluation process and in our case. Part A of the figure shows that the generic configuration has focused on a "transfer" type of relationship between external evaluation guidance and program participants. In this configuration, the empowering relationship is bidirectional. For example, Fetterman (1994) has described how external evaluators can serve program participants through training, facilitation, and advocacy roles. Similarly, Linney and Wandersman (Chapter 12, this volume) developed a specific manual *(Prevention Plus III)* to provide external evaluation guidance to local programs.

Our case differed from this generic configuration by involving multiple parties with multiple perspectives. The results were collaboration; commitments; the interchange of program-focused and insiders' perspectives of program operations; and intellectual excitement, investment, and energy around the evaluation of community-based substance abuse prevention programs. Thus, part B of Figure 9.1 shows the interaction that took place among local evaluators, partnership directors, a cross-site evaluation team, and the federal agency staff over the evaluation ideas that were developed: a customized evaluation framework for the community partnership program.

Part B suggests that all of these ingredients are part of the empowerment process—operating with multidimensional and not just bidirectional interactions. Basically, we are able to describe the empowerment process as a group-oriented activity. In this configuration, empowerment means empowering all parties—not the implicit "technology transfer" model whereby one party (typically an external evaluator) is an adjunct in the empowerment of another (typically, the program participants). The successful empowerment in our case should enable both local and cross-site evaluators to complete their work in a collaborative fashion. The partnership directors should be able to envision or implement their local programs more effectively. Finally, the federal agency should have learned important programmatic lessons—to report the results of its program to Congress and to design newer and better programs.

The Evaluation Ideas: A Customized Framework for Evaluating Community Partnerships

A critical aspect of our case was the attention given to the specific evaluation ideas being communicated. Such attention has not always been explicit in the reports of other empowerment evaluation experiences. Yet all evaluation, including empowerment evaluation, needs to be intimately and constantly concerned with the quality and validity of the evaluation ideas.

Our case included a draft paper that captured the essence of a customized framework for evaluating community partnerships. The paper also included illustrative examples of key concepts. The initial group of participants—local evaluators, cross-site evaluation team, and CSAP staff—all considered these contributions to be collective ones, and did not try to attribute the work to single (or a small set of) individuals.[1] Figure 9.2 presents the framework, indicating eight major categories of variables set in a presumed causal flow. Appended to this chapter are a series of illustrative examples of the subcategories and items falling within these eight major categories. The framework is potentially an important contribution to the understanding of how community partnerships work— as well as a practical and usable model because of the inductive way in which it was developed. To this extent, we believe that the customized framework will be a contribution to the advancement of knowledge.

The basic components of the framework fall into a potential causal sequence, as depicted in Figure 9.2:

1. *Partnership characteristics* and
2. *partnership capacity* are presumed to lead to
3. *community actions and prevention activities,* which in turn produce
4. *immediate process and activity outcomes* and then
5. *alcohol, tobacco, and other drug (ATOD) prevention outcomes,* together with
6. *non-ATOD community outcomes,* and finally
7. *ATOD impacts,* all occurring in the presence of
8. *contextual conditions.*

At first, the framework appears to be merely a reiteration of other similar open-systems models (e.g., Katz & Kahn, 1978). The customized framework, however, actually builds on these other models and extends them with several critical features. These features were created

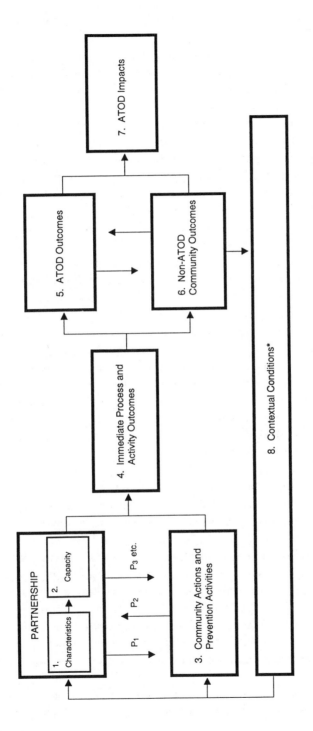

Figure 9.2. Customized Framework for Evaluating Community Partnerships

NOTE: P_1, P_2, P_3 = Phase 1, Phase 2, Phase 3.
*Other arrows from Contextual Conditions to all other components not shown.

193

collegially by the local evaluators—an important part of empowerment evaluation—and are the main basis for believing that the framework has advanced the state of the art.

A Dynamic Framework. First, there is a recursive or dynamic
interaction built into the framework, especially between the capacity
building of the partnership and the implementation of community
actions and prevention activities. This feature is reflected in the arrows
in the exhibit labeled "P_1," "P_2," "P_3," and so on, with "P" standing
for *phase.* Thus, partnerships may quickly start an action or activity
(P_1), may later increase their partnership capacity (P_2), and then may
repeat or expand the action or activity in a more enduring manner
than at first (P_3). Any sequence of partnership-building and activity
implementation phases may be followed by a partnership, but Figure
9.2 shows only the illustrative sequence just described.

The dynamic feature reflects the developmental, real-life evolution
of both the partnership's capacity and the strength and meaningfulness
of its actions and activities. Ultimately, this evolution is intended to
produce the *institutionalization or routinization* of partnership functions, reflected in part by the eventual survival and operation of the
partnership beyond the 5-year period of CSAP funding. In addition,
the nature of the immediate and other outcomes is expected to differ
(and to improve) as a function of the different phases through which
a partnership has progressed.

The time interval for a "phase" is not predictable or necessarily
uniform. Some partnerships may proceed through a phase quickly,
others slowly. What marks each phase, however, is that the partnership has successfully attained a new benchmark compared with an
earlier phase. If the new benchmark has not been attained, no new
phase has been attained, and development has not occurred. Concep-
tually, a benchmark would be agreement that an alcohol, tobacco, and
other drug (ATOD) prevention activity, once started on a transient or
temporary basis, might now be assumed to be enduring. As another
example, a benchmark would be the completion of the initial recruitment of key members, which would enable the partnership to proceed
with a continual (but normal) replacement and expansion process.

Partnership-Specific Immediate Outcomes. Second, the nature of
the immediate process and activity outcomes are highly specific to

partnership operations—improved coordination and reduced duplication of services. These immediate outcomes are, therefore, different than those normally found either in the generic open-systems models or in nonpartnership ATOD frameworks. The distinctiveness of these outcomes justifies the investment in the partnership strategy. Where these immediate outcomes do not occur, a partnership will have failed to demonstrate its value over and above investment in a traditional ATOD prevention activity in the absence of a partnership arrangement (thus, the central placement of the immediate outcomes in Figure 9.2).

Interactions With Contextual Conditions. Third, the framework recognizes that contextual conditions, such as school truancy and dropouts and poor or dilapidated housing, also exist in a dynamic relationship with the partnership and its actions and activities. Not only do the contextual conditions exert an influence on the partnership and its actions and activities, but these, in turn, can produce non-ATOD outcomes that represent changes in the contextual conditions (e.g., improvements in school or housing conditions). The proposed framework, therefore, explicitly shows how the partnerships can affect the environment within which they operate—a major goal of the partnership program.

A Broadened Understanding of ATOD Impact. Fourth, the customized framework reflects a richer and more accurate understanding of ATOD impact, far beyond the traditional measure of ATOD "use" as the main impact variable. As shown in Appendix 9.A, other relevant impacts include reduction of ATOD markets, demand reduction, deferral of use, and harm reduction.

Together, these components and the features of the customized framework have been developed to reflect actual conditions in the ongoing partnership programs. Although not apparent in the figure or this discussion, the conceptual strength of the framework is its inductive rather than deductive nature. The subgroups of local evaluators who created the framework did so on the basis of their ongoing experiences with their partnerships. This process assured that the framework was grounded in the reality of the partnerships' ongoing work. The complete empowerment process, as well as its early outcomes, are, therefore, the subject of the remainder of this chapter.

The Empowerment Process in Practice

Our empowerment process in practice focused heavily on the concern over the quality of empowerment evaluation. An important point of reference for this concern has been the interchange between David Fetterman and Daniel Stufflebeam regarding empowerment evaluation. Fetterman's presidential address at the American Evaluation Association's 1993 annual meeting—"Empowerment Evaluation"—called for evaluators to use evaluation concepts and techniques to foster self-determination among those being evaluated (Fetterman, 1994). According to Fetterman, evaluators can "teach people to conduct their own evaluations and thus become more self-sufficient" (p. 3). Optionally, "evaluators can serve as coaches or facilitators to help others conduct their evaluation" (p. 4) or even "serve as direct advocates—helping to empower groups through evaluation" (p. 6). Overall, as Fetterman (1994) summarizes, "Empowerment evaluation is explicitly designed to serve a vested interest—program participants" (p. 10).

Fetterman's call was then roundly challenged by Stufflebeam (1994). At worst, Stufflebeam (1994) noted: "What worries me most about Dr. Fetterman's portrayal of empowerment evaluation is that it could be used as a cloak of legitimacy to cover up highly corrupt or incompetent evaluation activity" (p. 324). This is because, according to Stufflebeam, empowerment evaluation gives over authority "to the client/interest group to choose criteria, collect data, and write/edit and disseminate reports, all in the name of self-evaluation for empowerment" (p. 324). Stufflebeam also found empowerment evaluation to fall short of meeting the program evaluation standards approved by the Joint Committee on Standards for Educational Evaluation (Joint Committee, 1994). One of Stufflebeam's recommended solutions was, therefore, to apply the "meta-evaluation" component of these standards to empowerment evaluation—subjecting empowerment evaluations to independent evaluations against the standards of the evaluation field.

A later rejoinder by Fetterman (1995) directly addressed many of Stufflebeam's concerns. A major clarification provided by Fetterman was that empowerment evaluation is not assumed to replace all forms of evaluation. In fact, empowerment evaluation and external evaluation are not only not mutually exclusive, they enhance each other. Fetterman also noted that the educational standards cited by Stufflebeam had not yet been accepted by the American Educational Research Association or by the American Evaluation Association. Thus, imposing the tenets of the standards was premature—but Fetterman nevertheless proceeded to illustrate how empowerment evaluation appears in many critical ways to meet the standards.

The ongoing debate and this collection provide much guidance to those who might want to use empowerment evaluation. Fetterman (1994, 1995) provides a long list of precautionary and quality-control measures to enhance the practice of empowerment evaluation. More work can be done, however, to further refine the quality of empowerment evaluation practice. The simple question is this:

- Has an empowerment evaluation (like any other evaluation) been conducted in an exemplary manner, with results that are of high quality?

Everyone knows that there is a world of difference between exemplary and mediocre external evaluations. Therefore, as a complementary strategy, empowerment evaluation also must continue to attend to this critical matter. The challenge is to attain the benefits of Fetterman's empowerment evaluation without risking the loss of the standards and quality control of concern to Stufflebeam and others. How we tried to accomplish this in our case experience is described next.

Implementation of Empowerment Evaluation. At the outset of the community partnership program, no attempt had been made to develop uniform design or instrumentation requirements for the local evaluations. Such requirements would have been premature and potentially unreflective of the diversity of local conditions and needs faced by the partnership grantees. At the same time, CSAP did commission the creation of *Prevention Plus III,* which provided general guidance to the local evaluators (as well as evaluators of other types of substance abuse prevention programs). Linney and Wandersman (Chapter 12, this volume) discuss and document the usefulness of this general guidance.

Nevertheless, when these 5-year grants were drawing to a conclusion, CSAP initiated steps to ensure, as much as possible, utilization and benefit from the data and findings of the local evaluation projects in concert with the cross-site evaluation. CSAP engaged both groups of evaluators for the purpose of establishing an evaluation framework to help guide the final identification, reporting, and utilization of all evaluation data and findings. The result was the customized framework that has been described earlier.

The implementation process involved several steps over an extended period of time. Key to this process was the involvement of an ever-widening circle of participants, eventually extending to the entire set of relevant parties. In hindsight, the process met or exceeded all expectations for participation that have been stipulated by Fetterman's description of empowerment evaluation.

Three major activities marked the implementation process. First, CSAP planned and convened a pair of meetings (held 2 months apart) of a subgroup of the local evaluators, together with the cross-site evaluation team. All told, 14 local evaluators, representing about 35 partnerships, participated (some local evaluators served more than a single grant). The local evaluators were asked to identify process and outcome variables, based on their ongoing experiences, and to provide operational definitions for these variables. The activity occurred as a group process over a 2-day period at each of the meetings. The cross-site evaluation team facilitated the meeting, along with the CSAP staff.

These meetings provided the opportunity for the evaluators as well as CSAP's technical staff to step out of their traditional roles and expectations, and to engage fully in a participatory and (eventually) mutually empowering exercise. The immediate result was the subgrouping of a large number of variables and the creation of a coherent framework (logic model) showing how a partnership might theoretically produce the desired substance abuse prevention outcomes and impacts. The process enabled the framework to reflect the peculiar features of the partnership program—hence, it was considered a "customized" framework. Following the meetings, a written draft of the framework was developed and shared with all of the original participants, whose comments were used to write a redraft.

Second, the redrafted framework was presented and discussed at a second meeting, 6 weeks later, involving another set of nearly 20 local evaluators—but ones who had not participated in the earlier pair of meetings. The meeting also involved the participation of nearly 20 partnership directors. The discussions emphasized the earlier process that produced the framework and its tentative nature as a draft. At this time, an early sign of empowerment was already revealing itself: The framework was not regarded as belonging to CSAP or to any individual party. Rather, the framework evoked a mutual perspective of co-ownership, paralleling the experience of most of the contributors in this collection.

Third, 2 months later, a further draft was distributed to all of the remaining 250 partnership directors and local evaluators, who were given 1 month to review the draft before convening at a plenary meeting (annual conference). The annual conference then provided the opportunity for extensive discussion of the framework among subgroups of local evaluators, in relation to specific, ongoing evaluation activities. Although CSAP sponsored the conference, the six separate "breakout" sessions were facilitated by the original set of 14 local evaluators, not CSAP staff or the cross-site evaluation team.

The result of the annual conference was the further development of the framework. The array of issues and comments at the conference had been recorded and also were shared widely.

Adherence to Evaluation Standards. To demonstrate our continuing concern with the quality of the empowerment process and evaluation ideas, we assessed this implementation process as just described according to the four major categories of the educational evaluation standards (Joint Committee, 1994). Briefly, these categories are as follows:

1. *Utility* standards are intended to ensure that an evaluation will serve the information needs of intended users.
2. *Feasibility* standards are intended to ensure that an evaluation will be realistic, prudent, diplomatic, and frugal.
3. *Propriety* standards are intended to ensure that an evaluation will be conducted legally, ethically, and with due regard for the welfare of those involved in the evaluation as well as affected by its results.
4. *Accuracy* standards are intended to ensure that an evaluation will reveal and convey technically adequate information about the features that determine worth or merit of the program being evaluated.

Of course, the final attainment of these standards can only be assessed when the entire experience has been completed. The implementation process, however, already appeared to match the needed requirements in the following ways.

The process was congruent with the utility standards in that the broadest array of potential users was involved in the process. Four groups of stakeholders were identified: local evaluators, the cross-site evaluation team, partnership directors, and federal staff. Consequently, the process was successful in highlighting multiple perspectives—for instance, providing useful evaluation information for local partnerships to seek further funding where necessary, but also providing cross-site information for CSAP's program review purposes.

The process followed the feasibility standards because a minimum of new burden was placed on the participants. The framework was a way of organizing existing data and did not dictate the use of any particular instruments or evaluation design. Throughout the process, people found the concepts and ideas of the framework to be revealing and helpful in organizing their thinking about existing partnership activities. To this extent, the framework represented an ideational or cognitive breakthrough that in fact created rather than dissipated energies—congruent with Fetterman's "illumination" stage of empowerment evaluation.

Propriety standards were relevant in that the empowerment process also had to deal with data sharing among all the parties, as well as with the partnerships being evaluated. The patterns of sharing were worked out through extensive discussions, which included the large number of local evaluators and the partnership directors. These discussions were held throughout different sessions at the annual conference.

The accuracy standards were assured through the creation of a specific, reviewable product—the customized framework. Moreover, the framework was subjected to extensive peer review. Some of the peer review was reflected through interchange of data and drafts among the participants. Additional peer review occurred through an external expert panel that provided technical assistance to the cross-site evaluators. Further peer review occurred where results were submitted for presentations or publications.

In summary, the implementation process was lengthy and involved extensive investments of time and effort. The CSAP staff, numerous local evaluators, and the cross-site evaluation team all committed many days of their time in engaging in the process. All members of the group appeared to be invested in the process, and the immediate outcomes are noted next.

Empowerment Outcomes

Within weeks and months of the empowerment evaluation experience, a variety of empowerment outcomes appeared evident. Although no one to our knowledge has systematically tracked these outcomes, we have become aware of 13 outcomes (enumerated below) through our informal contacts and our own professional activities. Assessing these outcomes—together with the presentation of the customized framework and the discussion of how the empowerment process adhered to evaluation standards—represents the third step in our overall concern for the quality of empowerment evaluation. Reporting about these outcomes, however, first requires a brief discussion of how we have conceptualized them.

The possible outcomes of empowerment evaluation appear initially to resemble the concepts related to evaluation utilization (e.g., Patton, 1986; Weiss, 1978) in which an overall impact—"utilization"—may be reflected by multiple outcomes (e.g., dissemination, communication, and utilization; Ball & Anderson, 1977). In this vein, empowerment appears to be reflected by at least two outcomes: action and validation. The action outcome could consist of (a) dissemination of the customized framework, (b) communication to others regarding the structure and function of the framework, and (c) utilization of the framework in a variety of evaluation activities. The second outcome— validation of the empowering ideas—is equally important, to assure that action is taking place around valid ideas. Both outcomes interact and are dynamic. For the sake of clarity, however, we discuss the action and validation outcomes separately.

Action Outcomes. In our case, one form of action outcome was the active adaptation and adoption of the framework. In Outcome 1, we have been told or have direct knowledge of several instances in which evaluators and their staffs quickly used the framework by applying it to new topics such as organizational development—not just partnership evaluation (and not just evaluation). We also know of one instance in which the framework was incorporated into the research design for a new evaluation of a new partnership (Outcome 2).

Other action outcomes derived from the partnership directors, and not the local or cross-site evaluators. In Outcome 3, newly developing partnerships have claimed that the framework has provided important insights into coalition-formation and implementation strategies. The framework has been one way of implementing a strategic planning process. Similarly, directors of mature partnerships have remarked how helpful the framework would have been if it had existed during the early stages of their planning processes, because the framework appears to represent accurately their subsequent experience (Outcome 4).

An originally planned action outcome also has occurred: The federal agency has integrated the framework into grantee final reporting requirements (Outcome 5). Because of the empowerment process, the federal staff also have assimilated the framework into their thinking about program development (Outcome 6). Finally, federal staff have incorporated the framework as the design for the local evaluations of an entirely new program, whose awards were only made in the fall of 1995 (Outcome 7).

Validation Outcomes. Among the validation outcomes, we have documented instances of the framework having been presented at a national meeting of an evaluation association (Outcome 8). It is proposed as a panel topic for another national meeting of evaluators (Outcome 9) and as the design framework for conducting a new cross-site evaluation of coalitions in a large western state (Outcome 10). We also know that the framework has been incorporated into a manuscript on evaluating local partnerships, just submitted for publication (Outcome 11), and that it has been used to analyze the early data in a local evaluation (Outcome 12). In the national, cross-site evaluation, the framework has been used to analyze preliminary cross-site data, with the results already reported at a national workshop of about 200 evaluators (Outcome 13). These all are considered validating outcomes because of their appearance in evaluation research circles—exposing the framework and its applications to peer review and expert scrutiny. Such review can only reinforce and improve the basic ideas implicit in the framework.

Empowerment Impact. The total empowerment impact is embedded in the overall pattern of individual outcomes just enumerated. First, all relevant parties to the original empowerment evaluation process (federal agency staff, local evaluators, and cross-site evaluators) have been involved in some action or validating outcome. In fact, key persons (partnership directors) not originally involved in the empowerment process also have been part of the outcomes. Second, these involvements have occurred in the spirit of "co-ownership." Each party has considered the framework to be his or her own, rather than something produced externally. To this day, for instance, the framework is not referenced as a "CSAP" or federal framework.

Summary

Our case has contributed in several ways to the evolving concepts and practice of empowerment evaluation. First, it reinforces the conception that the empowerment process involves groups of persons becoming empowered together, not the linear transfer of assistance or ideas from another party (an evaluator or technical assistance expert).

Second, our case highlights a critical ingredient—the substantive ideas that are the subject of evaluation (in our case, the customized framework). These substantive ideas must be scrutinized as one way of assuring quality over the empowerment evaluation process.

Third and most important, our case has integrated a concern for quality and validity throughout the empowerment evaluation process, not just in discussions of the customized framework. Also important has been the demonstration that the process can be assessed according to formal evaluation standards (such as the four major categories of educational evaluation standards)—and that the outcomes of the entire process can and should be tracked.

Future work on this topic can only lead to further improvements and refinements of the knowledge and tools for empowerment evaluation. A focus on additional outcome measures of "empowerment" should continue, if we are to strengthen practice. Another topic meriting further investigation is the development of a better sense of the anticipated dynamics of the empowerment process—for instance, how long the diffusion of ideas and empowerment outcomes can be expected to last. Additional case studies of empowerment evaluation, especially those representing replications (Yin, 1994), also would be invaluable. These and other refinements will sharpen a craft that can continue to be practiced well.

APPENDIX 9.A

Examples for Customized Framework

1. *Partnership Characteristics*

Eligibility rules
Number of partners
Governance structure
Organizational structure
Staff size and diversity
Age, ethnic, and racial diversity of partners and population

2. *Partnership Capacity*

Human resources:	The ability to recruit and mobilize people
Organizational resources:	The ability to create a viable organization
Planning:	The ability to develop and implement responsive plans
Internal and external communication:	The ability to communicate internally and externally
Managerial capability:	The ability to make decisions and diffuse conflicts (when appropriate)
Institutional knowledge of the "system":	The ability to move an issue through, or make a change in, the external environment

3. Community Actions and Prevention Activities

Incentive activities:	Aimed at increasing participation and visibility
Strategic activities:	ATOD activities of substantive duration
Policy and legislative changes:	Changes in a community's rule system, related to ATOD prevention theory
Outreach activities:	Aimed at maintaining or increasing support for the partnership as well as awareness of ATOD problems and issues
Community development:	Aimed at changing community conditions that affect ATOD in the long run

4. Immediate Process and Activity Outcome Variables

Coordination and collaboration not present prior to the partnership
Spin-offs (from the partnership) of new services
Reduction in the duplication of existing services
Promotion of appropriate, comprehensive mix of multilevel services
Noncompetition with existing services
Managerial effectiveness

5. ATOD Outcome Variables

Changes in risk perception
Changes in perceived norms and beliefs
Increases in community protective factors
Increases in community resilience
Implemented policy changes
Mobilization on ATOD issues
Changes in ATOD prevention and treatment services
Increased knowledge and attitudes about ATOD abuse
Intentions or pledges not to use

6. Non-ATOD Community Outcome Variables

Sociopolitical condition
Socioeconomic conditions
Community infrastructure
Community health conditions (including violence)

7. ATOD Impact Variables

Community indicators of ATOD impacts:	Drug-related arrests and emergency-room cases
	ATOD-related crime reduction
	Workplace drug use
	Per capita consumption of alcohol
	Youth or adult surveys of ATOD abuse
Reduction of ATOD markets:	Availability, accessibility, and price of drugs
	Existence of open-air drug dealing
	Availability and sales of alcohol and tobacco
Reduction of demand:	Frequency of selling or being offered drugs
	Frequency of heavy drinking or feeling drunk or high
	Tobacco use
	Ability to purchase alcohol
Deferral of use:	Deferral of first use of illegal drugs
	Drugs used by underage persons
	Tobacco and alcohol sales to underage persons
Reduction of harm:	Distribution and monitoring of clean needles
	Lower incidence of drunk and/or drugged driving
	Reduction in ATOD-related gangs and violence

8. Contextual Conditions

Sociopolitical conditions
Socioeconomic conditions
Community infrastructure
Community health conditions (including violence)

Note

1. The local evaluators were Judd Allen, Edgar Butler, Gladys Baxley, Ester Cadavid-Hannon, David Chavis, William Hansen, Wayne Harding, Nancy Jacobs, Gilbert Robledo, Wendy Rowe, Glenn Solomon, Fred Springer, and Abraham Wandersman. A few other consultants also participated in these meetings but were not local evaluators.

References

Ball, S., & Anderson, S. B. (1977). Dissemination, communication, and utilization. *Education and Urban Society, 9,* 429-450.

Fetterman, D. M. (1994). Empowerment evaluation. *Evaluation Practice, 15,* 1-15.

Fetterman, D. M. (1995). In response to Dr. Daniel Stufflebeam's: "Empowerment evaluation, objectivist evaluation, and evaluation standards: Where the future of evaluation should not go and where it needs to go." *Evaluation Practice, 16*(2), 321-338.

Joint Committee on Standards for Educational Evaluation. (1994). *The program evaluation standards.* Thousand Oaks, CA: Sage.

Katz, D., & Kahn, R. L. (1978). *The social psychology of organizations.* New York: John Wiley.

Patton, M. Q. (1986). *Utilization-focused evaluation.* Beverly Hills, CA: Sage.

Springer, J. F., & Phillips, J. L. (1994). Policy learning and evaluation design: Lessons from the community partnership demonstration program. *Journal of Community Psychology* [Special issue], pp. 117-139.

Stufflebeam, D. L. (1994). Empowerment evaluation, objectivist evaluation, and evaluation standards: Where the future of evaluation should not go and where it needs to go. *Evaluation Practice, 15,* 321-338.

Weiss, C. H. (1978). Improving the linkage between social research and public policy. In L. E. Lynn, Jr. (Ed.), *Knowledge and policy: The uncertain connection* (pp. 23-81). Washington, DC: National Academy of Sciences.

Yin, R. K. (1994). *Case study research: Design and method* (2nd ed.). Thousand Oaks, CA: Sage.

10

Evaluation and Self-Direction in Community Prevention Coalitions

JOHN F. STEVENSON
ROGER E. MITCHELL
PAUL FLORIN

For several years, the Community Research and Services Team has been evaluating community-level prevention programs in Rhode Island. In that context, we have employed a variety of methods intended to increase the control by local program participants over the community conditions that affect them. Over time, our experience has led to our own way of conceptualizing empowerment and integrating that objective with our evaluation work. In this chapter, we will provide a brief introduction to the specific context in which our ideas and methods have evolved; a discussion of some of the issues, benefits, and complications in linking conceptualizations of empowerment to program evaluation; some central examples of relevant methods we have developed; and our current perspective on the promises and limitations of these methods.

COMMUNITY COALITIONS AS A CONTEXT FOR EVALUATION

We believe that an understanding of our ideas and methods will require a grounding in the context in which they have evolved. We do not see *empowerment* or *empowerment evaluation* as standing outside the actual situations in which the terms are applied.

Community coalitions intentionally constructed to bring about reduction of alcohol and other drug abuse are part of a broader movement in the field of prevention. Although evidence for their effectiveness as a policy approach is still limited, these local voluntary groups have been proposed as an appropriate mechanism for attacking a wide variety of difficult community problems, including crime, violence, substance abuse, and delinquency (Butterfoss, Goodman, & Wandersman, 1993; Farquhar et al., 1990; Jacobs et al., 1986; Kaftarian & Hansen, 1994; Pentz et al., 1989).

Experimentation with these efforts is based on the premise that a multifaceted approach that alters the fabric of communities can have powerful preventive effects (Hawkins, Catalano, & Miller, 1992; Pentz et al., 1989; Perry, 1986). The language used to justify coalition building as a prevention strategy is permeated with an empowerment motif (Butterfoss et al., 1993; Chavis & Florin, 1990; Fawcett, Paine, Francisco, & Vliet, 1993). The logic for the utility of this strategy posits a community development process that builds confidence, competencies, and social connections among participants. By developing broad-based citizen participation, coalitions can increase local ownership, which in turn can lead to better access to existing resources and increased commitment to maintaining prevention activities over time. The rationale envisions a systemic change process, moving beyond the individual level to influence key decision makers, extend interorganizational networks, and develop new prevention and health promotion policies within the community.

We believe evaluators have a critical role to play not only in judging the success of this policy strategy but also in identifying the conditions and support structures that may enhance or impede the effectiveness of the strategy (Florin, Mitchell, & Stevenson, 1993). Evaluation itself is clearly part of the support structure, and like many evaluators we conceive of our work as building the learning capacity of the organizations we are evaluating (Fetterman, 1994a, 1994b; Forss, Cracknell,

& Samset, 1994). Although we have not used the term *empowerment evaluation* to describe our work, we do see a close connection between some aspects of our approach and the description of empowerment evaluation offered by Fetterman (1994a, 1994b, and this volume). Many of the issues and methods associated with the "empowerment" label have been present in the evaluation field for decades, and we appreciate the opportunity to reflect on our experience with some of these issues and methods in the current volume.

Conceptual and Practical Issues in Empowerment Evaluation

Our roles as evaluators in projects that attempt to empower participants have led to pointed discussions on a number of difficult questions: If these organizations have empowerment as an objective, how should we conceptualize and measure it? Must the evaluation be carried out in a manner that is consistent with the "empowerment" values of the projects themselves? If the answer is no, what are the implications for our ability to work effectively with program participants and staff to collect data and to capture a fair picture of the "empowerment process" presumably under way? If yes, to what extent are we abdicating our "objective" roles as arbiters of program quality? Which of the many players in the complex, multilevel systems with which we work are to be empowered? To what extent is the evaluation successfully building the capacity of the participants to use evaluation findings in their own self-monitoring and improvement processes? Can one really "give evaluation away"? We will introduce and discuss some of these issues here, and return to them later. Our point in introducing these issues is to provide the reader with a sense of the questions we encountered as we developed the methods and tools described in this chapter. We think this provides a clearer context for understanding how our work evolved.

DEFINING EMPOWERMENT

In the context of our community coalition research, we have developed and empirically investigated a multilevel, social-action-oriented view of empowerment (McMillan, Florin, Stevenson, Kerman, & Mitchell, in press). As we have already noted, coalitions are by their nature concerned with empowerment issues. Moving from broad conceptual brush strokes to concrete operationalizations, however, has challenged us to sort out definitional ambiguities. Our work suggests important distinctions about *what* it means to be empowered and *who* is intended to be empowered.

First, our work recognizes that a variety of definitions have been used for the content of empowerment, referring alternatively to values, processes, or outcomes (Zimmerman, in press). Such definitions have also varied in the extent to which they view empowerment as a general or situation-specific competency. Our own approach has been to view psychological empowerment as including knowledge, skills, perceived competencies, and expectancies for individual and group accomplishments in the specific arena of their coalition work. Second, we have taken a multilevel approach to empowerment, positing three levels at which changes in power may occur: (a) the individual level, at which *psychological empowerment* takes place; (b) the intraorganizational level, at which the *empowering organization* may make possible the collective empowering of its members; and (c) the extraorganizational level, from which relevant social systems can be judged more or less *organizationally empowered,* that is, successful in influencing their environment (McMillan et al., in press; Swift & Levin, 1987; Zimmerman, in press). It is not certain that changes at one level will necessarily be associated with changes at another level.

Consider the following scenario. An attempt has been made to create a group that is representative of multiple community sectors and that is committed to developing local solutions to alcohol and other drug abuse problems. The individuals involved become increasingly close-knit and committed. They change in the following ways: becoming more knowledgeable about prevention, more confident in their skills, and more expectant that they are and will continue to make contributions. At the same time, however, their outreach efforts have slowed. Significant sectors of the community that were missing at the outset are still missing, with lessening energy and attention devoted to broadening the constituencies represented by the coalition. Is this empowerment? At the individual level perhaps yes, at the organizational level perhaps no.

To examine the links among different levels of empowerment, we have employed a multilevel data analysis method (Kenny & LaVoie, 1985) to identify the characteristics associated with coalitions that "collectively empower" their members and that are "organizationally empowered" in that they achieve system-level goals. For example, McMillan et al. (in press) found modest, positive relationships between coalitions that "empowered" their members and those that had been able to affect the policy decisions and resource allocations of other influential community institutions (i.e., organizational empowerment). The important lesson for this discussion is that there may be very different conclusions about what to do and what is happening depending on one's definition of empowerment.

EMPOWERMENT AMBIGUITY AND EVALUATION CONFUSION

How does our work on empowerment inform thinking about empowerment evaluation? Fetterman (1994a) has defined *empowerment evaluation* as "the use of evaluation concepts and techniques to foster self-determination" (p. 1). As our discussion of empowerment demonstrates, there are many ways of interpreting this statement, with differing implications for evaluation practice. Because the programs with which we work intend to increase self-determination at both the individual (coalition member) and the collective (coalition) levels, we have been sensitized to the importance of distinguishing between the two. As we have learned over the course of our experience, the nature of power, and the means by which evaluation may influence it, differ substantially for these alternative levels (see also Fetterman, Chapter 1, this volume). Failure to resolve the ambiguity about what one means by *empowerment* is likely to lead to confusion in the evaluation arena as well.

EMPOWERMENT EVALUATION AND ROLE CONFUSION

Responding to the interests of multiple stakeholders in the interests of empowerment can lead to confusion regarding the evaluator's role. Each stakeholder represents another possible target for increased self-determination. The prevention coalitions we have studied mount a variety of actual prevention activities, for example, each of which has an intended target population within the community. A particular activity may be directed at all fifth and sixth graders in the community, another at Latino families, another at alcohol vending establishments ("servers"), and another at public housing residents. Are all of these groups to be empowered through the conduct of the evaluation? And what of the local agency directors and staff who may be providing much of the prevention programming—are they to be empowered as well?

O'Neill (1989), for example, describes a consultation to a battered women's shelter that attempts to empower them in using needs assessment and evaluation data to assist them in advocating for needed services. The consultant helps to collect needs assessment data that help to secure needed funding from the local town council. What happens, however, when subsequent data reveal that these funds are being used to support deserving clients but not necessarily the target of the initial appeal? The failure to clarify role responsibilities leads to an ethical dilemma as to whether the evaluator's allegiance belongs to the program staff, program clients, or the town council that provided program funding. This is another challenge for the well-meaning evaluator.

EMPOWERMENT EVALUATION AND
THE REALITIES OF "GIVING EVALUATION AWAY"

A central theme of this chapter (and indeed of this collection) is increasing local control by building individual and organizational capacity for evaluative activities. To what extent, though, do the people to whom we want to "give evaluation away" have the interest or technical skills to do evaluation competently? And how credible is the work of program participants as an objective picture of program effectiveness? When the results of an evaluation may be used to justify resource allocation, one should expect strong challenges from vested interests regarding the adequacy and interpretation of any evaluative findings. These are challenges that an "amateur" evaluator may not be able to meet very easily.

In one of our projects, our request for coalition volunteers to collaborate in designing and conducting a small-scale evaluation of an after-school tutoring program for minority youth led to an angry meeting between evaluators and indigenous volunteers. There was suspicion about whether this was a "setup," because "amateurs" at evaluation certainly couldn't defend their program as well as a professional evaluator might be able to. There was also questioning of the fairness of expecting community volunteers to do work for which they believed the evaluators were being paid. Are we doing participants a disservice, as well as compromising the quality of evaluations, when we involve them in evaluation enterprises for which they may not have the skills or the time?

EVALUATION STANDARDS AND EMPOWERMENT

In what sense is our work with community coalitions "evaluation"? There are a variety of definitions of evaluation, reflecting alternative theories of the central task of evaluation (Shadish, 1994; Shadish, Cook, & Leviton, 1991). These differing definitions lead to disparate views of whether and how *empowerment,* defined in any of the ways we have discussed, can be an integral part of evaluation. For Lincoln (1994), the future of the social sciences and of evaluation lies in joining a revolutionary change process that will undo inequalities currently maintained, in part, by the rhetoric and research of positivist social science. For Stufflebeam (1994), on the other hand, extant standards for ethical practice in evaluation appear to call for an "objectivist" stance to the independent determination of program value.

We prefer to stay closer to our own evaluation praxis, in contexts where we believe we can justify our choices. It is to examples of our practice that we turn next; we will return to the broader issues in the final section of this chapter.

Evaluation Methods for Empowerment

Our own thinking about these issues has evolved with our experience. Our research group first used the title "Giving Evaluation Away" in a presentation on coalition evaluation at a conference of the American Evaluation Association in 1991 (Stevenson, Florin, & Mitchell, 1991). In a 1993 conference sponsored by the Center for Substance Abuse Prevention (CSAP), we presented a utilization-focused, stage-based model for improving the utility of evaluation by channeling it into a set of relevant tools for planning and self-monitoring by community prevention coalitions (Stevenson, Florin, & Mitchell, 1993). Since that time, we have continued to experiment with a variety of ways of learning about and working collaboratively with such organizations.

EVALUATION UTILITY

One way in which evaluators can empower their clients is a very traditional one: designing, conducting, and reporting evaluations in ways that give the client(s) greater control through improved usability of evaluation. We have focused on use in the context of a developmental model for evaluation feedback and self-correction—including evaluation tools and procedures in the early organizational stages of coalition development, evaluation in the planning process, and evaluation for refinement and institutionalization. If evaluation is to provide a useful mirror, feedback must be organized by a user-friendly conceptual map, so that attention is effectively focused on aspects of the organization and its environment that can lead to "small wins" and steady progress.

DEVELOPING AND APPLYING AN
ACCESSIBLE FRAMEWORK FOR FEEDBACK

We believe that reflection about the difficult issues in utilization can lead to increased participant power as well as improved quality of the evaluation, both in terms of its conceptualization and in terms of the quality of the data that are collected. In our work, this has come about through a gradual process of refined thinking about evaluation objectives and sharpened focus for evaluation methods in action. One example of this interplay is our development of a framework for coalition evolution.

Our initial work with coalitions occurred in 1988 when we began a process evaluation of 35 newly formed substance abuse prevention task forces (or coalitions) that were supported through state funding. The contract called for a 9-month, one-shot, process evaluation, and we provided what typically results from such efforts: a lengthy concluding report that contained data on scores of constructs that previous literature had suggested were important for the development of grassroots organizations (Florin, Chavis, Wandersman, & Rich, 1992). In these beginning days of research on coalitions, however, there was little theory regarding mediating processes of coalition functioning to guide integration of voluminous data from multiple sources (e.g., member mail surveys, key informant telephone surveys, face-to-face leader interviews, ratings of intervention plan quality, etc.). As we subsequently became evaluators of a 5-year coalition-based project sponsored by CSAP, we realized we would have to provide much more brief, focused, and "accessible" reporting of results if we expected to influence these coalitions' formative development. We began to ask ourselves the following questions: What do these coalitions really need to know from the data we are collecting? Are some types of information more important to them at some times rather than others? How do we encourage an investment in evaluation that would make consideration of formative evaluative results more likely?

Our struggle with these questions led us to examine literature on stages of change among individuals (Prochaska & DiClemente, 1992) and to try to apply such concepts to coalitions. We constructed a developmental framework for coalition evolution that helped us to organize a variety of our measures as reflecting crucial tasks that coalitions must address, representing perhaps a series of developmental stages (Florin et al., 1993). The developmental tasks presented in Table 10.1 offer a potential schema for organizing the feedback presented to task forces. At the very beginning of coalition formation, attention might most productively focus on the adequacy of Initial Mobilization (e.g., appropriate membership and constituency representation on the coalition) and of Establishing Organizational Structure (e.g., formally designated roles and procedures, the members' perceptions of organizational process, their view of the costs and benefits of participation). Building Capacity for Action included a range of individual-level capacities (e.g., member skill development and efficacy expectations) as well as organizational-level capacities (e.g., interorganizational networking and linkages). Planning for Action (including needs assessment), Implementation, Refinement (via internal evaluations of program activities), and Institutionalization of successful undertakings round out our series of developmentally staged feedback topics. Although these stages are not invariant, and some are likely to

TABLE 10.1 Stages and Tasks of Coalition Development

Stage of Coalition Development	Examples of Tasks Associated With Each Stage
Initial mobilization	Recruitment of critical mass of active participants
	Engagement of key community constituencies or sectors
Establishing organizational structure	Establish structure for working group that clarifies roles and procedures
	Adequately address both task and maintenance functions of the group
Building capacity for action	Member-level capacity: Orient members to concepts and provide skill building
	Organizational-level capacity: Establish interorganizational linkages with other important players in the community
Planning for action	Assess needs as perceived by community constituencies; prioritize and clearly state coalition goals and objectives
	Select an array of intervention strategies based upon literature about program effectiveness
Implementation	Develop a sequential work plan that sets time lines, allocates resources, and assigns responsibilities
	Implement activities in a manner that involves key organizational players, networks, and broad citizen participation
Refinement	Use process evaluation data for specific program refinements that incorporate community reactions
	Target strategies that build toward a comprehensive and coordinated array of programming strategies across community sectors
Institutionalization	Member-level institutionalization: Establish processes for leader succession and recruitment of new members
	Organizational-level institutionalization: Integrate functions into ongoing missions of existing organizations

SOURCE: From Florin, Mitchell, and Stevenson (1993). Reprinted by permission of Oxford University Press.

be revisited, they provided a series of benchmarks to guide our evaluation feedback about task accomplishment at each stage.

We found this framework useful not only because it helped us to present findings in a consistent and comprehensive manner but also because it began to serve as a useful tool for self-assessment. For example, staff responsible for providing technical assistance to these coalitions used this framework to develop consultation forms (or "contracts") that would help specify and narrow the kinds of individualized assistance and training that individual coalitions might be in need of developing. This framework then began to be a "common language" that individuals used to talk about problems and prospects in their work. As we continued to collect data, this heightened attention to "stages" and "tasks" caused us to ensure that we were assessing each of these areas adequately in our multiple assessment methods and sources. Theoretically, we are now at the point of exploring whether significant outcomes can be predicted on the basis of one's functioning with regard to these tasks at early stages of development.

DEVELOPING A TYPOLOGY OF PREVENTION ACTIVITIES

We have incorporated a number of more focused utilization-oriented methods into our stage-based approach. One good example is the development of a typology of prevention activities. Linked to the Planning for Action stage, this typology development process resulted in both a more comprehensive means for evaluating prevention plans but also in a usable tool for self-evaluation that assisted coalitions in their own planning activities.

In our initial work with coalitions in 1988, one of our most urgent questions was this: What were these groups actually doing or planning to do to prevent alcohol and other drug abuse? Each coalition was required by the state Department of Substance Abuse to develop a detailed 3-year plan of their proposed prevention activities. Across these proposals were over 900 individual activities, including such diverse activities as "publish results of drug use survey," "provide communication, conflict management training," "offer substance-free prom night activities," "develop workplace policies," and "make liquor license renewals contingent on participation in alcohol server training." How could we characterize these proposed activities in a way that would both serve the purposes of our evaluation and serve as a helpful planning tool for coalitions?

We examined initial efforts in the literature at categorizing prevention activities that confirmed our sense of the importance of this task. Tobler (1986), for example, had characterized programs (rather than activities) as part of a meta-analysis of school-based research on alcohol and other drug abuse prevention. Linney and Wandersman

(1992) had developed a broader set of categories as part of their planning and evaluation tool, *Prevention Plus III*. Such typologies can serve as an important point of intervention by encouraging community-based groups to consider more deliberately and systematically the full range of intervention options available to them. Such frameworks can also broaden perspectives by highlighting strategies that have not typically been used or by encouraging use of familiar strategies in new settings. For example, our informal sense of the material was that system change strategies and policy change strategies were receiving little attention. There had not been a system whose reliability had been tested, however, and our efforts to use existing schemas clearly indicated the complexity involved.

Over several iterations, the result of our efforts was the development of the following categories: Increasing Knowledge/Raising Awareness, Building Skills/Competencies, Increasing Involvement in Drug-Free/ Healthy Alternative Activities, Changing Institutional or Organizational Policies, Increasing Attention to Law Enforcement and Regulatory Practices, Building Coalition/Partnership Capacity, Building General Institutional/Community Capacity, and Treatment/Early Identification and Referral. We were able to develop a relatively reliable system across several different kinds of raters (for a full description of reliability and validity issues, see Mitchell, Stevenson, & Florin, in press). In the context of our developmental framework, we used this typology to organize our reports about where local coalitions were putting their efforts and how this matched with both the emerging prevention literature and what seemed important. For example, an aggregate profile across recently formed coalitions suggested that they were putting a great deal of their effort into promoting awareness-raising activities, when recent literature was indicating the limits of such strategies alone.

As we began to use this tool, several interesting things happened. First, we received feedback about a few areas where the typology may have been unclear and where certain activities may have been unclear. Second, we received some positive feedback about the usefulness of the information we provided. Third, and most revealing, we relearned the lesson that much more is needed than simple feedback of evaluation results to have a tool or process adequately disseminated. Despite presenting summary feedback about a lack of policy and community change initiatives in the original 1989 plans, little actual change occurred in their subsequent efforts. So, we began to focus on the questions of where in their planning processes could such a tool be used, and what kind of infrastructure would be needed to support integration of a planning tool that should assist them in looking more broadly at intervention strategies. Multiple efforts with varied constituencies in the use of this typology ultimately created the kind of critical synergy needed to facilitate dissemination.

Ultimately, this typology was incorporated into several strong support systems relevant to coalitions. There were several arenas in which it began to be used as a way of organizing discussion and planning: first, at our "Program Evaluation for Prevention" workshop series, which was open to any and all coalition participants; second, at intensive planning sessions that were held for individual coalition groups by a CSAP-funded training and technical assistance consortium to help them assess more systematically the comprehensiveness of their prevention activities and strategies; third, as part of a resource bank that categorized "promising programs," this typology was used as one way of categorizing efforts to enable the coalitions to get greater access to programs that might fill gaps in their current programming; and, finally, we worked with the Rhode Island Department of Substance Abuse in revising the material it required from coalitions in order to receive an annual allotment of state funds. We suggested that this typology be used as a way for local coalitions to describe what they were intending to do in their yearly plan of proposed efforts. Thus, the typology is in use at the very time when these coalitions are most intensively involved in their planning processes.

All the informal feedback we have received is that the use of a common typology across different settings has sensitized coalition members to important distinctions among strategies and made it easier to communicate during the planning process.

TECHNICAL ASSISTANCE AND TRAINING TO GIVE EVALUATION AWAY

Another method we have found useful in evaluating community-based coalitions is an evolving system of workshop training and follow-up technical assistance designed to enhance the ability of local nonprofessionals to collaborate in the evaluation of their programs. This method is relevant for what we term the Refinement stage of coalition development.

Our evaluation context creates a special set of challenges for the evaluator, challenges that are not unique to community coalition evaluation but are particularly salient there. One important challenge is created by the tension between the need for standardized interventions and evaluation designs, contrasted with the recognized value of promoting local control and ownership. The logic model underlying our community prevention approach proposes that locally identified needs are the best guide to action and that "community empowerment" is itself part of the solution for such problems as crime and substance abuse (Chavis & Florin, 1990; Wallerstein, 1992; Zimmerman & Rappaport, 1988). The apparent contradiction between externally imposed research designs and spontaneous development of indigenous programs makes political tension inevitable. Evaluation may be experienced as a means to compel obedience to external authority; local ownership of the prevention program may be weakened, and useful evaluation may be undermined as well. The imposed design may not address locally important outcomes. The process of data collection may be sabotaged. Reports may be seen as irrelevant to local concerns. When local program staff and participants are members of disenfranchised minorities, these problems may be especially corrosive.

Major funding sources for prevention understandably want both local control (including representatives of populations most at risk, organically developed community programs, and a less dominating role for the "outside experts") and rigorous evaluation designs (with universally accepted measures of program impact that will identify effective programs and justify the expenditures being directed toward these programs). The need for standardized, cross-site study is very clear (Klitzner, 1993; Peterson, Hawkins, & Catalano, 1992; Shinn, 1990).

Is there a way to allow relatively autonomous community prevention efforts yet impose reasonable and useful evaluation requirements? The challenge is too complex for simple solutions, but part of the answer may be to expand the traditional evaluation role to include the provision of training and technical assistance to teach local prevention personnel and community activists to understand, use, and even conduct local program evaluations. Inspired by the pioneering work of Linney, Wandersman, and McClure, later revised and reissued as *Prevention Plus III* (Linney & Wandersman, 1992), we have been trying to give evaluation away since 1990 (Stevenson et al., 1991).

Here is one case example of our application of this approach. The stimulus for this application was described above. We received a contract to evaluate a statewide legislative initiative that created a "substance abuse task force" in every municipality in the state. We collected extensive data on the process of implementation in every community. Many of these community coalitions had only been in existence for a short period of time, and we looked for early indications of potential future success or problems. One of these indications was the quality of prevention plans submitted by the local coalitions to the state. One of the two weakest sections of the task forces' proposals for community action was their plan for evaluating their activities. In our final report (Florin, Mitchell, & Stevenson, 1989), we emphasized the need for ongoing technical assistance to these local efforts, including evaluation capacity building. Following up this recommendation, we have collaborated in developing a technical assistance system to support the local task forces (funded by the Center for Substance Abuse Prevention). The feature of this system that is relevant here is the provision of two workshops and follow-up technical assistance (under our supervision) to impart evaluation skills to coalition members and leaders. We have used a similar mechanism to build evaluation capacity in three other federally funded projects.

In the development of this intervention, we identified several goals. The first goal was *demystification*. If distrust is at least partly based on lack of knowledge and lack of a face-to-face relationship with evaluators, then all three can be improved by skill-building workshops. When fear of the unknown is reduced, a more positive attitude toward the potential utility of evaluation may emerge. Evaluators can also benefit when local activists share their expertise and their knowledge of local conditions.

A second major goal was to *improve the rigor* of the project evaluation. The quality of data collected may be improved, and ultimate conclusions about the processes and outcomes of the project may have more ecological validity (Gibbs, 1979), because they are better grounded in the evolving intentions of the participants. Small-scale program evaluations resulting from the workshops can become a part of the cumulative record of the project's effects, and shared instrumentation, where appropriate, can facilitate aggregation and cross-site comparisons, thus meeting the needs of the external audience of funding agencies and professionals.

A third goal was *increased local use of findings.* Starting early in the life of the project to link evaluation methods to project objectives can help to shape both the evaluation and the project itself. As time goes by, this is likely to enhance the utility of the results for local application. Project staff and volunteers who are involved with the evaluation from the outset, who understand it as a collaborative enterprise, and who participate in the development of reported conclusions are likely to be more ready to use the results and recommendations that emerge.

A fourth goal was to *improve individual and organizational capacity for self-direction.* As we have said, we are committed to enhancing the learning capacity of the system we are evaluating, and we believe this intervention improves both the program planning process and the evaluation process. In planning prevention programs, community activists may be impatient with the seemingly obscure and distracting requirements of evaluation. The value of "baseline data" as one means of needs assessment, the usefulness of having a system for keeping precise records about program participants and activities, and especially the need to define measurable and achievable outcome objectives during the planning phase, however, can all become part of increased capacity for effective planning. Beyond the contribution to planning, these interventions actually may increase the capacity necessary to conduct small-scale program evaluations, building increased control at both the individual level (self-efficacy and credentials) and the organizational level (better programs and new funding).

The actual intervention consists of several elements. Here we provide brief descriptions of each of these.

For each project in which we conducted workshops, we engaged in extensive preworkshop planning. This typically included a needs assessment questionnaire mailed to those who planned to attend. We also selected appropriate task group assignment principles—typically planning to bring together working groups that would continue to function after the workshops. We used this preliminary step to clarify for ourselves what the objectives of the intervention were for the particular circumstances (e.g., how much emphasis to place on individual vs. organizational capacity building).

Workshops were 2 to 3 hours in length. The first workshop began with introductions and ice-breaking. The icebreaker dealt with fears and hopes regarding evaluation, participants' expectations for the workshop, and our expectations. Then we solicited sample questions the participants thought evaluation might help them answer. We went on to introduce the four-step model provided by *Prevention Plus III,* classifying the questions asked by participants into the four steps. We often conducted an exercise/quiz at this point. After following one example we had prepared through the four steps, we briefly reviewed our own evaluation and placed its components within the four-step framework. In one variation of this review, we spent some time brainstorming how to decrease the costs and increase the benefits of our evaluation. Next we divided the participants into agency- or coalition-affiliation groups and asked them to identify one high-priority program suitable for outcome evaluation in the coming year. We asked for discussion of the objectives of this program, working with each task group on this problem. We closed with an attempt to support the planning of next steps. We also administered pre- and postworkshop questionnaires to assess changes in knowledge, attitudes, and behavioral intentions.

Between the two workshops, members of our evaluation staff visited local groups or agencies to work with those who were most interested in actually designing program evaluations. This was more challenging for our staff than we had anticipated. A lack of well-defined objectives and clearly bounded programs proved to be the rule rather than the exception, and requests for help with a variety of other organizational problems and needs were given higher priority than evaluation (Chadwick, 1993).

At the second workshop, we aimed for smaller, more work-focused groups. We began again with an icebreaker, this time one that turned the tables by asking workshop participants to evaluate the presenters. In small groups, they rated our wearing apparel and selected the best and worst features of each outfit. This strange reversal of power was made easier by our intentional choice of some unorthodox adornments (one particularly hideous tie remained a topic of discussion with coalition members a year after the event!). We used this exercise to stimulate a discussion of the different perspective when playing the evaluator role, the variety of influences on judgments of worth (including ethnic differences), and the value of getting more specific about what is working and what is not. After reviewing the four-step model, we divided participants into working groups and focused once more on writing good objectives. An exercise/quiz checked understanding of what constitutes a good objective. Group-level problem solving allowed these exercises to involve all of the participants in a less threatening way and also built team collaboration. Next, the groups wrote objectives for their high-priority program and designed outcome evaluations, linking the objectives to potential outcome measures. Finally, we discussed next steps.

Despite the workshop's focus on preparation for actual evaluation, we found that little evaluation would actually take place without follow-up technical assistance from our staff. In fact, there was a tendency for our staff to be drawn into more extensive consultative roles, such as helping with program planning and evaluating proposals from vendors for prevention programs.

We have some evidence for the effectiveness of this method. Our participant questionnaires have indicated some knowledge gain, transient effects on self-efficacy, strengthened belief in the general utility of evaluation, and increased readiness to undertake evaluation activities. With insistent follow-up technical assistance, we have nurtured a number of small-scale program evaluations by grassroots and community coalitions. In terms of improved *planning* of evaluations, preliminary findings from a quasi-experimental design indicate a significant benefit from follow-up technical assistance. We are continuing to experiment with this approach.

Wrestling With the Challenges

As we have developed the methods described above, we have been confronted by the issues that were raised in our opening discussion of conceptual and practical challenges. We return now to these issues, with some commentary on how we have chosen to deal with them.

WHERE IS POWER TO BE NURTURED?

First, how have we dealt with the conceptual issue of deciding the level at which to target empowerment efforts? Take, for example, our efforts to use workshops to build the capacity of local task forces to understand, use, and perhaps undertake evaluation activities. The local Department of Substance Abuse offered its support for this activity and asked for our recommendations about how to structure these workshops. We were interested in "giving evaluation away"— but to whom? An individual model of empowerment initially led us to consider opening up the workshops to whomever was interested. We began to wonder, however, to what extent changes in individual participants' interest in evaluation would subsequently be supported within their task forces. After all, we were really interested in increasing the capacity of *organizations* to act in a self-directing and effective way with regard to evaluation activities. Ultimately, we requested that each task force formally designate a person (or persons) with responsibility for evaluation issues to attend the workshop. This was intended to ensure that every participant had at least some endorsement from their task force to bring back information and put evaluation issues on the table. In addition, we required that at least two members from a task force register for the workshop. If there was a critical mass of interested individuals from each task force, we reasoned, it would be more likely that something would happen and that evaluation responsibilities would be dispersed beyond one person. Our measure of "capacity-building" success was not only change in participants' skills and knowledge but also changes in the task forces' subsequent activities. Although one can argue with the merits of these strategies, one point is clear: Our self-conscious efforts to "give evaluation away" at the organizational level led us in a different direction than if we had focused on individual-level change.

If empowerment can mean different things to different people, so can "empowerment evaluation." It is important that evaluators be as relentless in clarifying their own purposes, objectives, and outcomes of success as they typically are in their work with others (see also Fetterman, 1984, 1986; Stevenson, 1981; Stevenson & Ciarlo, 1982; Stevenson, Longabaugh, & McNeill, 1979).

CLARIFYING ROLE DEFINITIONS

Although we try to follow sound evaluation practice in incorporating the views of multiple stakeholders in our work, we do not explicitly set out to make our evaluation self-determining for every stakeholder. Our experience has been that it is particularly important to establish the ground rules early on regarding the role of the evaluator vis-à-vis the coalition and the evaluator's responsibilities to various stakeholders. Although this is certainly a good practice in any evaluation, it becomes especially important in our context of unclear evaluator obligation and complex power issues with multiple stakeholders.

In one project, we have attempted to deal with this issue by constructing a detailed statement of when and how evaluation results would be disseminated. Rather than serving as a sterile contractual document, this statement was designed to stimulate discussion regarding who "owned" the evaluation and what our responsibilities were to the various stakeholders involved. As professional evaluators, we certainly had our own position. Opening up this dialogue, however, allowed us to achieve clarity for all to whom we were accountable, as well as the limits of that accountability. The ultimate agreement details how quickly evaluation results will be distributed after data collection, who will get the first opportunity to provide comments, which stakeholders will receive final copies of interim results, and so on. The agreement also makes clear our responsibilities to the funding source.

EMPOWERMENT AND ORGANIZATIONAL CAPACITY BUILDING

Is it realistic to expect indigenous community residents, activists, and agency staff to become professional-quality evaluators? We never had that goal, and we do not think it is a realistic one. As we have worked with community groups, we have learned from experience and revised our objectives along with our intervention, emphasizing the collaborative role and program planning to a greater extent.

In our initial attempts to use workshop formats to train individuals to use *Prevention Plus III* as a planning and evaluation tool, we greatly underestimated the resources and skills needed in two areas: (a) the individual skill and knowledge involved in applying these concepts and (b) the degree of organizational infrastructure needed to support knowledgeable individuals in following through. Over time, we have come to conclude that the degree of expertise needed can vary considerably depending upon the evaluation task under consideration. Different kinds and amounts of capacity building are needed to address different kinds of evaluative issues. We feel that real "empowerment evaluation successes" can occur at different levels of expertise.

For example, after attending one of our workshops and after having received ongoing individualized technical assistance around interpretation of evaluation results regarding their coalition, one group decided they wanted to take additional steps. They proposed using a pre-post control group design to construct an evaluation of a specific program in which they were interested. They were able to conceptualize the problem, formulate a general design, and indicate a commitment to using evaluative findings in their subsequent funding decisions. As members of a consortium of substance abuse prevention task forces, they also had a professional evaluator available to provide individualized assistance around the sampling and measurement issues. The feedback about this dialogue was that involvement in the evaluation planning process alone helped coalition members to be much clearer about what the outcomes they were hoping to have achieved by the program under consideration. In addition, their commitment to formulating evaluation on their own clearly indicated an increased sense of responsibility and control in determining whether the programs they were promoting really worked. If this is an "empowerment evaluation" success, we feel that contributing factors were a clarity at the outset about the level of commitment and expertise needed to make the evaluation happen, and the availability of a support system to plug gaps in the areas of needed expertise.

Because we work with complex, multilevel systems, the issue of "capacity" can be elusive. Capacity for performing important organizational functions, including various kinds of evaluation, may reside at different levels of the system. Although grassroots coalitions may be encouraged to conduct small evaluations with high-intensity technical assistance, outcome instruments shared across local sites may allow a higher-order evaluation that is quasi-experimental or even truly experimental, with an interest in learning more about how the overarching support structure is doing in nurturing the grass roots. We continue to work at both levels, building evaluation capacity as we work to provide formative feedback for broader organizational development.

WHEN, HOW, AND WHY TO SHARE CONTROL OR GIVE IT AWAY

What justification can we offer for "giving evaluation away" and working to make evaluation more relevant for self-determination? Clearly, the Center for Substance Abuse Prevention, our ultimate funding source for the projects we have described, wants objective evidence regarding program effectiveness. Over the past several years, their calls for proposals have become ever more specific about the design and measurement requirements they expect proposed evaluations of demonstration projects to meet. At the same time, however, CSAP has endorsed the role of evaluation as a formative tool for self-change. In recently promulgated principles, CSAP (1993) endorsed the value of evaluation in a variety of forms and levels, the use of evaluation as a "tool for improvement," and the "need to invest in capacity building at the community level in order to develop the skills needed for communities to design and carry out their own evaluations" (p. 2).

We accept and attempt to accomplish both of these broad goals. When evaluators become a part of the program they are evaluating, classic objectivity requirements are compromised and the intervention becomes inextricably entwined with the evaluation process. We believe this is a healthy state of affairs for the complex, struggling social innovations we are attempting to study. We do not believe this relationship precludes the attempt to provide relatively accurate, unbiased feedback or the possibility of higher-order evaluation that takes the embedded, formative role of local evaluation into account. We do not prescribe this approach for all evaluation activities, but in our context the gains outweigh the risks.

There clearly are real risks. The evaluation standards cited by Stufflebeam (1994) make a strong statement about the professional responsibilities of the evaluator, and they also make clear the basis for credibility of evaluation: professional skills; established methods; and an objective, impartial stance. To abandon this basis for credibility is to give up a well-defined identity. We wish to stake a claim to a middle ground where there are standards of evidence, guidelines for reducing bias, and the intention to construct externally relevant stories of program effects, but these are coupled with a recognition of the special needs of particular clients and contexts. Are traditional methods and measures for "professional-quality" evaluation wholly satisfactory in alternative cultural settings with legitimately suspicious and hard-to-reach populations? Will our imposed designs alienate the activists we must work with if real community change is to take place? Can we be confident that measures developed elsewhere will be valid for a particular culture and setting? We certainly see a crucial role for professional evaluators, but we believe it will include learning some new skills as well as teaching some new skills in order to work in the collaborative environment we have been exploring.

At this point, it should be abundantly clear that we are not arguing for the universal application of "empowerment evaluation," nor are others who advocate its use (e.g., Fetterman, 1994a). We believe there are conditions in which sharing control is an effective evaluation strategy, and we believe that the coalitional context in which we have been working is a good example of those conditions. There are multiple levels in the evaluation design, and control is not being given away in the same sense at all levels. There is a special emphasis on empowerment as an aspect of the program being evaluated, making this an important norm for all involved. There are complex, emergent social processes that do not present a static, narrowly circumscribed target for evaluation. The clients for the evaluation, both local and national, are supportive of evaluation methods consistent with an empowerment approach. We enjoy working in that kind of an environment, and we believe the conceptualization and tools we have evolved are appropriate to the context.

References

Butterfoss, F. D., Goodman, R. M., & Wandersman, A. (1993). Community coalitions for prevention and health promotion. *Health Education Research, 8*, 315-330.
Chadwick, K. (1993, November). *Providing technical assistance for community program evaluation.* Paper presented as part of a panel, Empowering Stakeholders in

Community Prevention Evaluation, held at the annual meeting of the American Evaluation Association, Dallas, TX.

Chavis, D., & Florin, P. (1990). *Community development, community participation, and substance abuse prevention.* San Jose, CA: County of Santa Clara, Department of Health.

CSAP. (1993). *CSAP evaluation agenda: Ten principles.* Unpublished document. (Available from CSAP, Substance Abuse and Mental Health Services Administration, Rockville, MD 20857)

Farquhar, J. W., Fortmann, S. P., Flora, J. A., Taylor, B., Haskell, W. L., Williams, P. T., Maccoby, N., & Wood, P. D. (1990). Effects of community wide education on cardiovascular disease risk factors: The Stanford Five-City Project. *Journal of the American Medical Association, 264,* 359-365.

Fawcett, S. B., Paine, A. L., Francisco, V. T., & Vliet, M. (1993). Promoting health through community development. In D. Glenwick & L. A. Jason (Eds.), *Promoting health and mental health: Behavioral approaches to prevention* (pp. 233-255). New York: Haworth.

Fetterman, D. M. (1984). Guilty knowledge, dirty hands, and other ethical dilemmas: The hazards of contract research. In D. M. Fetterman (Ed.), *Ethnography in educational evaluation* (pp. 211-236). Beverly Hills, CA: Sage.

Fetterman, D. M. (1986). The ethnographic evaluator. In D. M. Fetterman & M. A. Pitman (Eds.), *Educational evaluation: Ethnography in theory, practice, and politics* (pp. 21-47). Beverly Hills, CA: Sage.

Fetterman, D. A. (1994a). Empowerment evaluation [Presidential address]. *Evaluation Practice, 15*(1), 1-15.

Fetterman, D. A. (1994b). Steps of empowerment evaluation: From California to Cape Town. *Evaluation and Program Planning, 17*(3), 305-313.

Florin, P., Chavis, D., Wandersman, A., & Rich, R. (1992). A systems approach to understanding and enhancing grassroots organizations: The Block Booster Project. In R. Levine & H. Fitzgerald (Eds.), *Analysis of dynamic psychological systems* (pp. 215-243). New York: Plenum.

Florin, P., Mitchell, R., & Stevenson, J. (1989). *A process and implementation evaluation of the Rhode Island Substance Abuse Prevention Act.* Report submitted to the RI Department of Substance Abuse and the RI General Assembly, Providence, RI.

Florin, P., Mitchell, R., & Stevenson, J. (1993). Identifying training and technical assistance needs in community coalitions: A developmental approach. *Health Education Research: Theory and Practice, 8*(3), 417-432.

Forss, K., Cracknell, B., & Samset, K. (1994). Can evaluation help an organization to learn? *Evaluation Review, 18*(5), 574-591.

Gibbs, J. (1979). The meaning of ecologically oriented inquiry in contemporary psychology. *American Psychologist, 34,* 127-140.

Hawkins, J. D., Catalano, R. F., & Miller, J. Y. (1992). Risk and protective factors for alcohol and other drug problems in adolescence and early adulthood: Implications for substance abuse prevention. *Psychological Bulletin, 112*(1), 64-105.

Jacobs, D. R., Luepker, R. V., Mittelmark, M. B., Folsom, A. R., Pirie, P. L., Mascoili, S. R., Hannan, P. J., Pechacek, T. F., Bracht, N. F., Carlaw, R. W., Kline, F. G., & Blackburn, H. (1986). Community-wide prevention strategies: Evaluation design of the Minnesota Heart Health Program. *Journal of Chronic Disease, 39,* 775-788.

Kaftarian, S. J., & Hansen, W. B. (1994). Improving methodologies for the evaluation of community-based substance abuse prevention programs. *Journal of Community Psychology, 22,* 3-6 [CSAP Special Issue, Community Partnership Program].

Kenny, D. A., & LaVoie, L. (1985). Separating individual and group effects. *Journal of Personality and Social Psychology, 48,* 339-348.

Klitzner, M. (1993). A public health/dynamic systems approach to community-wide alcohol and other drug initiatives. In R. C. Davis, A. J. Lurigio, & D. P. Rosenbaum (Eds.), *Drugs and the community: Involving community residents in combating the sale of illegal drugs* (pp. 201-224). Springfield, IL: Charles C Thomas.

Lincoln, Y. S. (1994). Tracks toward a postmodern politics of evaluation. *Evaluation Practice, 15*(3), 321-338.

Linney, J. A., & Wandersman, A. (1992). *Prevention Plus III.* Rockville, MD: Center for Substance Abuse Prevention.

McMillan, B., Florin, F., Stevenson, J., Kerman, B., & Mitchell, R. E. (in press). Empowerment praxis in community coalitions. *American Journal of Community Psychology.*

Mitchell, R., Stevenson, J., & Florin, P. (in press). A typology of prevention activities: Applications to community coalitions. *Journal of Primary Prevention.*

O'Neill, P. (1989). Responsible to whom? Responsible for what? Some ethical issues in community intervention. *American Journal of Community Psychology, 17,* 323-341.

Pentz, M. A., Dwyer, J. H., MacKinnon, D. P., Flay, B. R., Hansen, W. B., Wang, E. Y., & Johnson, C. A. (1989). A multicommunity trial for primary prevention of adolescent drug abuse. *Journal of the American Medical Association, 261,* 3259-3266.

Perry, C. L. (1986). Community-wide health promotion and drug abuse prevention. *Journal of School Health, 56,* 359-363.

Peterson, P. L., Hawkins, J. D., & Catalano, R. F. (1992). Evaluating comprehensive community drug risk interventions: Design challenges and recommendations. *Evaluation Review, 16*(6), 579-602.

Prochaska, J. O., & DiClemente, C. (1992). In search of how people change: Applications to addictive behaviors. *American Psychologist, 47,* 1102-1114.

Shadish, W. R. (1994). Need-based evaluation: Good evaluation and what you need to know to do it. *Evaluation Practice, 15*(3), 347-358.

Shadish, W. R., Cook, T. D., & Leviton, L. C. (1991). *Foundations of program evaluation: Theories of practice.* Newbury Park, CA: Sage.

Shinn, M. (1990). Mixing and matching: Levels of conceptualization, measurement, and statistical analysis in community research. In P. Tolan, C. Heys, F. Chertok, & L. Jason (Eds.), *Researching community psychology* (pp. 111-126). Washington, DC: American Psychological Association.

Stevenson, J. (1981). Assessing evaluation utilization in local human service agencies. In J. Ciarlo (Ed.), *Utilizing evaluation: Concepts and measurement techniques* (pp. 35-57). Beverly Hills, CA: Sage.

Stevenson, J., & Ciarlo, J. (1982). Enhancing the utilization of mental health evaluation at the state and local levels. In J. Stahler & W. Tash (Eds.), *Innovative approaches to mental health evaluation* (pp. 367-386). New York: Academic Press.

Stevenson, J., Florin, P., & Mitchell, R. (1991, November). *Giving evaluation away.* A panel presented at the annual meeting of the American Evaluation Association, Chicago.

Stevenson, J., Florin, P., & Mitchell, R. (1993, February). *Understanding dissemination and implementation issues: Maximizing the usefulness of evaluation.* Paper presented at the New Dimensions in Prevention conference (sponsored by the Center for Substance Abuse Prevention), Washington, DC.

Stevenson, J., Longabaugh, R., & McNeill, D. (1979). Meta-evaluation in the human services. In H. Schulberg & J. Jerrell (Eds.), *The evaluator and management* (pp. 37-54). Beverly Hills, CA: Sage.

Stufflebeam, D. L. (1994). Empowerment evaluation, objectivist evaluation, and evaluation standards: Where the future of evaluation should not go and where it needs to go. *Evaluation Practice, 15*(3), 321-338.

Swift, C., & Levin, G. (1987). Empowerment: An emerging mental health technology. *Journal of Primary Prevention, 8,* 71-94.

Tobler, N. S. (1986). Meta-analysis of 143 adolescent drug prevention programs: Quantitative outcome results of program participants compared to a control or comparison group. *Journal of Drug Issues, 16,* 537-568.

Wallerstein, N. (1992). Powerlessness, empowerment, and health: Implications for health promotion programs. *Health Promotion, 6*(3), 197-205.

Zimmerman, M. (in press). Empowerment theory: Psychological, organizational and community levels of analysis. In J. Rappaport & E. Seidman (Eds.), *The handbook of community psychology.* New York: Plenum.

Zimmerman, M. A., & Rappaport, J. (1988). Citizen participation, perceived control, and psychological empowerment. *American Journal of Community Psychology, 16*(5), 725-749.

11

Fairness, Liberty, and Empowerment Evaluation

DENNIS E. MITHAUG

Empowerment evaluation applies to individuals, organizations, and societies or cultures (Fetterman, Chapter 1, this volume). The purpose of empowerment evaluation on the individual level of analysis is individual self-improvement through methods that compare Person with herself in attempts to improve Person's capacity to fulfill her mission in life in ways that are consistent with her needs, interests, and abilities.

At the organizational level, empowerment evaluation is a powerful tool that helps participants in an organization connect their needs, interests, and abilities with the means and ends that define the organization's activities and purpose. The purpose of empowerment, on this level of analysis, is organizational self-improvement through methods that compare the organization with itself in an effort to improve organizational capacity to fulfill its mission in ways that are consistent with the needs, interests, and abilities of its participants.

The concept of empowerment evaluation is as fundamental to assessing societal fairness in liberty for all as it is to evaluating

AUTHOR'S NOTE: Portions of this chapter also appear in a book manuscript titled *Equal Opportunity Theory: Fairness in Liberty for All.*

organizational effectiveness through the voluntary participation of all. To make this connection between fairness and effectiveness, we need only remember that in both cases people feel empowered to the extent they are self-determined. Empowerment evaluation at the societal and cultural levels is implicated in a process that determines the extent to which all members of society have a fair chance of pursuing those self-defined ends in life that are most fulfilling.

Empowerment evaluation at the program or organizational level is the most highly visible form of this approach (Fetterman, 1994a, 1994b, and Chapter 1, this volume). It is an approach to evaluating social programs that permits those carrying out the work of the organization to gather information on progress toward programmatic ends so as to increase a sense of control, to improve a sense of involvement in the mission and vision of the enterprise, and to reach beyond themselves in the pursuit of excellence. Empowerment evaluation empowers as it evaluates by reversing the order of control and rectification. Rather than imposing evaluation and remediation from outside, empowerment evaluation encourages self-evaluation and self-adjustment from inside. It creates user-friendly environments by motivating improvement rather than fomenting user-hostile environments that create defensiveness. It assumes goodwill and plays on that assumption so those with insider knowledge about how things work will use what they know to correct or adjust what they discover through self-evaluation does not work.

In addition to these parallels between individual, organizational, and societal or cultural uses of empowerment evaluation, there is always the question of which organizations and societies are most likely to adopt empowerment evaluation. The argument in this chapter is that empowerment evaluation is most likely in societies that value fairness in liberty for all because, when the premise of fairness in liberty for all is acknowledged, then empowerment evaluation is the logical method for monitoring progress toward that ideal.

Equal Opportunity Theory and Empowerment Evaluation

Equal opportunity theory helps us understand this connection between the ideal of fairness in liberty for all and the appeal of empowerment evaluation by focusing social redress on the discrepancy between the *right* of self-determination and the *experience* of self-determination, a condition in need of analysis and evaluation (Mithaug, 1995). Equal opportunity theory locates the cause of this discrepancy in two measurement targets—*individual capacity* and *social opportunity*. The theory claims that the discrepancy between the right and the experience of self-determination is due to the lack of capacity and lack of opportunity among individuals whose personal, social, and economic circumstances are beyond their control. By claiming that every member of society deserves an optimal chance of securing the good in life, the theory explains our collective responsibility for assuring fair prospects for all. The theory shows that when prospects for self-determination are distributed fairly, they are equally optimal for all. Although one person's pursuits will be different than another person's, prospects are nonetheless comparable because all individuals have roughly the same chance of pursuing or not pursuing, of fulfilling or not fulfilling, their own ends *over the long term*.

This means that social inequality is a problem of unequal prospects for engaging and succeeding in self-determined pursuits. When prospects for pursuing the individually defined good in life are not distributed equally among members of a society, the ideal of liberty for all is jeopardized. And in all countries of the world, including the United States, this is the case. Substantial numbers of individuals fail to engage and succeed in their own pursuits and as a consequence lose control over life's circumstances. The persistent pattern of failure they experience leads to a loss of hope and a growing sense of helplessness and despair that destroys the very basis of their self-respect. They become victims in a cycle of personal, social, and economic decline that debilitates and erodes their capacity for improving their own prospects for life. These individuals need and deserve help to reestablish their experience of self-determination.

Equal opportunity theory justifies social redress on behalf of less-well-situated individuals by claiming that (a) all persons have the right to self-determination, (b) psychological and social conditions of freedom cause some individuals and groups to experience unfair advantages in determining their future, (c) declines in self-determination prospects for the less fortunate are due to social forces beyond their control, and (d) as a consequence of these declines, there is a collective obligation to improve prospects for self-determination among less-well-situated groups. The collective action proposed by the theory is to optimize prospects for self-determination among the less fortunate by improving their *capacity* for autonomous thought and action, by improving their *opportunities* for effective choice and action, and by optimizing the match between individual capacity and social opportunity by eliminating obstacles and constructing opportunity that encourages more frequent expressions of self-determination.

Empowerment evaluation can play a central role in tracking progress toward the fairness in liberty for all ideal to the extent it focuses upon these key variables of self-determination—*capacity, opportunity,* and *engagement.* In the remainder of this chapter, I will present what I believe to be the essential components of a societal approach to empowerment evaluation. I will begin by reviewing recent gains toward fairness for students with disabilities and by describing a conception of person and a definition of opportunity that is consistent with equal opportunity theory and with the optimal prospects principle. I will end the chapter by describing a method of assessing levels of self-determination that employs these new conceptions of person and opportunity.

Gains Toward "Fairness in Liberty for All"

Today we believe every person deserves a reasonable chance of engaging and succeeding at pursuits in life that are commensurate with his or her interests, needs, and abilities. This is what we mean by fairness in liberty for all. It is an ideal that has become universal and, as a consequence, has become a standard against which we assess the moral worth of a society. One way to check our progress toward social justice is by noting how far our nation has come in extending the freedom experience to children and youth with disabilities. Whereas a half century ago there was little thought (much less consideration) of the opportunity they deserved for liberty under the Constitution, now there is serious discussion about how social policy and educational opportunity can improve their prospects for self-determination by the end of the century. The *National Agenda for Achieving Better Results for Children and Youth With Disabilities* articulated this expectation for the year 2000, which "begins with images of children and youth with disabilities having access to supports and services that lead to self-actualization, self-determination, and independence" (U.S. Department of Education, 1994, pp. 4-5). This is but a reaffirmation of what people around the world have come to believe is a condition of life in every democratic society—that all persons have a right to self-determination. Indeed, Article 1 of the International Covenant on Civil and Political Rights (ICCPR) adopted by the General Assembly of the United Nations in 1960 states: "All peoples have the right to self-determination. By virtue of that right they freely determine their political status and freely pursue their economic, social and cultural development" (Humana, 1992, p. 385).

This vision for children and youth with disabilities is consistent with its legitimating anchor in *Brown v. Board of Education* in 1954 and its enabling legislation in the Education for All Handicapped Children Act of 1975, the Individuals With Disabilities Education Act of 1990, and the Americans With Disabilities Act of 1991. It is also consistent with the claim that all societies should *promote the realization of the right to self-determination* among *all* members of society, as stated by the United Nations in Article 1 of the ICCPR:

> The States Parties to the present Covenant, including those having responsibility for the administration of the Non-Self-Governing and Trust Territories, *shall promote the realization of the right of self-determination,* and shall respect that right, in conformity with the provisions of the United Nations Charter. (Humana, 1992, p. 385, italics added)

This vision captures a special moment in human history, a time when one country—the United States—has come as close as any in realizing this right of self-determination for *all* members of society, including those least likely to succeed in life. Indeed, its policies in the treatment of children and youth with disabilities are a beacon of hope for advocates of fairness in liberty for all. Because if members from least-likely-to-succeed groups can find their way to fulfillment through pursuit of the self-determined life, then there is no reason members of every other excluded group cannot benefit from universal fairness as well. For arrayed against children with disabilities are widely held beliefs deeply grounded in arguments for the futility and waste of social redress on behalf of any person lacking substantially in capacity for living the self-determined life.

But in the United States, social redress on behalf of all left-out groups is based upon a set of premises about capacity and opportunity that challenges ancient conceptions of person and society on fundamental grounds. This new set of assumptions about what constitutes a person and what can be defined as an opportunity is captured in the central idea behind the *optimal prospects principle,* which presumes that every person is an individual with a special set of talents, interests, and needs and is deserving of a fair chance—*an equally favorable opportunity*—of expressing those unique attributes in pursuit of self-defined ends in life. Consequently, there can be no overarching social mechanism for sorting individuals into categories of deserving and undeserving when it comes to distributing access to the fair chance. Every person deserves a fair chance. This is a pervasive value in American life. Only when everyone has had a fair chance is it also fair and just to expect each person to live with the consequences produced by their own engagement of that fair chance.

The optimal prospects principle promotes fairness in liberty for all by focusing social redress on the essential *means* for being self-determined— the individual's *capacity* and *opportunity* to choose and to enact choice in pursuit of self-defined needs and interests. The principle is sensitive to variation in both capacity and opportunity and to the interaction between them. It reflects the empirical fact that when either capacity or opportunity to self-determine is diminished or constrained, the probability of self-determination diminishes, and when one's chances of engaging in self-determined pursuits decline, then fairness in liberty for all is threatened. The optimal prospects principle is based upon an understanding of how individuals interact with opportunity to improve their chances of getting what they need and want in life. When opportunities are *just-right challenges*—when they offer the right amount of risk for the gain expected—they will be pursued. *All persons* regardless of who they are, where they come from, or whether or not they have a disability or a disadvantaged background will think and act on just-right opportunities *repeatedly* to learn what they need to learn and to adjust what they need to adjust to reach the ends they most desire (Mithaug, 1993). In other words, all persons have the ability to regulate their thoughts, feelings, and actions in pursuit of ends that define themselves *as self-determining persons, that define themselves as being free.*

Application of the optimal prospects principle on behalf of individuals in need of social redress results in just-right matches between opportunity and capacity. These just-right matches, in turn, engage the thoughts and actions of those receiving that redress by empowering them to enhance their own capacity and to improve their own opportunities for living the self-determined life. It matters little if the persons empowered are disabled, nondisabled, impoverished, or enriched because the goal is the same for all, *increased engagement* in challenging opportunity to pursue desirable ends in life. Indeed, this has been the direction, if not the content, of compensatory policies emanating from equal opportunity programs designed to improve prospects for disadvantaged and disabled individuals in past decades. Court decisions and legislative mandates in the 1960s and 1970s focused upon building *capacity* for learning through Head Start for disadvantaged youth and through individualized educational planning and instruction for students with disabilities. They also attempted to restructure *social opportunity* by requiring capacity building to take place in desegregated schools for African American students and in least restrictive environments for students with disabilities. The intention was for these early experiences to improve prospects for pursuing adult opportunity after school. Schooling was to provide comprehensive social redress through capacity building and opportunity enhancement.

The lessons to be learned from these policies of social redress on behalf of the least advantaged members of society is that age-old conceptions of person and opportunity have changed, although our ways of measuring and assessing people and their opportunity perhaps have not. To understand how capacity and opportunity interact to produce variation in the experience of self-determination, we must conceive of Person as having *variable capacity* that functions to expand and contract over time, depending as it does on variation in the optimality or favorableness of opportunity—the social and physical circumstance in which choice and action must occur. We must move from thinking of people as *static* entities with distinguishing color, sustaining physical disability, enduring low IQ, and fixed economic condition, and consider them instead as dynamic entities capable of building capacity to make circumstance more favorable for their pursuits in life. We must conceive of Person as an entity with variable capacity that expands and contracts according to the opportunity she engages when expressing that capacity.

The Concept of "Person"

In the past, Person was defined at birth as attached to a station in the social structure specifying a lifetime of fixed opportunity and outcome. Capacity was a constant, fused with fixed opportunity and predictable outcome defining what it was to be a person. To know one's social position was to know everything else about that person. To some extent, this conception has endured to this day, even though it conflicts from time to time with belief about our unlimited potential for growth and accomplishment. Perhaps this is because individual capacity is often confused with individual potential, which has a longtime association with psychometric measures of cognitive ability and the popularization of the intelligence scores.

Nevertheless, there are conceptions of individual capacity more in keeping with what we know about social life and about the sociopolitical process affecting opportunity for self-direction in that life. Contemporary views of intelligence such as those reported in Sternberg's (1988) *The Triarchic Mind: A New Theory of Human Intelligence* and Howard Gardner's (1993) *Frames of Mind: The Theory of Multiple Intelligences* remind us that, stripped of its psychometric manifestations, intelligence is still reflective of adaptive functioning through self-regulated problem solving. It is intelligent adaptation to environmental circumstance that differentiates humans from each other and from other species, and this capacity is not easily captured in a single psychometric score of mental ability. Contrary to popular claims that intelligence scores are genetic markers for position in the social structure (Herrnstein & Murray, 1994), IQ correlates poorly with the real-life problem solving required in adapting to life's circumstances, as Sternberg (1988, p. 211) has pointed out.

The reason for these weak associations between intelligence test scores and human accomplishment is that intelligence is *more than* what can be measured by a single psychometric test. Human intelligence is, according to Sternberg, a kind of "mental self-management of one's life in a constructive, purposeful way." It is management of one's personal, social, economic, and technical resources that affects the pursuit of opportunity to secure from the environment those means that are necessary for reaching desirable ends in life. Human intelligence is the capacity to act and react to opportunity in ways that yield beneficial gain. According to Sternberg (1988):

> Intelligence involves the shaping of, adaptation to, and selection of your environment. . . . *There may be no single set of behaviors that is intelligent for everyone.* . . . What does appear to be common among *successful people is the ability to capitalize on their strengths and compensate for their weaknesses.* Successful people are not only able to adapt well to their environment but also to modify this environment in order to increase the fit between the environment and their adaptive skills. (pp. 16-17, italics added)

The reason psychometric scores on intelligence measures fail to reflect fully the range and variety of human engagement in opportunity for self-determined pursuits is that individuals vary in their capacity and in how they employ that capacity to determine their own ends in life. The failure of psychometric measures to assess this capacity should not suggest that such assessments are not possible, because this is not the case. In fact, we can track variation in a person's capacity to self-determine in several ways.

First, we can consider an individual's level of *self-knowledge* and *self-awareness* of personal needs, interests, and abilities because individuals vary in how acutely aware they are of themselves and, as a consequence, how effectively they will act in engaging opportunity to advance pursuits that are consistent with this awareness of self. Moreover, this variation in their sense of self affects their motivation to pursue *self-directed capacity building* that is consistent with their self-knowledge and self-awareness. This view is consistent with Nathan Branden's (1994) definition of *self-esteem* as

> confidence in our ability to think, confidence in our ability to cope with the basic challenges of life, and confidence in our right to be successful and happy, the feelings of being worthy, deserving, entitled to assert our needs and wants, achieve our values, and enjoy the fruits of our efforts. (p. 4)

It is also consistent with the Supreme Court's decision in *Brown v. Board of Education* (1954) regarding the effects of segregated schooling on the self-esteem of black children and how lowered self-esteem diminishes their capacity to benefit from schooling.

A second way we can track variation in capacity is by assessing the skills, knowledge, and motivation individuals use *to manage their resources effectively and efficiently* to get what they need and want in life. Those who are skilled self-managers, who have acquired information about when and how to use their skills effectively, and who are motivated to apply those skills and knowledge persistently and continuously are usually more successful than those deficient in these respects. People in business and industry have recognized the importance of these skills since Napoleon Hill first described them more than 30 years ago in *Think and Grow Rich* (1960), which identified five steps to securing desirable ends in life: (a) Choose a definite goal to be obtained, (b) develop sufficient power to attain that goal, (c) perfect a practical plan for attaining the goal, (d) accumulate specialized knowledge necessary for attaining the goal, and (e) persist in carrying out the plan (p. 157). Sternberg (1988) has identified a similar set of self-management skills that he considers essential for success:

> My own research and that of others suggests that they are among the most important, because they must be used in the solution of almost every real-world problem. The processes are: recognizing the existence of the problem, defining the nature of the problem, generating the set of steps needed to solve the problem, combining these steps into a workable strategy for problem solution, deciding how to represent information about the problem, allocating mental and physical resources to solving the problem and monitoring the solution to the problem. (p. 79)

The third way of tracking variation in capacity is by assessing the resources available to Person for engaging different opportunities. These resources can be classified as personal, social, economic, and technical. *Personal resources* include the excess time, energy, and behavior Person has to engage a given opportunity; *social resources* are the time, energy, and behavior of other people that are available to her; *economic resources* are the exchangeable currency and capital she has at her disposal to access resources she is lacking; and *technical resources* are the specific means over which Person has personal control for engaging specific types of opportunities.

Individuals with sufficient personal resources to engage specific opportunities have the time and behavior necessary to alter social and physical circumstances defined by that opportunity in ways that increase their chances of getting what they need and want. Individuals with social resources have access to other people's resources in altering environmental circumstances for their own benefit. Individuals who have economic resources can purchase resources they lack in order to optimize environmental circumstance for a given pursuit, and individuals who have technical resources have personal control over the means of changing an environmental circumstance simply by applying their knowledge, skills, behaviors, and time to get the results they want when they want them.

By tracking these three dimensions of capacity—self-knowledge, self-management, and resource access—we can assess how different patterns of growth and constraint can affect and be affected by changes in environmental circumstance at any point in time. Variation in capacity will occur within and across the three domains according to the level and type of capability present: Different individuals will have different interests, needs, and abilities; different strategies for managing themselves; and different resources for engaging opportunity. They will also have different experiences of empowerment due to variation in interaction among these dimensions and with those changing environmental circumstance we call opportunity.

This conception of Person suggests a different strategy for measuring capacity to self-determine. It suggests an approach that assesses an individual's self-knowledge, self-management, and resource availability. It also suggests a method that determines how these three domains interact to produce an effect on the environment—a change in the favorableness or optimality of opportunity for producing a desirable gain. Finally, this conception of Person suggests a different conception of "opportunity" to self-determine.

The Concept of "Opportunity"

According to this analysis, opportunity is an occasion in time and space that is more or less favorable for pursuit of a given end. And the degree of this favorableness is expressed as an "optimality." Hence, the greater the optimality of opportunity for a given pursuit contemplated by an actor, then the more favorable the environmental circumstance for producing gain by engaging that opportunity with action on those circumstances. Optimalities of opportunity are simply *probabilities of success* inferred by actors evaluating the circumstances surrounding the actions they plan to take to reach a desirable end. Person *infers* this prospect or probability by assessing her capacity for effective action based upon her self-knowledge, self-management, and resource availability and by evaluating the favorableness (optimality) of her social and physical circumstances. The inference drawn from these assessments of capacity and circumstance constitutes the optimality of opportunity that will affect her decision to act or not to act on that set of circumstances.

If she decides to act, she will attempt to alter those social and physical circumstances so she can reach her own ends and, depending upon the consequences of her actions, she will have altered the probability of success for subsequent endeavors. She will have changed the optimalities of opportunity by making her circumstances more favorable for subsequent action, or she will have failed to produce the desired effect and perhaps made subsequent opportunity less favorable. So from one engagement episode to another, Person's optimalities of opportunity will change, reflecting as they will different conditions of capacity and different conditions of environmental demand.

According to this conception of opportunity, it is difficult logically and empirically to separate the conception of person from the conception of opportunity because the two are interdependent. Person's self-knowledge, self-management, and resource access affect her inferences about environmental circumstances, and the optimalities of opportunity she infers from assessing those circumstances affect her decisions to act in pursuit of various ends in life. Moreover, Person's success or failure in changing environmental circumstances to improve her chances—to increase optimalities of opportunity—will feed back to affect her subsequent levels of self-knowledge, self-management, and resource availability. In other words, Person's capacity and opportunity interact in various ways to expand capacity and optimize opportunity or to contract capacity and suboptimize opportunity. The only way to know what is cause and consequence is by knowing Person's capacity and inferred opportunity because it is Person and no one else who assesses the favorableness of a given opportunity prior to engaging those circumstances that lead to a self-determined end in life.

So to claim that opportunities among a population of actors should be equal at some imaginary starting point or that they should yield equal outcomes at some imaginary ending point is to ignore the empirical and contingent reality that there is no basis for equivalence. Every individual has a different capacity and every opportunity a different inference of optimality because optimalities of opportunity depend upon an *individual's* assessments of capacity, circumstance, and the correspondence between the two. To expect equivalence of capacity and opportunity is to expect individuals to be identical and to expect the circumstances for their actions to be equivalent. All we can expect by way of equality, then, is that *prospects* or *probabilities* for self-determined action be roughly equivalent across individuals.

The phrase *equality of opportunity* often obfuscates this essential meaning of opportunity. Viewed simply as favorable occasions for action, opportunities can be equal only if they are equally favorable for the pursuits envisioned, only if the opportunities for engagement have comparable *optimalities*. They are comparable but not equivalent because each opportunity depends upon the purpose Person has in mind for altering circumstances to achieve some end. Therefore, opportunities for one purpose may have one optimality valence, whereas opportunities for another purpose will have a different optimality valence. The idea of equal opportunity is meaningless without some consideration of the optimality Person has inferred after examining a set of circumstances and after evaluating her capacity to deal with those circumstances to produce the outcome she desires.

The conclusion to draw from this analysis is that an individual's prospects for living the self-determined life are *specific to that individual* depending as they do upon that person's capacity to self-determine and the optimalities of opportunity she believes will affect her capacity to be successful. Therefore, given that (a) individuals vary in their capacity to self-determine and (b) individuals vary in their judgment about the favorableness of opportunity for the expression of that capacity, then (c) individuals vary in their prospects or chances of engaging in self-determined pursuits. So as capacity to self-determine declines, Person's assessments of her optimalities of opportunity may decline as well, and as these two condition become suboptimal, Person becomes less likely to self-determine; she is less likely to engage opportunities and initiate pursuits of her own choosing. The converse relationship is also possible, as are many other combinations of capacity-opportunity interaction. The point is that when the same people experience declining prospects for self-determination, then there is justification for social redress to improve their prospects, to make their chances comparable to the optimalities of opportunity other people experience.

The focus of social redress suggested for this problem of declining prospects for self-determination is twofold—to increase Person's capacity and to improve her opportunities. And the likely method of pursuing these policy objectives to reverse the declining prospects of the least advantaged members in society is through *educational interventions that build capacity to optimize opportunity*. And to be effective, these interventions will have to accomplish three objectives. First, they will have to improve Person's *understanding of her own needs, interests, and abilities* so she will be able to pursue opportunities consistent with those needs, interests, and abilities. Second, they must provide Person with the *specific personal, social, and technical knowledge and skills* required to optimize a full range of opportunities in the social and physical world. And, third, they must help Person learn to *manage those resources* effectively and efficiently when she engages opportunity to achieve her own ends in life.

The expected outcome of these educational interventions is *increased engagement* in self-determined pursuits. As Person engages opportunity more frequently, she is more likely to improve her social and physical circumstances for action leading to ends in life she values. And as these increased engagements succeed in improving her circumstances, they also enhance her capacity to engage more challenging opportunity. This, in turn, improves Person's chances of becoming a lifelong learner as she adjusts more optimally to her circumstances, making them more favorable for her own pursuits wherever and whenever she can. In this sense, Person experiences self-determination.

Levels of Self-Determination

The implication of this analysis for measuring progress toward fairness in liberty for all is that the *process* by which Person becomes self-determining is as important for assessing prospects for self-determination as are any outcomes produced by various pursuits. Because, for prospects for self-determination to improve, capacity must increase, opportunity must become more favorable, and Person must engage that opportunity. Indicators of progress toward fairness would, therefore, include (a) measures of individual capacity, (b) measures of the optimalities of social opportunity with respect to an individual's capacity, and (c) measures of the individual's engagement of opportunity for self-determined ends. These three variables provide a yardstick for monitoring progress toward fairness. When capacity indicators indicate growth over time, when optimalities of opportunity are judged to be increasing by those in need of social redress, and when engagement in personal pursuits is increasing for those least likely to succeed in life, then the discrepancy between the right and the experience of self-determination is declining for individuals least well situated in society.

This measure of freedom departs from those used by the United Nations, which focus on the sociopolitical circumstances or rights guaranteeing opportunity in an array of domains defining social life in different countries around the world, and they are different than measures of socioeconomic gain associated with different groups and reflected in income gaps between the most and least advantaged in society. Although these traditional measures of inequality validate the existence of the problem of unequal prospects for freedom by describing the discrepancy between the right and the experience of self-determination, they do not measure the *process* by which individuals gain and lose capacity and opportunity for self-determination nor do they measure how these variations affect the engagement of opportunity and the pursuit of self-defined ends in life. The *process* indicators of freedom—capacity, opportunity, and engagement—reflect the *means* of producing the self-determination experience while the *outcome* indicators used by the United Nations reflect the *distribution of advantages and disadvantages* among different members who vary in their experience of freedom. In this sense, the process indicators of freedom provide an answer to some of the major measurement problems Tim Gray (1991) identified in his book *Freedom:*

> Another conclusion is that more research is required if we are to make progress in applying the concept of freedom. In particular, considerable

analytical and empirical work is necessary in order to clarify and refine our understanding of the ways in which freedom can be measured and distributed. The current literature abounds with discussion (though sometimes aridly combative in nature) of the meaning of freedom, and there is increasing interest in the subject of the justification of freedom. But precious little effort has been devoted to the issues of the measurement and distribution of freedom. Yet comparative judgments concerning both the extent of freedom in different countries and the impact of governmental intervention upon the overall level of freedom and its distribution within a society are made confidently day after day, with little real grasp of the complexity of some of the issues involved. Since it seems that these judgments do play a role in important domestic and foreign policy decisions made by governments, it is a matter of some urgency that research is undertaken in order to place such judgments upon a sounder footing than at present. (p. 173)

The process indicators are also similar to those employed in research projects funded by the U.S. Department of Education's Office of Special Education Programs and charged with developing a definition and measure of self-determination appropriate for use with children and youth with disabilities. The working definition of *self-determination* agreed to by those projects was "choosing and enacting choice to control one's life to the maximum extent possible, based upon knowing and valuing one's self and in pursuit of one's needs, interests, and values."[1] The assumption undergirding this initiative to assess levels of self-determination was that students with disabilities needed to be more self-determined if they were to achieve their goals in life. Some of the projects developed measures of self-determination that included a school component for *capacity building* through more effective self-regulated problem solving to meet personal goals in life as well as an *opportunity* component for the expression of that capacity through pursuit of vocational training, job placement, and independent living.

The intended use of the self-determination data produced by these instruments was to guide educators in teaching the skills and behaviors necessary for students to become more self-determined at school, home, and in their community. The assessment instrument, developed by the American Institutes for Research and Teachers College, Columbia University (AIR-TC), assessed students' capacity and opportunity to self- determine, with scores on these two dimensions indicating the students' levels of self-determination.[2] This approach permitted variations in capacity and opportunity to yield comparable levels of (prospects for) self-determination across different students. For example, some students with greater capacity and lesser opportunity than other students scoring lower on capacity but more favorable on opportunity could have equivalent levels of (prospects for) self-determination.

The capacity component of the AIR-TC instrument focused only on two of the three capacity domains—*self-knowledge* and *self-management*. The *resource* component was not included. The self-knowledge and self-management components included items on thinking, acting, and adjusting to opportunity for self-determined pursuits. Students responded to the items by indicating how often they

- Identified and expressed their own needs, interests, and abilities
- Set expectations and goals to meet those needs and interests
- Made choices and plans to meet goals and expectations
- Took actions to complete plans
- Evaluated the results of actions
- Altered plans and actions as necessary to meet goals more effectively

The opportunity component of the assessment required students to indicate how often and to what extent they experienced circumstances at home and at school that permitted or encouraged them to perform each of these six behaviors.

The AIR-TC project tested and validated this self-determination instrument on 484 students with and without disabilities ranging in age from 6 to 25 years. The results indicated that (a) male and female students did not have significantly different scores on levels of self-determination; (b) students who were African American, Hispanic, or white did not have significantly different scores on self-determination; (c) economically disadvantaged students (i.e., those enrolled in free-lunch programs) had significantly lower scores on self-determination than other students; (d) students enrolled in special education programs had significantly lower scores on self-determination than students not enrolled in special education; and (e) students with mild disabilities had significantly higher scores on self-determination than students with moderate to severe disabilities (Wolman, Campeau, DuBois, Mithaug, & Stolarski, 1995, pp. 43-44).

Although the results yielded expected differences between advantaged and disadvantaged students and between students with and without disabilities, an additional finding indicated that older students with disabilities had higher scores on self-determination than younger students with disabilities. This finding is consistent with the claim articulated earlier that capacity and opportunity to self-determine are not fixed entities functioning like psychometric measures of intelligence. Instead, they interact to produce different prospects for self-determination. Gain toward the self-determined life is not simply a function of academic performance, although improvements here no doubt will have positive effects later.

Still, academic gain is but one of several areas of capacity enhancement that help students optimize opportunity to get what they need and want in life. Also, the fact that the AIR-TC instrument only measured two of the three capacity domains—self-knowledge and self-management—limits its ability to reflect a student's prospects for self-determination accurately. Had data been collected on resource availability in terms of students' personal, social, economic, and technical assets, then estimates of prospects for self-determination would have been based upon the relationship among all three capacity components—self-knowledge, self-management, and resource access.

This suggests a fruitful line of research for assessing the long-term benefits of educational intervention on behalf of capacity and opportunity. The idea that if schools do not improve students' academic achievement they have failed may be misleading because access to mainstreamed, public education can produce other capacity enhancements not evident in achievement scores. One enhancement that occurs as a consequence of attending mainstreamed, desegregated schools is socialization and social networking. Amy Wells and Robert Crain, for example, report a study on the long-term effects of school desegregation that suggests that attending desegregated schools helps break the perpetual cycle of segregation and isolation by providing black students with social contacts with white students that predispose them to access similar social networks when they seek to advance themselves after they leave school. Wells and Crain (1995) conclude: "There is a strong possibility . . . that when occupational attainment is dependent on knowing the right people and being in the right place at the right time, school desegregation assists black students in gaining access to traditionally 'white' "jobs (p. 552). In other words, attending a desegregated school functions to expand the *social resource* base of minority student for later use. And this, in turn, improves their prospects for self-determination later in life.

A complete assessment of an individual's capacity to self-determine would include information on the range and depth of one's re-sources—personal, social, economic, and technical. This information, combined with assessments of the individual's self-knowledge and self-management skills, would provide a substantial basis for deter-mining an individual's capacity to self-determine. It would also provide a basis for identifying the range of options—those *just-right* opportu-nities—that are most likely to engage Person in a self-determined pursuit. In other words, information on capacity and just-right oppor-tunity based upon assessed capacity would identify the *optimal pros-pects condition most likely to engage Person in the self-determination experience*—the pursuit of desirable ends that are consistent with Person's needs, interests, and abilities.

Although nothing like this now exists, the parts to the puzzle are known and are available for assembly for this purpose. For example, the AIR-TC instrument has demonstrated an approach to gathering information on self-knowledge and self-management but not on re-source access, whereas efforts like Michael Sherraden's "assets approach" offer guidelines for assessing Person's resources. In *Assets and the Poor: A New American Welfare Policy,* Sherraden (1991) uses the term *asset* instead of *resource* to evaluate Person's capacity for preparing for future action. His "theory suggests that assets are not simply nice to have, but yield various behavioral consequences such as enabling people to focus their efforts, allowing people to take risks, creating an orientation toward the future, and encouraging the development of human capital" (Sherraden, 1991, p. xiv).

Included in his assessment of a person's tangible and intangible assets are examples of the personal, social, economic, and technical resources described in this chapter. Sherraden's *tangible assets* include (a) money savings; (b) stocks and bonds; (c) real property, earnings from capital gains; (d) hard assets other than real estate with earnings in the form of capital gains (or losses); (e) machines, equipment, and other tangible components of production with earnings in the form of profits on the sale of products plus capital gains (or losses); (f) durable household goods with earnings in the form of increased efficiency of household tasks; (g) natural resources such as farmland, oil, minerals, and timber with earnings in the form of profit on sale of crops or extracted commodities plus capital gains (or losses); and (h) copyrights and patents, and earnings in the form of royalties and other user fees.

His *intangible assets* include (a) access to credit; (b) human capital such as intelligence, educational background, work experience, knowledge, skill, and health (also energy, vision, hope, and imagination); (c) cultural capital in the form of knowledge of culturally significant subjects and cues, ability to cope with social situations and formal bureaucracies, including vocabulary, accent, dress, and appearance; (d) informal social capital in the form of family, friends, contacts, and connections sometimes referred to as a "social network"; (e) "formal social capital or organization capital," which refers to the structure and techniques of formal organization applied to tangible capital; and (f) political capital in the form of participation, power, and influence, with earnings in the form of favorable rules and decisions on the part of the state and local government (Sherraden, 1991, pp. 101-105).

Given this array of indicators, it is reasonable to expect substantial differences in the level and type of resource capacity available to different actors, and it is also reasonable to expect that these differences will help explain why some members of a population have better prospects for pursuing what they need and want in life than others. Moreover, when these indicators are combined with indicators of self-knowledge and self-management, it is reasonable to expect that we can track the process by which enhancement or contraction of Person's overall capacity affects her prospects for self-determination.

In other words, empowerment evaluation measures that reflect the three capacity indicators—self-knowledge, self-management, and resource access—have the potential of providing information on progress toward the ideal of fairness in liberty for all. And because self-determination is a cornerstone of empowerment evaluation, this refinement of our understanding about just-right opportunities may help us improve our empowerment evaluation practice. It may also provide the type of information policymakers need to direct social redress toward the construction of those just-right opportunities that provoke the *desire as well as the experience of self-determination* among those least-well-situated members of society.

Conclusion

The purpose of this chapter was to show that empowerment evaluation reaches to the core of deeply held values in many cultures regarding the rights of individuals to participate in their own self-development, whether that development occurs within a context of personal improvement, organizational restructuring, or societal reform. The fundamental principle linking these deeply held values is that individuals have a right to participate in any change affecting their lives: *They have a right to self-determination.*

Empowerment evaluation respects this right by operating from the premise that evaluation data should be employed to *enhance the capacity* of individuals, organizations, and social institutions to pursue those ends in life they deem most worthy. It provides information people can use to adjust their sights on those just-right challenges that energize and empower change *within* themselves, their organizations, and their institutions. Empowerment evaluation is a mechanism for setting optimal expectations for meeting the challenge of adaptation during unremitting change. It empowers users to regulate their personal, social, and institutional problem solving toward those goals and missions worth pursuing. And in the process of guiding the problem-solving talent and energy of all, it creates the sense of community that is necessary for maximizing gain toward a common good.

Notes

1. The definition was developed at a Washington, D.C., meeting sponsored by the Department of Education, Office of Special Education Program, and attended by directors of five research projects charged with developing and validating methods and materials for assessing levels of self-determination in children and youth with disabilities. The five grantees conducting this research were the American Institutions of Research and Teachers College, Columbia University, the University of Minnesota, the Virginia Commonwealth University, Wayne State University, and the Association for Retarded Citizens—Texas.

2. David Fetterman contributed significantly to developing the framework and indicators for the Self-Determination Scale.

References

Branden, N. (1994). *Six pillars of self-esteem.* New York: Bantam.

Brown v. Board of Education, 347 U.S. 483 (1954).

Fetterman, D. M. (1994a). Empowerment evaluation [American Evaluation Association presidential address]. *Evaluation Practice, 15*(1), 1-15.

Fetterman, D. M. (1994b). Steps of empowerment evaluation: From California to Cape Town. *Evaluation and Program Planning, 17*(3), 305-313.

Gardner, H. (1993). *Frames of mind: The theory of multiple intelligences.* New York: Basic Books.

Gray, T. (1991). *Freedom.* Atlantic Highlands, NJ: Humanities Press.

Herrnstein, R. J., & Murray, C. (1994). *The bell curve: Intelligence and class structure in American life.* New York: Free Press.

Hill, N. (1960). *Think and grow rich.* New York: Fawcett Crest.

Humana, C. (1992). *World human rights guide* (3rd ed.). New York: Oxford University Press.

Mithaug, D. E. (1993). *Self-regulation theory: How optimal adjustment maximizes gain.* Westport, CT: Praeger.

Mithaug, D. E. (1995). *Equal opportunity theory: Fairness in liberty for all.* Unpublished manuscript.

Sherraden, M. (1991). *Assets and the poor: A new American welfare policy.* Armonk, NY: M. E. Sharpe.

Sternberg, R. J. (1988). *The triarchic mind: A new theory of human intelligence.* New York: Penguin.

U.S. Department of Education. (1994). *The national agenda for achieving better results for children and youth with disabilities.* Washington, DC: Author.

Wells, A. S., & Crain, R. L. (1995). Perpetuation theory and the long-term effects of school desegregation. *Review of Educational Research, 64*(4), 531-555.

Wolman, J. M., Campeau, P. L., DuBois, P. A., Mithaug, D. E., & Stolarski, V. S. (1995). *AIR Self-Determination Scale and user guide.* Palo Alto, CA: American Institutes for Research.

WORKSHOPS, TECHNICAL
ASSISTANCE, AND PRACTICE

Empowering Community Groups With Evaluation Skills

The Prevention Plus III Model

JEAN ANN LINNEY
ABRAHAM WANDERSMAN

During the last half of the 1980s, responsibility for community problem solutions and initiation of change efforts shifted from the federal government to state agencies and local community groups. Reductions in federal budgetary support for social programs were certainly one important factor contributing to this shift, but other changes were occurring in local communities that solidified the change in locus of intervention activity. During this time, there was renewed attention to the potential of local community groups and the success of individual community efforts at dealing with persistent problems like crime, school improvement, and environmental concerns. As frustration increased with the limitations of broad-scale social change efforts, community psychologists, sociologists, and other social analysts were recognizing the need for an enhanced "sense of community" and growing evidence of the desire of citizens to become involved in efforts to improve their quality of life (Kelly, 1990; Linney, 1990).

The slogan *think globally, act locally* not only applied to environmental issues but also began to describe implied policy and activity for a broad range of community problems.

Social change theorists and social activists among others have long recognized the importance of community ownership and stakeholder involvement in activities directed at change in communities (Alinsky, 1971; Fairweather, 1972; Seidman, 1983). The value of citizen participation and community collaboration in terms of enhancing the success of interventions and psychological outcomes as well as general satisfaction and quality of life for participants has been clearly documented (Heller et al., 1984; Prestby, Wandersman, Florin, Rich, & Chavis, 1990). Thus, the shift in action responsibility that occurred during the late 1980s, although politically and economically stimulated, was not without substantial theoretical and empirical support predicting its success in terms of positive outcomes for the individuals involved and the community problems addressed.

One area in which the shift from national to local responsibility has been most apparent is the problem domain of alcohol, tobacco, and other drug (ATOD) abuse. A substantial portion of the monies available for ATOD prevention programming has been directed toward the creation of local community partnerships or coalitions involving a broad spectrum of professionals, service providers, and citizens for the development of prevention initiatives tailored to the needs and circumstances of the local community (Butterfoss, Goodman, & Wandersman, 1993). Both the Robert Wood Johnson Foundation and the federal Center for Substance Abuse Prevention have focused considerable financial investment in this model of programming development. Likewise, the U.S. Department of Education Drug Free Schools and Communities initiatives have focused on distribution of monies to local communities for activities selected by community steering committees.

Along with the shift from national to local, from broad-brush approaches to local choice, has been increasing concern with accountability and documentation of the effect of the particular strategies and activities implemented. Thus, beginning in 1990, recipients of federal grant monies for prevention efforts were required to provide evaluation of their efforts. This requirement created a series of new problems for local citizen groups now empowered to work toward the prevention of ATOD abuse in their communities. In general, these local groups were likely to have both limited funds and little or no experience with evaluation. For example, a school might be awarded $1,000 to plan and implement ATOD prevention activities for a year, and to evaluate their success. Oftentimes, the parent-teacher association or another volunteer organization takes the lead in the ATOD prevention effort at the school level. By design, other community efforts are expressly intended to involve nonprofessionals (i.e., citizens other than the usual ATOD service providers, academics, and social service professionals).

How can an evaluation be completed when the financial resources available are not sufficient to hire an evaluator and it is unlikely that such expertise is available within the planning committees? How does a local planning group, primarily composed of citizen volunteers, begin to address the task of evaluation? The needs and dilemmas presented by this situation stimulated the development of *Prevention Plus III* (Linney & Wandersman, 1991), a four-step approach to assessing school and community prevention programs designed as an evaluation tool for use by nonprofessionals. *Prevention Plus III (PP-III)* has as its goals to demystify the process of evaluation and to provide local community groups with an evaluation tool that is not only relatively simple and straightforward but also readily applicable to local initiatives of varying complexity.

The *PP-III* model and the materials included in the workbook reflect the concepts of empowerment evaluation (Fetterman, 1994) in seeking to enhance the ability of local groups to plan, implement, evaluate, and redirect prevention program activities following a set of guidelines and practices widely accepted among professional evaluators. The workbook distills evaluation to four essential components: (a) question formation, (b) implementation/process evaluation, (c) assessment of proximal effects, and (d) assessment of broader impact. The *PP-III* materials provide a modifiable blueprint designed to help local program people not only to improve their programs but also to develop new skills.

Assumptions and Needs
of Nonprofessional Evaluators

Nonprofessionals in the position of conducting an evaluation often comment that they have no idea how to do an evaluation, do not know where to begin, and thus become paralyzed from taking the first step. In the initial design of *Prevention Plus III*, we sought to develop a tool that would provide a blueprint for the logic of evaluation, thus offering both a starting point and a mechanism to guide action. The development of *PP-III* was guided by considerable experience in working with community groups and recognition of a set of common problems experienced by nonprofessional evaluators. Two commonly observed initial problems are difficulty in the formulation of evaluation questions and translation of concepts into project-specific indices and evaluation activities. The literature on program implementation has documented the evaluation problems that stem from insufficient program planning and poor documentation of program activities. Inadequate planning and documentation are especially likely to hamper evaluation efforts of nonprofessional community groups. Another widespread misconception about evaluation is that evaluation is done after a program is over. Unfortunately, without attention to evaluation prior to and during program implementation, very little useful evaluative knowledge can be gathered at the end of the program. The *PP-III* materials were designed to reduce these barriers to evaluation specifically for nonprofessional community evaluation teams.

From experience in working with community groups not trained in evaluation methods, it is apparent that identification and delineation of evaluation questions is frequently a significant barrier to the evaluation process. Papineau and Kiely (1994) have reported similar experiences in their work with community groups in Montreal, Canada. The most common evaluation question identified by local groups is, "Well, did it work?" Despite genuine interest in this question, most nonprofessionals are not prepared to articulate a plan for determining whether a local initiative "worked." There are issues of clarifying what is meant by *working,* identifying indices that would provide evidence of an effect, and delineating the set of evaluation questions appropriate to stages of implementation.

As professional evaluators, we are trained to think causally and recognize that any intervention or prevention activity is based on an underlying model of the causal factors contributing to the problem to be prevented, and that if the activity or program is not implemented adequately, the effect is unlikely to occur; if the program doesn't affect the intervening variables, then the desired effect is unlikely to occur;

or the underlying assumptive model may be incorrect. For social scientists, this type of thinking becomes so automatic that we forget it is not a universal mode of analysis. A common community prevention activity might be to sponsor a Red Ribbon Awareness Campaign. The local group wants to effect a reduction in ATOD use by getting citizens to display red ribbons. Why would wearing a red ribbon lead to reductions in ATOD? Recognition that awareness of the hazards of alcohol use (which are intended to be associated with the red ribbon) is an intervening variable that may affect individual decision making or stimulate a friend to intervene to prevent another from driving while intoxicated is a style of critical thinking not necessarily typical outside of evaluator circles. Thus, an important goal of *PP-III* was to facilitate this type of causal thinking and, as much as possible, to provide the assumptive causal model and associated process and outcome evaluation questions for common community prevention initiatives. We sought to design materials that would stimulate analytical thinking about the ways in which prevention programs might affect ATOD use, realistic thinking about the effect of any one preventive effort, and careful planning for implementation.

A second major barrier to evaluation for nonprofessionals centers on measurement and instrumentation, specifically translating abstract constructs into identifiable, measurable variables and locating relevant instruments. Many of the regional Centers for Drug Free Schools and Communities have reported that the most common request for technical assistance is the identification of evaluation instruments and measures. Here again, as evaluators, we tend to think in terms of operationalizing abstract constructs and have an extensive inventory of surveys, questionnaires, and other assessments at our fingertips. In *PP-III*, we sought to provide assistance in identification of specific instruments and variables relevant to the goals and desired outcomes identified in the assumptive causal models for the specific activities being implemented. The *PP-III* volume includes a set of suggested instruments and references to other sources for relevant measures. Each of the *PP-III* activity worksheet modules includes specific reference to measures for each proximal outcome identified for that activity. The set of worksheet modules and measures included in the workbook is only a starting point for community groups and is intended to provide enough structure and direction to get evaluation activity started. The materials further prompt the user to adapt the modules and tailor them to his or her specific circumstances and program characteristics consistent with the general evaluation procedures described.

Failures in implementation and inadequate preprogram planning may account for a considerable portion of the variance in program failures. These shortcomings present a major limitation for evaluation as well. For volunteer or inexperienced program planners, enthusiasm and commitment to the long-term goals of their activity often short-change the planning phase and attention to implementation fidelity. *PP-III* was designed to reduce these problems by facilitating implementation planning and documentation of the process of implementation through structured worksheets addressing plans and implementation realities. We have found that this focus on process and implementation appears to have reduced the incidence of premature evaluation (i.e., evaluating the outcomes of a program before it has been implemented adequately), focusing greater attention on the mechanics of implementation and success in reaching the targeted audience.

Perhaps the most frustrating barrier to evaluation of community-initiated programs is the myth that evaluation begins after the program has been implemented. To many of the lay public, evaluation is something mysterious and magical done by an outsider after a program has been completed. There are a number of unfortunate outcomes of this kind of thinking. One common effect of this myth is inadequate documentation of program activities and participation, reducing the feasibility of a useful process or outcome evaluation, which then undermines the usefulness of any evaluation and further undermines belief in the value of subsequent evaluations. Professional evaluators know that the most useful evaluations are those that address the questions of greatest interest to those who have worked with the project and provide feedback for program improvement. In *PP-III,* we sought to convey the interconnectedness and importance of planning, documentation of process, and data collection regarding effects. The separate worksheets for each of the four steps of the evaluation process include references to each other such that information for one worksheet comes from another, structurally highlighting the interconnectedness of the several phases (e.g., planning, implementation, and outcomes).

Within the narrative and supporting materials, the *Prevention Plus III* volume briefly addresses some of the common fears people have about evaluation (e.g., "It's so boring." "I've never been good at math."). The volume provides a step-by-step, user-friendly set of materials that summarizes the logic of evaluation toward the goal of empowering nonprofessionals to conduct elementary evaluations of local program initiatives. We have adopted the assumption that, once people understand the logic of evaluation and have some structured success conducting an evaluation, they will become increasingly interested in more sophisticated designs and methods.

Structure of the *Prevention Plus III* Workbook

Prevention Plus III presents a simplified four-step model for program assessment and evaluation:

Step 1: Identification of Goals and Desired Outcomes
Step 2: Process Assessment
Step 3: Outcome Assessment
Step 4: Impact Assessment

Worksheets were developed for each of these steps, individualized for more than 50 common ATOD prevention activities covering the broad domains of efforts to raise awareness; increase knowledge; change norms and expectations about ATOD use; enhance parenting and family influences; enhance student skills; increase involvement in school and in healthy, legal alternatives; increase support services; and promote deterrence through regulatory and legal action.

A sample set of worksheets for assessment of a teacher in-service training program illustrates the *PP-III* format. The Step 1 worksheet prompts the local evaluator to identify goals, target groups, and desired outcomes. As in the sample shown in Figure 12.1, the worksheet for each program activity is preprinted with goals and desired outcomes commonly identified for that activity with additional space on the worksheet for the local team to add their own goals. In the example shown here, common goals for teacher in-service programs include increasing teacher knowledge of ATOD issues, enhancing their commitment to prevention, and encouraging teachers to integrate ATOD issues into other instructional activities. Each of the

Step 1: Identify Goals and Desired Outcomes

In-Service Programs for Teachers

Part A: Make a list of the primary goals of the program.
Ask yourself: "What were we trying to accomplish?" Of the goals listed, check the ones that apply to your program and add any others on the lines provided.

_____	increase teachers' knowledge and awareness of AOD problems
_____	encourage teachers to include AOD issues in their instructional activities
_____	enhance teachers' commitment to AOD use prevention
_____	_____
_____	_____

Part B: What groups did you want to involve?
Ask yourself: "Whom were we trying to reach?" Of the groups listed, check the ones that apply to your program and add any others on the lines provided.

	Target Group	**How many did you want to involve?**
_____	all staff	_____
_____	health teachers	_____
_____	counseling staff	_____
_____	physical education staff	_____
_____	_____	_____
_____	_____	_____

Part C: What outcomes were desired?
Ask yourself: "As a result of this program how would we like the participants to change? What would they learn, what attitudes, feelings, or behavior would be different?" Of the outcomes listed, check the ones that apply to your program and add any others on the lines provided.

_____	increase teachers' knowledge about AOD use
_____	increase awareness among teachers about AOD use
_____	change teacher attitudes toward use
_____	change student attitudes about AOD use
_____	increase teachers' willingness to intervene with students using AODs
_____	increase referrals for counseling from teachers
_____	increase enforcement of school AOD policy
_____	increase use of AOD materials in the curriculum
_____	_____
_____	_____

Figure 12.1. *Prevention Plus III* Worksheet for Identifying Goals and Outcomes

primary goals listed is translated into at least one outcome preprinted in Part C of the Step 1 worksheet, thus linking aims and goals to measurable proximal outcomes. Each part of each worksheet includes an everyday-language version of the relevant evaluation question to enhance understanding and stimulate directed thought in each area.

The Step 2 worksheet (Figure 12.2) addresses implementation and process evaluation with attention to specification of timetables, planned activities, service delivery objectives, products to be delivered, and barriers to implementation including attention to how well the target population was reached. The process assessment also focuses attention on "dosage" issues such as how much actual intervention time is directed at the individuals constituting the target population. This is a critically important issue for community groups who may spend a great deal of time working on some prevention activity that in fact represents only a minor dose of intervention for any one individual. Disappointing effects can discourage future initiatives, but if more attention had been focused on just how much intervention would be felt by the individuals being measured, modifications in the intervention or expected effects might have occurred. For example, a local group might spend a great deal of time preparing public service announcements without fully recognizing that only those who hear the announcement can be affected by this intervention and even for those the intervention itself may last only 30 seconds. The *PP-III* worksheets are structured to focus thinking on how much intervention is being delivered and how much of an effect can realistically be expected.

The Step 2 process worksheets are intended both for implementation planning and for evaluation of the adequacy of implementation. They are structured to focus attention on both accomplishments and failures, such as how many people were reached and who in the target group was not included. The information in the Step 2 worksheet provides the base for accountability reports including the who, what, and how-much types of information describing the project. Attention is further directed to how the activity might be improved in future implementation. Thus, there is a direct link made between the evaluation data recorded and efforts toward program improvement. Establishing the linkage between evaluation and programming should enhance the value of the evaluation activities for local program groups.

Step 2: Process Assessment Worksheet

In-Service Programs for Teachers

Part A: What activities were planned?
(Include a brief description of the components of the program. Ask yourself: "What did we actually do to prepare for this and implement it?" Form a chronology of events constituting this program and a quantity indicator for each.)

Activity in-service presentations on:	Date	Quantity Planned	Quantity Actual
_____	_____	_____	_____
_____	_____	_____	_____
_____	_____	_____	_____
_____	_____	_____	_____
_____	_____	_____	_____

Quantity Totals:

number of sessions _____ (s) length of time for each _____ (hr)
total hours of activity (s x hr) _____

	What written materials were available?	Total distributed
_____	manuals, brochures	_____
_____	other	_____
_____	_____	_____

Total other services delivered:

_____
_____
_____

What topics or activities were planned but not covered?
What happened that these were not accomplished?

Activity	Problem
_____	_____
_____	_____
_____	_____
_____	_____
_____	_____

Figure 12.2. *Prevention Plus III* Worksheet for Process Assessment

Part B: When was the program actually implemented (dates of activities, length of time for each) and who were the participants?

Date	Materials purchased	Quantity

Total number	Percentage of goal	Total number

Who was missing that you'd hoped to have participate in the program?

What explanations can be offered for the discrepancy between the projected and the actual participation?

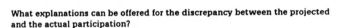

_____ *schedule conflicts*

_____ *competing programs/activities*

_____ *teachers felt they didn't need the program*

_____ *program uninteresting to teachers*

_____

_____

_____

Part C: How did participants evaluate the materials purchased?

_____ *Source of evidence: Participant Assessment Form (M2)*

_____

_____

_____

Part D: What feedback can be used to improve the program for the future?

Figure 12.2. Continued

Step 3: Outcome Assessment Worksheet

In-Service Programs for Teachers

1 Desired Outcomes List the desired outcomes from Step 1-Part C	2 Measure/Indicator Indicate the type of evidence you have for each outcome	None	3 Observed Scores Project Group Before	After	Comparison Group Before	After	4 Amount of Change Before vs After the Project	Comparison Group vs Project Group
1. increase teacher knowledge about drug use	1. knowledge test (M3, M4)							
2. increase teacher awareness of use	2. awareness measure (M9)							
3. change teacher attitudes toward use	3. AOD attitudes measure (M7)							
4. change student attitudes toward use	4. AOD attitudes measure (M7)							

Figure 12.3. *Prevention Plus III* Worksheet for Outcome Assessment

Outcomes are assessed with the Step 3 worksheet (Figure 12.3). The desired outcomes identified in Step 1-C are transferred to the Step 3 worksheet. The outcome assessment worksheet is structured to highlight the value in examining program effects as compared with another time point or a comparison group. The supporting text provides the logic for pretest-posttest comparisons and the need for comparison groups. The fill-in-the-blank format used in this step is formatted to match comparative evaluation designs (e.g., pre-post comparison group design). If the evaluation design is a one-shot post-only assessment, then the Step 3 worksheet will have many blank spaces. We have found that as local groups begin to gather information systematically about program effects, they begin to see the value of comparison points and work toward that in subsequent efforts.

A unique feature of *PP-III* is the inclusion of suggested measures for the preidentified outcomes in each module. These suggested instruments or information about their availability are included in one section of the volume. Each outcome that is preprinted in the Step 3

Step 4: Impact Assessment

Student AOD Use and Risk Indicators

1	2	3	4	5	6	7
Impact	Measure or Evidence	Program Group Before	Program Group After	Comparison Group Before	Comparison Group After	Amount of Change
1. Reduction in youth AOD use	Drug use survey					
2. Delay of onset of youth AOD use	Drug use survey					
3. AOD-related traffic crashes involving youth drivers	Number in 12-month period					
4. Decrease in DUI arrests among youth	Number of DUI arrests 12 months before and 12 months after activities					
5. Decrease in youth AOD-related arrests	Number of arrests 12 months before and 12 months after					
6. Decrease in youth AOD-related hospital emergencies	Number of drug-related hospital emergencies					
7. Change in number of student disciplinary actions for AOD offenses	Number in school or district					
8. Change in number of youth admissions for AOD treatment	Number of admissions in region in 12-month period					

Figure 12.4. *Prevention Plus III* Worksheet for Impact Assessment

worksheets is labeled to match a specific instrument included in the volume. Every effort was made to include suggested instruments that are easy to use, readily obtainable, and meet accepted standards for reliability and validity.

The Step 4 Impact Assessment (Figure 12.4) provides a worksheet intended for use with all of the program modules. The user is prompted to examine impact or distal outcomes for eight indicators of ATOD use at the school and community level. Here also, spaces are included in the worksheet for pre-, post-, and comparison group levels. Because *PP-III* was designed for use with ATOD prevention programs, the indices of impact are specific to use of alcohol and other drugs (e.g., actual use by youth, age of first use, ATOD-related traffic accidents, arrests for driving under the influence, drug treatment admissions) but this model can be applied to other problem domains with an alternative list.

Intended and Unintended
Outcomes of *Prevention Plus III*

Initial feedback and critique of *PP-III* from the professional evaluation community focused on the limitations of the simplified process-outcome model of evaluation. Some suggested that the materials set too low a standard for evaluation and that in fact the model should not be called evaluation. This feedback resulted in use of the word *assessment* instead of evaluation throughout the volume. There was concern from some that the simplified approach grounded in a quantitative and comparative model of evaluation would reduce expectations regarding the utility of evaluation among program staff. There was criticism that in fact evaluation is much more complex than presented in *PP-III* and that as professionals we should not mislead the lay community about what good evaluation looks like.

These may be valid concerns. The primary goal of the *PP-III* materials, however, was and remains to be a tool to enhance the capacity of local groups to conduct more useful program assessments and rudimentary evaluation, to increase their appreciation of the merits of evaluation, and to improve program planning and implementation via systematic attention to process, causal sequences, and outcomes. There is growing feedback from users that the materials have in fact resulted in these effects.

The *PP-III* materials were first disseminated in a series of train-the-trainer workshops conducted in 1989 by the Southeast Regional Center for Drug-Free Schools and Communities based in Atlanta. Training workshops were conducted statewide for drug-free schools coordinators and local school and community teams. Subsequently, the Office for Substance Abuse Prevention printed a revised version of the volume, which is distributed through the National Clearinghouse for Alcohol and Drug Information and its related networks. We know that more than 50,000 volumes have been distributed.[1] Although we have not systematically recorded inquiries and requests for the materials, we know that they are being used and modified by professionals and community groups across the country.

The most common use of the *PP-III* workbook seems to be in program planning and program development. Numerous groups have reported that as they began to plan an evaluation, they realized that they had not agreed on goals, adequately planned or implemented activities, or did not have a plan of interventions sufficient to effect change. Much of this realization occurred as the local group worked with the *PP-III* evaluation modules. The inclusion of both a narrative summary of prevention models and more than 50 specific prevention activities within the *PP-III* workbook stimulated enhanced programming in many places. Groups have told us that they expanded their prevention efforts after learning of new activities from their review of the *PP-III* modules. Local groups have used the *PP-III* materials in preparing grant applications, with the workbook helping a planning group to choose interventions and then to describe not only the causal models underlying their program but also a plan for evaluation.

Capacity building and empowerment of local community groups were primary goals in the design of *Prevention Plus III*. In work with community coalitions, we have seen how the materials have effected these outcomes. We know of ATOD community partnership projects across the country using the worksheet modules and have seen the development of enhanced planning skills, more sophisticated thinking about indices of outcomes and impact, and greater attention to comparative thinking in examining program effects. With the use of the *PP-III* materials, evaluation has become an ongoing part of the activities of local groups, allowing for midcourse corrections when implementation problems are identified early in the process and more stimulating, critical, and creative thinking about new directions for intervention. For example, in the evaluation of a theatrical production highlighting teens with drug addictions, the planning group realized early in their planning that without connecting this production to other school activities, the probability of an intervention effect would be quite small because the production itself would be only 75 minutes isolated from other facets of the lives of the teen audience. In working with the *PP-III* materials prior to implementation, the planning group decided to add teacher and curricular components to the intervention, in effect boosting the dose of intervention and, it was hoped, enhancing the likelihood of effect (Griffin & Kennerley, 1994).

Because of the simplicity of the *PP-III* materials, community groups are more likely to attempt to use them. We have found that, despite the concern from professionals that the materials are too simple, quite a number of community groups have needed assistance in getting started. Working with community partnerships in Rhode Island, Stevenson, Florin, and Mitchell (1993; Stevenson, Mitchell, & Florin, Chapter 10, this volume) have adapted the *PP-III* modules and prepared an interactive computer-assisted learning package. They have found that most community groups need to learn Steps 1 and 2, then cover Steps 3 and 4 in a subsequent session. The feedback from local groups is clear, however—once the process has begun, as members begin to understand the logic of the process, its utility for them, and the timeliness of feedback, enthusiasm, and investment in systematic evaluation grows. As we had hoped, local groups have become increasingly sophisticated in their questions and designs, and in some cases have sought collaboration with professional evaluators.

In at least three states in different parts of the country, efforts are under way to standardize a *PP-III* worksheet module for use by similar programs across the state (see Dugan, Chapter 13, this volume, for one example). This approach offers the possibility of generating a common database across sites, with local additions and individualization. Such a plan has the potential to capitalize on service initiatives across many localities to generate a common database that might be used to test broader theoretical models and contribute to a scientific knowledge base as well.

PP-III in its current form is specific to alcohol and other drug prevention efforts, but the approach and generic versions of the worksheets make it applicable to many social program domains. We are aware of efforts to develop *PP-III*-like modules for prevention efforts in child abuse, HIV/AIDS, and teen pregnancy. The approach has been used successfully with local mental health groups for the assessment of new program initiatives involving community groups and mental health consumers (Anderson & Linney, 1994). In these instances, we worked with program staff in identifying program goals and desired outcomes, drawing initial statements from their program descriptions and funding applications. The process worksheets were modified to reflect the specific activities to be accomplished and their service delivery objectives with special attention to barriers faced in the initial implementation phases. The Step 2 process worksheets were used as the base for monthly reports to the funding agency. After initial consultation in the development of the *PP-III* worksheets, the local staff were able to maintain ongoing program monitoring and evaluation activity. The longevity in evaluation achieved from the involvement by local program staff is an important benefit of the *PP-III* approach.

As local communities have become responsible for program initiation and implementation, they have also been expected to provide accountability and evaluation reports. *Prevention Plus III* provides a useful tool to initiate this process, empowering local groups with the skills needed to improve programming and determine effectiveness. *PP-III* represents a starting point, a set of materials to stimulate a process. Professional evaluators and those concerned with the development of programs to address community problems can further facilitate the process by working with community groups helping them to develop more sophisticated skills in statistical analysis (e.g., Sirkin, 1994; Weitzman & Miles, 1995), techniques of qualitative evaluation (e.g., Fetterman, 1988, 1989, 1993; Miles & Huberman, 1994; Patton, 1980; Yin, 1994), and cost-effectiveness analysis (e.g., Levin, 1983). Given that the overwhelming majority of local programs are conducted with virtually no program assessment or evaluation, we believe that efforts toward even very rudimentary evaluations should be championed.

Note

1. Copies of *Prevention Plus III* are available free from NCADI and can be obtained by calling 1-800-SAY-NO-TO.

References

Alinsky, S. (1971). *Rules for radicals*. New York: Random House.
Anderson, C., & Linney, J. A. (1994). *Evaluation of the South Carolina Mental Health Partnership: Final report* (Unpublished report). Columbia: South Carolina Mental Health Partnership.

Butterfoss, F., Goodman, R., & Wandersman, A. (1993). Community coalitions for prevention and health promotion. *Health Education Research: Theory and Practice, 8,* 315-330.

Fairweather, G. W. (1972). *Social change: The challenge to survival.* Morristown, NJ: General Learning Press.

Fetterman, D. M. (1988). *Qualitative approaches to evaluation in education: The silent scientific revolution.* New York: Praeger.

Fetterman, D. M. (1989). *Ethnography: Step by step.* Newbury Park: Sage.

Fetterman, D. M. (1993). *Speaking the language of power: Communication, collaboration, and advocacy: Translating ethnography into action.* London: Falmer.

Fetterman, D. (1994). Empowerment evaluation. *Evaluation Practice, 15,* 1-15.

Griffin, L., & Kennerley, M. (1994). Evaluating a media-based strategy for alcohol and other drug prevention. *Community Psychologist, 28*(1), 4-5.

Heller, K., Price, R., Reinharz, S., Riger, S., Wandersman, A., & D'Aunno, T. (1984). *Psychology and community change: Challenges for the future.* Homewood, IL: Dorsey.

Kelly, J. G. (1990). Changing contexts and the field of community psychology. *American Journal of Community Psychology, 18,* 769-792.

Levin, H. (1983). *Cost effectiveness: A primer.* Beverly Hills, CA: Sage.

Linney, J. A. (1990). Community psychology into the 1990s: Capitalizing opportunity and promoting innovation. *American Journal of Community Psychology, 18,* 1-18.

Linney, J. A., & Wandersman, A. (1991). *Prevention Plus III: Assessing alcohol and other drug prevention programs at the school and community level (A four-step guide to useful program assessment)* (DHHS Publication No. [ADM] 91-1817). Washington, DC: Department of Health and Human Services.

Miles, M. B., & Huberman, A. M. (1994). *Qualitative data analysis: An expanded sourcebook.* Thousand Oaks, CA: Sage.

Papineau, D., & Kiely, M. C. (1994). Participatory evaluation: Empowering stakeholders in a community economic development organization. *Community Psychologist, 27*(2), 56-57.

Patton, M. Q. (1980). *Qualitative evaluation methods.* Beverly Hills, CA: Sage.

Prestby, J. E., Wandersman, A., Florin, P., Rich, R., & Chavis, D. (1990). Benefits, costs, incentive management and participation in voluntary organizations: A means to understanding and promoting empowerment. *American Journal of Community Psychology, 18,* 117-150.

Seidman, E. (1983). Unexamined premises of social problem solving. In E. Seidman (Ed.), *Handbook of social intervention* (pp. 48-67). Beverly Hills, CA: Sage.

Sirkin, R. M. (1994). *Statistics for the social sciences.* Thousand Oaks, CA: Sage.

Stevenson, J., Florin, P., & Mitchell, R. (1993, June). *Building the capacity of community prevention efforts with evaluation skills.* Presentation at the 4th Biennial Conference on Community Research and Action, Williamsburg, VA.

Weitzman, E. A., & Miles, M. B. (1995). *Computer programs for qualitative data analysis: A software sourcebook.* Thousand Oaks, CA: Sage.

Yin, R. K. (1994). *Case study research: Design and methods.* Thousand Oaks, CA: Sage.

13

Participatory and Empowerment Evaluation

Lessons Learned in Training and Technical Assistance

MARGRET A. DUGAN

In the spring of 1991, I received a phone call from a colleague asking for advice. Her nonprofit organization needed to assess the effectiveness of its curriculum-based support groups (CBSGs)[1] designed to delay and prevent alcohol, tobacco, and other drug use in at-risk youth. She wanted to develop a replicable, easy-to-use evaluation process for programs using the CBSG model. "How can we do this," she asked, "when we have no money to pay a consultant and no one on staff trained in evaluation techniques?" At the time, I certainly did not consider myself an expert in the field. I had, however, begun to believe that evaluation was not the exclusive domain of highly trained and expensive consultants but was probably a process that people could learn to do with support.

"If I were you," I told her, "I'd find someone to teach my staff how to conduct an evaluation."

"OK, when can you start?" she asked.

This brief conversation was the beginning of a 3-year collaboration that produced results neither of us could have anticipated. What follows is an account of how we developed a participatory and empowerment evaluation that went far beyond simply proving program effectiveness.

The First Attempt: A Traditional Evaluation

Our first effort in 1992 was a crushing failure. Out of six sites participating in a traditional evaluation process statewide, we received only 16 completed pre-/posttests with no comparison group scores. It seemed that almost a year's planning and work were down the drain.

But in learning how we had gotten off track, we were able to begin the process anew. In the end, our best teachers were our clients. We asked them where we had gone wrong, and they told us: We want to be part of an easy-to-use process from the beginning. We want to be trained in small groups so there can be plenty of feedback. We want weekly contact with the project coordinators and ongoing technical assistance. We want one-stop shopping when we need help. We want simpler materials specific to our needs. We want to be scientific, but not detached; valid in our findings and assured—in fact, guaranteed—that we will be the authors of our own development. We want less emphasis on rigorous data collection and more on our critical abilities as a collective.

The Heart of the Matter: Whom Should Evaluation Serve?

As we analyzed what went wrong, a dichotomy emerged—the needs of the participants versus those of the funder. To see the difference, we created a hypothetical evaluation environment schema (see Figure 13.1).

This activity clarified the problem. We had tried to complete a funder evaluation in a context demanding participation. Community psychologist Jim Kelly (1970) seems to capture the problem from the perception of the participants:

> The most arrogant guys around are often [the] professionals who analyze, position, reflect, study, commission, postpone, garble, intrude, and play with, but rarely play out, the cross-currents of community events. It is

Participatory Evaluation	Funder-Driven Evaluation
• Shared responsibility (doing with the community)	• Professional responsibility (doing for the community)
• Power residing with participants	• Power vested in agencies
• Participants seen as experts	• Professionals seen as experts
• Planning and services implemented on the basis of program needs assessment	• Planning and services responsive to each agency's mission
• Leadership develops shared vision, broad support, and participatory problem solving	• External leadership based on authority, position, and title
• Appreciation of ethnic diversity	• Indifference to ethnic diversity
• Emphasis on cooperation, collaboration, and shared resources	• External linkages limited to networking and coordination
• Inclusive decision making	• Closed decision-making process
• Accountability to participants	• Accountability of the agency
• Evaluation to document program development and improvement	• Evaluation primarily to determine funding .
• Maximum community involvement at all levels	• Community participation limited to providing feedback

Figure 13.1. Contextual Contrasts

[their] quiet and sometimes folksy and affable arrogance that can interfere with . . . opportunities to adopt tentative explorations and offbeat enterprises that are an integral part of psychology. . . . Few . . . have been trained to cathect to a locale. I am confident that few I have been taught to worry about our communities, and still fewer . . . have given . . . time to see the promotion of a civic cause fulfilled. (p. 524)

Kelly (1986) also helps provide an answer that we began to explore:

An antidote for coping with the regal quality of our scientific heritage and the discomfort about not being able to reduce the barriers between research and practice is to develop an investigative style that makes inquiry not just a right or privilege but makes community research a genuinely collaborative process. Without such collaboration our ideas run the risk of being sterile and our impact puny and shallow. With such collaboration researchers and citizens have the opportunity to generate a style of work that will contribute to our own personal development and to the evolution of our social settings. (p. 589)

At the Crossroads: Validity Anxiety

A friend of the project pointed out "that failure is just as important as success in any worthwhile endeavor. It allows us to find out what does not work" (J. Funkhouser, 1995, personal correspondence). As we began to understand why we were so unsuccessful in our first evaluation, we looked for direction. We turned to those who seemed to have greater understanding of participation (e.g., community psychologists and naturalistic and empowerment evaluators) for an answer. We found significant support for context-specific empowerment and participatory problem-solving approaches (Florin, Chavis, Wandersman, & Rich, 1992; Florin & Wandersman, 1984; Kelly, 1986; Rappaport, 1987; Wandersman, Chavis, & Stucky, 1983; Zimmerman, Israel, Schulz, & Checkoway, 1992; Zimmerman & Rappaport, 1988). We became convinced that there is what Patton (1979) calls "slippage between an . . . ideal rational model [and the] . . . day-to-day, incrementalist, and conflict-laden realities of program implementation" (p. 328). To account for this slippage, we decided a heavier evaluation emphasis on program development and operation (i.e., process) was needed. To accomplish this, we adopted a responsive, more empowering approach described as naturalistic evaluation (Stake, 1975) or, more recently, empowerment evaluation (Fetterman, 1994). This suited our need to honor the rich diversity of the participants by emphasizing the pluralism in values and viewpoints that characterized each program setting as well as the consequent belief that rarely is there one truth that all can accept.

The evaluation teams volunteered from 24 sites statewide: 16 sites were from urban settings, 6 sites were within 30 miles of an urban center, and 2 sites were rural or more than 30 miles from an urban center. Participants were school or community based. The majority of team members were trained educational or mental health professionals. In addition to the professionals, there were also community volunteers participating in the evaluation and training. No one in the project, either professional or volunteer, had any evaluation training, and very few had actively conducted any evaluation activities. The selection criteria for each site team included (a) an identified target population of at-risk youth between the ages of 8 and 15, (b) a support group program conducted around an approved curriculum, and (c) a willingness to attend training and collect evaluation data.

We had two objectives: (a) Rather than attempt another objective verification of goals and outcomes, we wanted to respond to a wider range of issues and concerns held by stakeholders. (b) In the process of illuminating their multiple perspectives, we wanted to build their evaluation capacity. Nevertheless, we were uneasy about validity and participant objectivity. We were apprehensive that somehow our expert judgment would be considered less valid by our stakeholders than other, more scientific approaches. We did encounter experts who typified Fetterman's (1995) viewpoint that "empowerment evaluation's fluidity is frightening for external evaluators who have not lived with program participants" (p. 187). We agreed with Fetterman that the kind of evaluation we wanted to do felt more "natural for those [of us] who work in and with social programs on a daily basis." Stake (1975) says that this type of approach "is an attempt to respond to the natural ways in which people assimilate information and arrive at understanding" (p. 2). Our design was emergent—one that could evolve though the evaluation process itself in response to the needs of the stakeholders and their issues as they developed—a process that Fetterman (1995, p. 196) suggests is indicative of an evaluative folk culture.

Our concern that the participants could not evaluate because they could not be objective and had too much program ownership, was simply unfounded. Although the participants were concerned about what would happen to their funding if the evaluation did not prove program effectiveness, these fears dissipated once they understood the evaluation process and were assured that the funder expected program improvement, not perfection, from the participants. They were able to look at their findings with a collective intelligence. This group analysis served as a check and balance that proved not only to be objective but more analytical than expected. Fetterman (1995) describes this process of collecting credible data as a highly (often brutally) self-critical process in which disenfranchised people and programs "ensure that their voice is heard and that real problems are addressed" (p. 183).

Improving our evaluation design was not the only barrier we faced. We had significant time and money issues—little or none of both. Our only hope was to empower the participants to help us gather and interpret data. We, in turn, would help them develop a logic model and analyze the outcome. We found support in the literature for this type of cooperation, which empowered participants in program planning and decision making (Fetterman, 1994; Linney & Wandersman, 1991; McLaughlin, 1987; Zimmerman & Rappaport, 1988).

Once the participants' needs were understood, we began our research. As a result, we now understand that the study of evaluation had its origins in the field of educational research and testing that began in the United States in the 1930s (Stufflebeam & Shinkfield, 1985). Since that time, the field has developed into a recognized profession with a clear set of standards for practice that we could employ (Cook & Shadish, 1987; Patton, 1987; Shadish & Reichhardt, 1987; Stufflebeam, 1994; Stufflebeam & Shinkfield, 1985). In the early stages of development, evaluation, like most other social sciences, looked to logical positivism as a guide for methodological choices (Cook & Shadish, 1987). *Hard data* was the watchword of the day. We also discovered that we were not the first to question the traditional methodology of quantitative experimental models, their underlying assumptions, and their relevance and applicability to real social settings (Cook & Shadish, 1987; Fetterman, 1982, 1994; Rossi & Freeman, 1989). For example, Cook and Shadish (1987) express their concern regarding possible shortcomings of traditional practices that use program goals to formulate hypotheses that could be tested experimentally:

> This strategy assumes that programs are relatively homogeneous, have goals that are totally explicit, postulate effects that can be validly measured, and can be assessed using feasible experimental designs that rule out all spurious interpretations of a treatment effect. All these assumptions have come under attack, not only in evaluation, but in science at large. (p. 55)

We now felt comfortable to move away from the more quantitative, experimental approach emphasizing measurement and prediction of social phenomena to a more naturalistic, real-world approach using qualitative evaluation methods that emphasize process description and understanding of social phenomena (Patton, 1980). Altman (1986) and others (Fetterman, 1989, 1994; McGraw et al., 1989; Patton, 1987) supported our shift of direction, suggesting that process-oriented evaluation promotes understanding of program implementation, the causal events leading to change, and the specific program components that most influence change. The principle of grounded discovery was central to our type of design, as themes or lessons learned emerged from participatory actions rather than being formulated prior to beginning data collection (Cook & Shadish, 1987; Glaser & Strauss, 1967).

Some projects begin as participatory and evolve to a more empowered approach, as our work did. Empowerment evaluation assumes that (a) the evaluation is used explicitly to contribute to the process, and (b) participants not only are involved but also control the process (Fetterman, 1995). In general, participatory work follows a continuum from limited participation to an ideal of full control. Empowerment evaluation begins closer to the end point of participatory work.

Because our perspective changed dramatically, we solidified our philosophical approach with a set of core beliefs and values. This was different than our original approach, because we tried to describe values rather than make value judgments. We felt it our ethical responsibility to the participants to ask, "How can we be of service?" Stake (1980) says, "It is an approach that sacrifices some precision in measurement, hopefully to increase the usefulness of the findings to persons in and around the program" (p. 76). Unlike others (Scriven, 1983), we do not think our actions can be value free. Experience has shown us that it is impossible to make choices in what Shadish, Cook, and Leviton (1991) refer to as "the political world of social programming" (p. 455) without values generated with participants. The following are the shared values produced from our evaluation:

1. The most meaningful changes are those that occur in the people themselves and that reflect an increased capacity for initiating and carrying out social change.
2. Change occurs best when greater emphasis is placed on the process for change rather than on the results of change.
3. The definition of *human capacity* is more concerned with self-sufficiency, self-determination, and empowerment than with changes that can be statistically measured.
4. Success is measured by the extent to which people are able to identify their own problems and form a consensus to propose appropriate solutions.

From these beliefs, a set of values emerged. The following are the shared values from the evaluation process:

1. Evaluation is something ordinary people can do.
2. Evaluation is guided by three key terms: *participatory, systematic,* and *simple methodologies.*
3. Evaluations should strive to show key issues, not prove hypotheses.
4. Knowledge that transfers benefits from one situation to another should be described in simple lessons-learned statements.
5. People are responsible for their own development, not agencies or experts.

We were now ready to leave the library and return to the real world. Linney and Wandersman (1991) provided the tool we needed to put our participatory plans into action: *Prevention Plus III: Assessing Alcohol and Other Drug Prevention Programs at the School and Community Level (A Four-Step Guide to Useful Program Assessment).* Using their materials, we concentrated on three validity-increasing principles: (a) a systematic process for all participants to follow, (b) leaving a paper trail, and (c) training and supporting all participants. We decided that some methodological expertise was needed to maintain validity, but that experts did not need to drive the evaluation simply for the sake of relieving validity anxiety. We refined and broadened our role as evaluation experts. Our role became one of facilitator, advocate, trainer, coach, mentor, and occasional expert. We were no longer detached observers but began to build trust and work closely with program personnel to implement an objective evaluation design.

The Second Attempt: Their Needs, Not Ours

From what seemed a failure came lessons that shaped our current, successful strategies. We learned to place greater emphasis on the process by which change occurs rather than on the result of that change. As evaluators, our job is to increase the ability of people to progressively take greater control over their own program development. Our self-assessment design is not contradictory to scientific rigor. Fetterman (1994) is correct when he advocates "educating others to manage their own affairs in areas they know (or should know) better than we do" (p. 10).

Keeping in mind our three key terms—*participatory, systematic,* and *simple methodologies*—we worked to keep the process simple, scientifically valid, and low cost, using only tested public domain materials. We collected nearly 1,000 ($N = 960$) matched pre-/posttests and 275 matched comparison group surveys from 24 rural and urban sites that were completed and analyzed in less than 6 months. During that time, focus groups were conducted, along with facilitator and program surveys, producing a qualitative and quantitative database for analysis. The process and outcome data were analyzed collectively. The evaluation project produced (a) a set of agency program improvements now being implemented, (b) a plan for advanced and ongoing evaluation training statewide, and (c) a user-friendly, four-step self-assessment process for field sites. At the same time, the evaluation (a) enhanced the agency's capacity to attain nearly $300,000 in state-level funding for technical assistance, including evaluation and related services statewide; (b) created an evaluation "cookbook" to explain how agencies can develop their own evaluation process; (c) added a statewide pool of experienced, trained lay evaluators who can identify and solve their own problems; and (d) increased the agency's capacity to write and implement competent, reliable evaluations for themselves.

Although some programs may be able to perform their own evaluation adequately without assistance, the majority simply cannot. The willingness to complete a satisfactory participatory and empowerment evaluation is most likely to occur if the evaluation is based on a developmental model. Therefore, we felt our first priority was to break down the evaluation process into successive stages to build systematically evaluation capacity of our on-site partners.

Using a model developed by Paul Florin (Florin, Mitchell, & Stevenson, 1993) and the lessons we learned, we created a developmental technical assistance framework. To complete this framework, we observed participants and categorized common evaluation characteristics, tasks, and technical assistance needs as they related to the role of the evaluator.

Figure 13.2 describes five stages of development, the needs associated with each stage, and the level of technical assistance that will meet those needs.

STAGES	PARTICIPANT'S TASKS	EVALUATOR'S TASKS	EVALUATOR'S ROLES
Stage One Organizing for Action **Knowledge** of own program strategies and of research to support them. **Belief** that what they do works. **Skill building** in designing and implementing their evaluation.	• Assign evaluation team members • Clarify roles, policies, and procedures • Participate in training • Assess program needs; identify gaps • Identify similar programs for support network • Identify technical assistance resources • Complete individual evaluation plan and time lines with evaluation facilitator and/or experts	• Demystify evaluation process, build trust • Get to know participants • Acknowledge fears • Clarify roles, policies, and procedures • Provide all evaluation resources available • Assign/train facilitators • Assist with needs assessment • Assist with goal/outcome identification • Mentor • Serve as advocate, if needed • Assist with individual evaluation plan • Be patient	• 20% Facilitator • 30% Mentor • 20% Advocate/Supporter • 20% Trainer • 10% Expert

Stage Two
Building Capacity for Action
	• Complete logic model
	• Crystallize goals and objectives
	• Concentrate evaluation effort on process development
	• Maintain link with support network
	• Utilize technical assistance
	• Maintain momentum, despite wanting to give up
	• Adopt multiple evaluation strategies to measure process & outcome
		• Provide feedback
		• Coach, especially when participants feel overwhelmed
		• Facilitate:
		  – logic model completion
		  – goal/outcome completion
		  – process data collection
		  – build evaluation capacity
		• Reinforce big picture frequently
		• Check in weekly
		• Encourage use of network
		• Use humor
			• 40% Facilitator
			• 20% Trainer
			• 20% Coach
			• 20% Expert

Stage Three Taking Action More specific evaluation task orientation within program; application of knowledge; beliefs and skill building from Stage Two.	• Collect process data • Utilize expert help; apply self-correction • Utilize network of resources; ask for help • Look for patterns in process data • Use lessons-learned technique • Complete outcome data collection • Analyze findings collectively • Write report	• Review big picture again • Provide feedback • Reinforce lessons-learned approach • Coach those not up to speed • Teach, if needed • Facilitate outcome design and data collection • Provide expert direction when needed • Support network, get them talking • Advocate as needed • Help disseminate results • Maintain momentum with those who think they are finished	• 40% Facilitator • 10% Advocate • 40% Coach • 10% Expert
Stage Four Refining the Action Less task orientation; more complex application of skill building & program refinements based on program evaluation.	• Apply lessons learned approach • Add strategies to build comprehensive program activities • Expand and diversify services • Disseminate findings • Train others in evaluation process • Publicize program effects • Incorporate evaluation into strategic plan & budget	• Build momentum • Wait for them to ask • Help participants understand how to refine program • Facilitate funding requests • Facilitate strategic planning • Advocate with the participants and for the participants	• 25% Mentor • 25% Advocate • 50% Expert
Stage Five Institutionalizing the Action Highly developed evaluation skills including self-efficacy, independence, problem solving, interdependence.	• Provide leadership, capable of mentoring • Maintain momentum • Build program resources and capacity; help others to do the same • Initiate wider program efforts into community, if appropriate • Continue to use evaluation data for program refinement • Implement a well developed plan for self-sufficiency • Form partnerships through cooperation & collaboration	• Maintain momentum • Mentor • Provide expert direction, if requested • Support cooperative efforts • Offer ideas for self-sufficiency • Publicize accomplishments • Recognize strengths publicly • Refer others in need to the participants	• 80% Mentor • 20% Expert

© 1994 Margret Dugan

Figure 13.2. Participatory and Empowerment Evaluation Framework

Stage 1: Organizing for Action

In Stage 1, participants organized themselves into evaluation teams for preplanning. Our role was that of mentor and facilitator. We wanted to level the playing field from the very beginning. We listened more than we talked. We took our time. We built trust that enabled our field-based partners to assimilate what for most of them was new and complex information with a very steep learning curve.

Training was the focus of this stage. We planned it very carefully. We did not repeat our mistakes. Instead of one large group lecture on "how to," we created small, interactive sessions where we trained two or three site teams at a time. Instead of setting up a central training location, we trained locally. Instead of a consultant show-and-tell, we used the training as an opportunity for agency staff to empower themselves as trainers and facilitators. Agency staff accompanied the evaluation consultant to training, gradually taking over more and more of the activities themselves. It should be noted that not all members of the agency felt comfortable training or were qualified to train. Nor were they required to do so. Some were excellent facilitators, but simply were not trainers. After a few training sessions, however, we developed a core of trainers and team facilitators. Fetterman (1995) says, "Training becomes a part of the evaluation process. In empowerment evaluation the first workshop generates a design and a preliminary assessment" (p. 182). As the project continues, workshops and training are provided on an as-needed basis (a cyclical and reinforcing process).

We slowed everything down to a pace all participants could handle. Our main goal was to demystify evaluation and build trust. We also worked to overcome the initial barriers we created in the first, more traditional attempt. We did what participants requested. We provided small, interactive training groups that allowed our on-site partners to work with their facilitator to develop a site-specific evaluation plan. This type of training required about 20% to 30% direct instruction from experts, and 70% to 80% of the training was devoted to small group, facilitated, skill-building activities.

There was no lectern, lecture, or lecturer. We sat at one table together and talked. For every 20 minutes of expert opinion, there was at least that much time for discussion and practical applications. Every concept developed in training was followed by an activity that had been evaluated by previous participants and revised as indicated. All activities were targeted to their sites exclusively. We incorporated whatever program materials they brought. We used slides, fill-in-the-blank forms, discussions, simulations, role-plays, and a great deal of humor to vary activities and reinforce concepts. We tried to model the thoughtful dialogue process necessary in participatory decision making. Agency facilitators sat with each team and helped clear up any misunderstandings on the spot. We acknowledged that evaluation was hard, but very attainable. Although this sounds like a painstaking process, the components—needs assessment, theoretical development, evaluation design, and action plans—were all covered in a day. The participants rated trainers, facilitators, and the content of each training as outstanding.

The outcome of this training is an evaluation plan with time lines. An additional outcome of this training is that evaluators can determine where each program and participant is developmentally. Evaluators can then allocate technical assistance, time, and talent based on a given level of development.

As we trained, we laid a foundation for a clear understanding of a needs assessment. This approach enabled evaluation teams and their assigned facilitators to identify gaps in service and to forecast appropriate programmatic actions and resources. Individual evaluation plans/time lines were completed. We then completed Step 1 of our evaluation process, identifying goals and outcomes. To help participants focus on specific outcomes, we had them ask themselves three simple questions: (a) What are we trying to accomplish? (b) Whom are we trying to reach? and (c) How would we like the participants to change as a result of this program? If teams were not ready for the next stage, we were patient and intensified our individual efforts to help them organize. After training, we assigned phone and fax buddies and made weekly contacts. At this point, we concentrated more on the people process than on the outcome of the evaluation. We provided names and numbers of all available free and low-cost resources.

Stage 2: Building Capacity for Action

Once each site was organized to complete an evaluation, we started to develop their skills and feelings of self-efficacy. Our goals at this level were threefold: (a) to develop knowledge and an understanding of how their program works, (b) to build a belief that what they do will work, and (c) to build their skills in designing and implementing a successful evaluation.

Because we wanted a theory-driven evaluation, and the participants agreed it was important to understand how their model worked, we developed a generic theory and logic model for their use. We taught them how to construct their own models or to adapt ours. Everyone was expected to have a one-page logic model. *We now know that understanding how their program works is the single most important evaluation tool for novices.* It gives them a benchmark to measure against and increases their confidence in their own ability to describe and measure what they do. It empowers them with their major stakeholders—funders and local decision makers. In this stage, our role as facilitators doing *with,* and not *for,* became more intense and frequent. We concentrated on logic model development, *PP-III* Steps 1 (goals and outcomes) and 2 (process), as well as outlining a plan for Step 3 (outcome). We checked in weekly and began coaching them. We used the framework to help the participants see where they were in the process. This seemed to relieve some of the competitiveness or anxiousness about "taking forever to get it."

After training, we went to the heart of a participatory and empowerment evaluation: formative assessment. We devised a simple, categorized questionnaire with the topics (a) program, (b) participants, (c) time frame, (d) activities, and (e) materials. We asked the participants to answer the following questions in each category to start their process data collection.

Program:
    (a) "What type of program do you have?"
    (b) "Where did the program take place?"
Participants:
    "How did participants get into your group?"
Time frame:
    "What time frame are you evaluating?"
Activities:
    (a) "What activities are planned for that time frame?"
    (b) "How many were actually provided?"

(c) "What other activities are planned?"

(d) "What did you have planned, but did not cover?"

(e) "What happened?"

(f) "Why were these goals not accomplished?"

(g) "What percentage of your goal did you reach?"

Materials:

"What written materials are available?"

To provide a structure for answering these questions, we simplified and adapted *PP-III* materials, helping sites tailor these materials to their own evaluation process. Unfortunately, the existing *PP-III* process forms were cumbersome. We found that the *PP-III* forms tended to limit creativity and stifle community problem solving (everyone thought they had to "stay within the lines"), so we collectively revised them—several times. We used how-it-might-look examples to help individuals visualize process data organization and measurement. We kept asking: (a) "What happened?" (b) "Why did it happen?" (c) "What lessons did you learn?" and (d) "What would you do differently?" We taught them to look for patterns in their data. This approach and the adaptations were labor-intensive and required frequent revisions, but have had long-term benefits. We now have forms and examples that can easily be tailored for site-specific use. We also have a set of the most commonly asked process questions and their answers. We used all of the data and forms we gathered, adapted, or created for the evaluation cookbook that was later developed.

At the same time that we were working with the evaluation sites, we advocated for participants with the state funder, asking the funder to accept a more process-oriented approach. The funders were receptive and helpful. Their support was a turning point in the evaluation. We trained state-level grant monitors, program and evaluation specialists, and other funding stakeholders in our evolving process.

This state-level training was the same as our field sites training, but it was shortened to a half day. The value of this training was that those who monitored and supported our evaluation teams clearly understood the empowered evaluation process. Not all state agency participants were overly enthusiastic about attending the half-day training, nor did everyone attending want any additional training in evaluation or feel that evaluation pertained to them. Nevertheless, the long-term benefits for both the state agency and our on-site partners turned out to be invaluable and worth a few disgruntled folks. We would urge anyone attempting the same type of statewide adoption evaluation to work with both the funder and the grantees. In a participatory process, the evaluator can serve both as advocate and as facilitator, helping to reduce program-level anxiety and promoting avenues for shared program improvement opportunities between the funder and grantee. Empowered grantees can use the evaluation as a means of defining and defending programming decisions, if necessary.

Participants began to feel part of a bigger effort as the funder publicly sanctioned their involvement in the project. The state funder fortuitously decided to require *PP-III* as a model for all projects they fund. This inspired participants who were already on board and empowered them even more. Some participants were able to assist other prevention programs' evaluation efforts. This transfer of knowledge gained from participants' own evaluation experiences was beneficial for all concerned. Participants realized and were empowered by the fact that they understood the process well enough to teach others. By sharing knowledge and experience, the participants began to create an informal, statewide resource network. Although not part of this study, these informal networks for sharing and transferring knowledge and skills need to be investigated as a potential asset for resource-poor agencies.

Stage 3: Taking Action

In Stage 3, participants crystallized their program's goals and outcomes and began to implement process data collection. Programs at this level needed facilitation as well as intense coaching. Participants often felt overwhelmed and wanted to give up. At this point, process data collection was ongoing, and this was particularly hard for the participants. Although participants were trained and given extensive assistance to collect process data, they (a) did not understand how to complete the forms, (b) did not have the time to complete them, or (c) lacked the commitment either to complete or to revise their work. We intensified our efforts and revised process forms according to their needs. Oddly enough, the participants could not articulate why or how the process was going astray. This inability was due mostly to a lack of experience and fear of making a mistake. Nevertheless, we kept asking questions and refining our approach. We increased contact with those who needed help with the process. We listened and changed directions as indicated. Our role was that of expert as well as facilitator and coach.

To start building the confidence needed to complete Step 3 (outcome), we encouraged people to look for patterns in their process data that would illuminate issues. Our goal was to convince them to trust their own process as much as statistically significant findings. We encouraged the lessons-learned technique: to develop articulate statements that facilitate transfer of knowledge from one experience to another. This incremental learning strategy is important because, as helpful as a participatory and empowerment model is in bringing about positive change, it will fail if the proper developmental steps are not taken.

Until they were competent, we relieved sites of outcome analysis and completed this step for them. We adapted survey scales and used a simple, statistical procedure to measure change. We provided each evaluation team with a very simple, two-page report detailing the output from this analysis. We taught participants to read and interpret this report so as to finish their own evaluation reports to the state funder. The teams we coached and trained can now collect and analyze outcome data—some with assistance, others on their own. Our role has reversed from contacting them on a regular basis, to expecting them to ask for help when they need it. When appropriate, we refer them to other participants in the project who have solved similar problems. This supports the informal network they built.

Stage 4: Refining the Action

Programs that have reached Stage 4 are more self-sufficient and will dictate how much assistance they need and/or want. Participants now turn to us for more technical information on outcome measures, focus group implementation, and other qualitative and quantitative methodology. They are just beginning to apply to their programs the lessons they learned. To help foster their independence, we created an evaluation cookbook. This simple, 40-page resource outlined the benefits of a good evaluation, explaining step by step how to complete a participatory or empowerment design. The cookbook defined terms and explained with detailed examples how to complete a needs assessment and logic model.

Process Assessment Form

Type of program:

CBSG

Where did program take place?

Community

School

How did participants get into your group?

Parent Requested

Self-Referred

School-Referred

What time frame are you evaluating?

July 94 to July 95 (1 Year)

What CBSG activities are planned for the time frame? Describe what activities are planned for your program:

ACTIVITY	DATE	HOW MANY PLANNED	HOW MANY ACTUAL	% OF GOAL REACHED
Find volunteers	July 94	8 volunteers	0	0%
Train volunteers	Sept 94	8 trained	0	0%
Locate site for group	July 94	4 sites	4	4/4 = 100%
Recruit & intake participants	Aug 94 – Sept 94	200 participants	155	155/200 = 78%
Prepare materials	Aug 94 – June 95	200 materials	155	155/200 = 78%
Complete groups	Sept 94 – June 95	29 groups	22	22/29 = 76%

How many group sessions are planned? Number of sessions __220*__ *(s) length of time for each* __1__ *(hr) total hours of activity (sessions x hr)* __220__ *(hours of group).*

*total number of groups held multiplied by number of sessions/group. For example: you facilitated 22 groups, 10 sessions each. 10 x 22 = 220.

Figure 13.3. Process Assessment Form: Part 1

We used only the forms, questions, answers, and examples that were tested and revised during training. With each evaluation step outlined, we provided a simple, fill-in-the-blank form adapted from *PP-III*. For example, the forms for Step 2 (process) are shown in Figures 13.3 to 13.5.

Process Assessment Form

What written materials are available	Total distributed
Handouts used in group	2015 handouts
13 stickers x 155 participants	1550 stickers
CBSG program brochures	3500 brochures

What other activities are planned?

Teacher In-Service

Parent Group

Summer Camp

Red Ribbon Week

PROGRAM IMPLEMENTATION

When was the program actually implemented (dates of activities, length of time for each) and who were the participants?

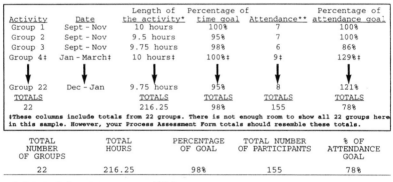

Activity	Date	Length of the activity*	Percentage of time goal	Attendance**	Percentage of attendance goal
Group 1	Sept – Nov	10 hours	100%	7	100%
Group 2	Sept – Nov	9.5 hours	95%	7	100%
Group 3	Sept – Nov	9.75 hours	98%	6	86%
Group 4‡	Jan – March‡	10 hours‡	100%‡	9‡	129%‡
Group 22	Dec – Jan	9.75 hours	95%	8	121%
TOTALS		TOTALS	TOTALS	TOTALS	TOTALS
22		216.25	98%	155	78%

‡These columns include totals from 22 groups. There is not enough room to show all 22 groups here in this sample. However, your Process Assessment Form totals should resemble these totals.

TOTAL NUMBER OF GROUPS	TOTAL HOURS	PERCENTAGE OF GOAL	TOTAL NUMBER OF PARTICIPANTS	% OF ATTENDANCE GOAL
22	216.25	98%	155	78%

 * Our goal from page one was 220 hours.
** We planned for 200 participants. We served 155. 155 divided by 200 = 75%.

TO BE ANSWERED AFTER THE POSTTEST:
What did you have planned, but did not cover? What happened that these were not accomplished?

ACTIVITY	PROBLEM
Recruit volunteers	Tried, but no one came
Summer camp	No funds available by end of year
Parent group	Canceled; only 2 parents came

Figure 13.4. Process Assessment Form: Part 2

To encourage their self-confidence, we publicized the network and their success statewide. Some have used their new skills to write grants and have been funded. All were able to comply with new state-level evaluation requirements with little or less effort than ever before.

Process Assessment Form

Who was missing that you'd hoped to have participate in the program?

Parents

Local community volunteers

Kids from south side of town

Donors for summer camp

What explanations can be offered for the discrepancy between the projected and the actual participation?

1) For the most part, attendance was at or above expectations; however, due to normal atttrition, some groups were less than expected.

2) We don't know how to market groups so working parents will want to come.

3) We need to work harder to foster relationships with south side agencies to locate sites and recruit kids.

4) Nobody on our staff had time to do fund-raising for summer camp.

How did participants evaluate the activities? (Questionnaires, surveys, focus groups, etc.)

1) IPFI-A pre/posttest

2) Process questionnaire about group (given after group)

3) Parent survey

4) Focus group

What feedback can be used to improve the program for the future?

1) IPFI-A Indicated dropouts need more help with resilience

2) Kids did not want written activities; preferred games and experiential activities

3) Parents want child care, dinner and a later start to come to parent group

4) Adolescent focus group indicated need for adolescent activities instead of "kid stuff"

Figure 13.5. Process Assessment Form: Part 3

They are able to disseminate their findings to a broader audience. They look to other experts for advice. They attend others' training to enhance their skills. They see us as mentors and loyal supporters.

Stage 5: Institutionalizing the Action

We are now at Stage 5. The goals at this level include not only self-sufficiency, self-determination, and empowerment but the idea of service to others. The nonprofit agency conducting the statewide evaluation is now self-sufficient, needing little expert guidance. It has received additional funding for statewide technical assistance and evaluation training. Evaluation is a routine part of its budgeting and its day-to-day operation. It mentors others statewide. It now can scan and analyze outcome data for others. Together with the state funder, it has published their findings. Few participants in the project are at this stage; the majority, however, are making steady progress. One said, "I never really thought I could do this. Never! And yet we wrote our whole evaluation by ourselves this year. I still can't believe it."

For the most part, the steep learning curve for evaluation has leveled off. As a collective, we created an evaluation learning community or evaluative folk culture (Fetterman, 1995) that did not previously exist—a community that sees social settings as rich resources for empowering people to exercise influence. The evaluation process has influenced the way people think about their capacity to bring about change important to them. Zimmerman et al. (1992) refer to these empowerment components as interactional and behavioral (e.g., between persons and their environment) mechanisms that enable them to master successfully their social or political systems. The results we have experienced in this simple evaluation design and implementation are consistent with this theory (Zimmerman & Rappaport, 1988; Zimmerman et al., 1992).

The Ideal Versus the Real

A caveat about our framework is needed. Ideally, evaluation teams would move through the developmental continuum in a linear fashion. Once a level is mastered, the participants would move to the next level. They would simply need training, facilitation, and coaching from experts to do so. It would seem that participants should continually move toward self-sufficiency and, once there, would remain self-sufficient.

In reality, it is not so simple. There is no clear linear movement. Participants may move back and forth along the evaluation continuum. Individual participants and/or the entire team may be at a different level at a different time. Participants need different levels and types of assistance at different times. The evaluator or facilitator must understand the developmental framework and must be absolutely flexible in its application. This process takes time and patience. For example, the evaluation teams we mentored and coached have been slow to internalize evaluation (Stages 4 and 5) as a natural part of their strategic planning. But slowly this is changing, particularly for those who seemed the most willing to make mistakes, learn from them, and get back on track.

In the End: Lessons Learned

The final product of this evaluation was not the 200-page report it generated but the lessons we learned. There are three primary lessons we can now articulate. The first of these lessons has to do with our belief that people can evaluate. The second deals with the importance of shared beliefs and values. The third describes the advantages and disadvantages we experienced. The following is our lessons-learned list:

LESSON 1

The supportive statement for Lesson 1 is that the most meaningful changes are those that occur within people themselves and that reflect an increased capacity for initiating and carrying out social change.

1. Evaluation is something ordinary people can do.
2. Participatory and empowerment evaluation should strive to show lessons learned, not prove hypotheses.
3. The evaluator's job is to build the evaluation capacity of the program participants.
4. Lasting program development is a process through which people grow in their ability to take control of their own lives and improve wherever necessary.

Participatory and empowerment evaluation is a systematic self-assessment process that occurs within the context of a program. Although participatory and empowerment evaluation relies on more inductive, qualitative measures, it also employs the quantitative. A well-thought-out logic model drives it. This requires a needs assessment and well-developed goals and outcomes. This evaluation approach has two major components—(a) process assessment and (b) outcome assessment—that seek to judge the ultimate value or merit of the endeavor from the participant's viewpoint. Of the two components, process is more important for people. They can use ongoing process data collection to begin to see patterns in their program. They can then write lessons-learned statements that can be analyzed in relationship to outcomes. Their judgments address not only the accomplishment of the program's goals and outcomes but other equally important formative factors. These include (a) examination of the chronological sequence of program planning and implementation; (b) analysis of program structure, components, and delivery systems; (c) a better understanding of contextual factors in which the program takes place; (d) determination of whether the outcomes and program design make sense in reality; (e) participation rates and participation characteristics; (f) perception of program participants; (g) levels of community awareness; (h) resources used for program operation; and (i) an analysis of unplanned results, as well as the planned ones.

LESSON 2

The supportive statement for Lesson 2 is that evaluators and participants should develop a set of written shared beliefs and values prior to making judgments.

Because it is vital to help people see the utility of evaluation and go beyond the description of activities and the measurement of merit, judgments must be made. Participants may feel threatened by judgments. This is a major barrier for decision makers because they do not know what standards will be used to judge their work or fund their program. The participants we worked with were uncertain and fearful about making any judgments. To dispel this fear, it was necessary to develop a set of shared core beliefs and values regarding the evaluation process.

These general statements become standards against which accomplishments or nonaccomplishments can be judged. These statements should be written and shared so that judgments can be applied uniformly. Evaluators and teams may need to reinforce these beliefs and values with the funder or other stakeholders.

Participatory and empowerment evaluations may fail from the lack of shared expectations, particularly if evaluators are brought in from the outside to consult. Outsiders seldom fully share the passionate values of grassroots participants—hence, judgments may be made or based on a divergent value system. Values, as well as accomplishments, should be brought under scrutiny throughout the evaluation process. All parties related to the program—funders, donors, clients, staff, evaluators, beneficiaries, and others—have a role, a right, and a voice in articulating these values. The evaluator may need to negotiate, or even become an advocate, for participants in this regard.

LESSON 3

The supportive statement for Lesson 3 is that there are certain advantages and disadvantages associated with participatory and empowerment evaluation. This list was generated with the help of Richard Krueger (1994) and Jean King (1994).

Advantages of Participatory and Empowerment Evaluation

1. Increases talent pool
2. Increases the quality and breadth of feedback for evaluation
3. Provides a systematic, flexible process for people
4. Increases the likelihood that evaluation will be undertaken and results will be used, because critical stakeholders designed and helped analyze results
5. Saves money in the long term—on-site travel, technical assistance, less training
6. Increases participants' capacity to initiate and carry out social change
7. Allows important program variables to emerge
8. Improves communication with audiences, because reports may be written in a form appropriate to the needs/interests of different stakeholders
9. Emphasizes improving local practices rather than funder-based expectations
10. Can produce context-specific materials

Disadvantages of Participatory and Empowerment Evaluation

1. Some rigor may be lost.
2. Training is essential to maintain control.
3. Time is needed to train and build empowering relationships.
4. Participants' skills and levels of commitment vary.
5. Evaluation facilitators must have strong process skills.
6. Motivation levels may vary among participants.
7. Difficulties may arise when abstraction is needed.
8. Participants may not be able to explain why their program produces few effects.
9. Highly technical reports are not typically available.
10. It involves high front-end costs.

ADDITIONAL LESSONS LEARNED

Finally, the following are lessons learned but not fully explored.

1. Some rigor may be lost. Using multiple measures, a systematic approach, and leaving a paper trail, however, help offset this loss.
2. Reporting problems and data collection idiosyncrasies will always exist, but the collective message assembled from a variety of areas across a variety of program indicators is undeniable.
3. Rigor improves as participants' skills and commitment increase.

No evaluation can be all things to all people. Participation warrants special attention, however, because it focuses on practicality and feasibility. Data collection is richer, more meaningful, and less difficult when participants are invested. Diversity and shared ideas stimulate creativity and expand the ability of those involved to see alternatives. Social change is not something imposed by external agents, but the prerogative of those on the front line who share the risks involved in decision making. Action researcher Jean King (1994) sums it up best: "If you want evaluation to make a difference, common sense and good evaluation practice tell us to involve people."

Note

1. During the past decade, over 150,000 high-risk elementary school children across Texas have participated in dozens of primary prevention and early intervention programs, all based on a single model. The model, Curriculum-Based Support Groups (CBSGs), was developed by Cathey Brown, executive director, Rainbow Days, Inc., a nonprofit agency based in Dallas, Texas. This model was first funded in 1983 by the Texas Commission on Alcohol and Drug Abuse (TCADA) to prevent substance use among children of alcoholics and other drug abusers. Since then, the model has been enhanced to meet the needs of at-risk youth in general and expanded to include programs for secondary school youth. Texas was the first state to fund this particular prevention strategy for youth, and TCADA continued to fund the expansion and enhancement of the model throughout the 1980s. Like most grassroots prevention strategies, CBSGs had been established over years of practice in a range of communities and schools but had not been systematically evaluated. Therefore, TCADA requested that a preliminary evaluation project be conducted to assess the CBSGs' effectiveness for at-risk youth and its replicability in different community settings.

To develop a simple, easy-to-use process that could be replicated, a complete review of the prevention literature was completed. The outcome of this research was the adoption of *Prevention Plus III (A Four-Step Guide to Useful Program Assessment)* as an evaluation model. This four-step process was adapted for the CBSG evaluation. During the summer and fall of 1992, the first pilot test of the model was conducted. Revisions were made to evaluation measures and processes to increase their usefulness and usability. After the pilot test, the evaluation was conducted in the summer of 1993. Then further revisions to survey instruments were made, as were significant changes to the collection of process data. Recruitment of those willing to volunteer for evaluation sites, training, and data collection began in the summer of 1993 and continued through the fall of that year. Data were collected and analyzed in the spring and summer of 1994. In the spring and summer of 1994, the evaluation process developed for CBSG programs was documented in a report to TCADA. An evaluation guidebook was developed as well.

References

Altman, D. G. (1986). A framework for evaluating community-based heart disease prevention programs. *Social Science and Medicine, 22*(4), 479-87.

Cook, T. D., & Shadish, W. R. (1987). Program evaluation: The worldly science. *Evaluation Studies Review Annual, 12,* 31-70.

Fetterman, D. M. (1982). Ibsen's baths: Reactivity and insensitivity (A misapplication of the treatment-control design in a national evaluation). *Educational Evaluation and Policy Analysis, 4*(3), 261-279.

Fetterman, D. M. (1989). *Ethnography: Step by step.* Newbury Park, CA: Sage.

Fetterman, D. M. (1994). Empowerment evaluation [American Evaluation Association presidential address]. *Evaluation Practice, 15*(1), 1-15.

Fetterman, D. M. (1995). In response to Dr. Daniel Stufflebeam's: "Empowerment evaluation, objectivist evaluation, and evaluation standards: Where the future of evaluation should not go and where it needs to go." *Evaluation Practice, 16*(2), 321-338.

Florin, P., Chavis, D., Wandersman, A., & Rich, R. (1992). A systems approach to understanding and enhancing grassroots organizations: The Block Booster Project. In R. Levine & H. Fitzgerald (Eds.), *Analysis of dynamic psychological systems* (pp. 215-243). New York: Plenum.

Florin, P., Mitchell, R., & Stevenson, J. (1993). Identifying training and technical assistance needs in community coalitions: A developmental approach. *Health Education Research, Theory and Practice, 8*(3), 417-432.

Florin, P., & Wandersman, A. (1984). Cognitive social learning and participation in community development. *American Journal of Community Psychology, 12,* 689-708.

Glaser, B. G., & Strauss, A. L. (1967). *The discovery of grounded theory: Strategies for qualitative research.* Chicago: Aldine.

Kelly, J. G. (1970). Antidotes for arrogance: Training for community psychology. *American Psychologist, 25,* 524-531.

Kelly, J. G. (1986). Context and process: An ecological view of the interdependence of practice and research. *American Journal of Community Psychology, 14*(6), 581-589.

King, J. A. (1994, November). *Evaluation that makes a difference: Strategies for involving others in evaluation studies.* Presentation at the meeting of the American Evaluation Association, Boston.

Krueger, R. (1994, November). *Evaluation that makes a difference: Strategies for involving others in evaluation studies.* Presentation at the meeting of the American Evaluation Association, Boston.

Linney, J. A., & Wandersman, A. (1991). *Prevention Plus III: Assessing alcohol and other drug prevention programs at the school and community level (A four-step guide to useful program assessment)* (DHHS Publication No. [ADM]91-1817). Rockville, MD: U.S. Department of Health and Human Services.

McGraw, S. A., McKinlay, S. M., McClements, L., Lasater, T. M., Assaf, A., & Carleton, R. A. (1989). Methods in program evaluation: The process evaluation system of the Pawtucket Heart Health Program. *Evaluation Review, 13*(5), 459-483.

McLaughlin, M. W. (1987). Implementation realities and evaluation design. *Evaluation Studies Review Annual, 11,* 205-223.

Patton, M. Q. (1979). Evaluation of program implementation. *Evaluation Studies Review Annual, 4,* 318-345.

Patton, M. Q. (1980). *Qualitative evaluation methods.* Beverly Hills, CA: Sage.

Patton, M. Q. (1987). *Creative evaluation.* Newbury Park, CA: Sage.

Rappaport, J. (1987). Terms of empowerment/exemplars of prevention: Toward a theory for community psychology. *American Journal of Community Psychology, 15,* 121-148.

Rossi, P. H., & Freeman, H. E. (1989). *Evaluation: A systematic approach* (3rd ed.). Newbury Park, CA: Sage.

Scriven, N. S. (1983). The evaluation taboo. In E. R. House (Ed.), *Philosophy of evaluation* (pp. 75-82). San Francisco: Jossey-Bass.

Shadish, W. R., Cook, T. D., & Leviton, L. C. (1991). *Foundations of program evaluation: Theories of practice.* Newbury Park, CA: Sage.

Shadish, W. R., & Reichhardt, C. S. (Eds.). (1987). The intellectual foundations of social program evaluation: The development of evaluation theory. *Evaluation Studies Review Annual, 12,* 13-29.

Stake, R. E. (1975). *Evaluating the arts in education: A responsive approach.* Columbus, OH: Merrill.

Stake, R. E. (1980). Program evaluation, particularly responsive evaluation. In W. B. Dockerall & D. Hamilton (Eds.), *Rethinking educational research* (pp. 72-87). London: Hodder & Stoughton.

Stufflebeam, D. L. (1994). Empowerment evaluation, objectivist evaluation, and evaluation standards: Where the future of evaluation should not go and where it needs to go. *Evaluation Practice, 15*(3), 321-338.

Stufflebeam, D. L., & Shinkfield, A. J. (1985). *Systematic evaluation: A self-instructional guide to theory and practice.* Boston: Kluwer-Nijhoff.

Wandersman, A., Chavis, D., & Stucky, P. (1983). Involving citizens in research. In R. Kidd & M. Saks (Eds.), *Advances in applied social psychology* (Vol. 2, pp. 189-212). Hillsdale, NJ: Lawrence Erlbaum.

Zimmerman, M. A., Israel, B. A., Schulz, A., & Checkoway, B. (1992). Further explorations in empowerment theory: An empirical analysis of psychological empowerment. *American Journal of Community Psychology, 20*(6), 707-727.

Zimmerman, M. A., & Rappaport, J. (1988). Citizen participation, perceived control, and psychological empowerment. *American Journal of Community Psychology, 16*(5), 725-750.

14

The Plan Quality Index

An Empowerment Evaluation Tool for Measuring and Improving the Quality of Plans

FRANCES DUNN BUTTERFOSS
ROBERT M. GOODMAN
ABRAHAM WANDERSMAN
ROBERT F. VALOIS
MATTHEW J. CHINMAN

This chapter focuses on the work of an evaluation team that collaborated with three community coalitions for alcohol, tobacco, and other drug abuse prevention to develop, test, and implement an empowerment evaluation method. What began as an external research evaluation project eventually incorporated the main tenets of empowerment evaluation (Fetterman, 1994a, 1994b, 1995). These included collaboration among coalition members, staff, and evaluators; improvement of the coalition and its plan of action; flexibility in solving problems and trying new approaches; and the use of both quantitative

AUTHORS' NOTE: We would like to thank staff from the South Carolina Community Partnerships in Fairfield, Newberry, Lexington, Spartanburg, and Union-Cherokee counties for their patience and helpful insights throughout the development of the PQI.

and qualitative methods. The role of the evaluation team evolved from a somewhat traditional, impartial role to that of a "coach or facilitator" (Fetterman, 1994a) helping to build group capacity for self-evaluation and empowerment.

The change in the role of the evaluators is reflected in the evolution of a new instrument, the Plan Quality Index (PQI). The index was developed and used for two distinct purposes: first, as a research tool to assess the quality of community plans and, ultimately, the effectiveness of the coalition; and, second, as a consultation tool to structure feedback to coalition staff and members to help them improve their community-based prevention plans and activities. This chapter illustrates how, based on active collaboration between the evaluators and the coalitions, the PQI evolved in its application. It was first used primarily as a quantitative measure of plan quality, and later became a qualitative tool to facilitate the improved functioning of the coalitions.

We begin by describing the rationale for a plan quality instrument in the context of the Community Partnerships programs, where our evaluation methods were developed and tested. Then, we describe the Plan Quality Index (PQI), an evaluation tool developed to measure the quality of coalition plans. Next, we describe the developmental stages of the PQI, first as a research tool and, later, as a consultation and feedback tool. This refinement was based on input from coalition staff who felt that the objective, score-based evaluation was not sensitive enough to evaluate the planning process fairly. As a result, a more narrative-based approach was collaboratively developed. Finally, we discuss the future development of the PQI and the implications of using a collaborative evaluation approach for the empowerment of coalitions.

Rationale for a Measure of Plan Quality

Coalitions of community agencies, institutions, and concerned citizens that plan and carry out programs to combat chronic health conditions have become a popular health promotion strategy. A community-based coalition is a group of individuals representing diverse organizations, factions, or constituencies within the community who agree to work together to achieve a common goal (Feighery

& Rogers, 1989). Many funding agencies require community-based coalitions to include a comprehensive planning phase (Davis, 1991; Klitzner, 1991; Sorensen, Glasgow, & Corbett, 1991; Steckler, Orville, Eng, & Dawson, 1992; U.S. Department of Health and Human Services, 1988; Wandersman & Goodman, 1991). A review of the literature enforces the fact that planning is essential for successful programs (Butterfoss, Goodman, & Wandersman, 1993). Literature and data that focus on coalition planning or the relationship of planning to outcomes are conspicuously absent, however.

Health promotion and disease prevention coalitions often have goals that take a long time to achieve. If program goals are ambitious, then careful planning is important for goal achievement and for the continued satisfaction of participants and funders (Hawkins & Catalano, 1992). Even though planning is essential, this component is often a hurried step in community-based programs with little attention paid to its process (Linney & Wandersman, 1991). Planning is equally challenging and difficult for trained staff, evaluators, and volunteers. Often, planning is initiated without the benefit of a complete needs assessment (Wandersman, Goodman, Butterfoss, & Imm, 1992). Evaluation of community-based chronic disease prevention programs, such as the Planned Approach to Community Health (PATCH) and the Community Chronic Disease Prevention Program (CCDPP), demonstrate that such projects need more training and technical support in *planning* over the life of the funding period to be successful and institutionalized (Goodman, Steckler, Hoover, & Schwartz, 1993; Green & Kreuter, 1992; Morton, 1992).

Plans are important intermediate outcomes of coalition work. Therefore, the *quality* of prevention plans and the planning process must be carefully assessed. Without careful planning, public health agencies may waste human and material resources (Goodman & Wandersman, 1994). One promising mandate for community involvement in planning comes from recent Center for Disease Control (CDC) guidance in the area of HIV prevention, which points to *participatory community planning* as an essential component of these programs. This type of planning is "evidence-based and incorporates the views and perspectives of the groups . . . for whom the programs are intended, as well as the providers of . . . services" (Centers for Disease Control and Prevention, 1993, p. 3). The U.S. Conference of Mayors also is currently involved in the evaluation of the quality, nature, and success of these community planning efforts for HIV prevention.

The Community Partnership Model

In the current example, community prevention plan development was examined in three Community Partnership programs funded by the U.S. Center for Substance Abuse Prevention (CSAP). The first partnership is composed of three counties that used a coalition of community members and social service agency representatives to plan and implement comprehensive strategies to prevent alcohol, tobacco, and other drug (ATOD) abuse. The second partnership involved one county and the third involved two counties.

In all three partnerships, the coalitions formed and developed according to the model illustrated in Figure 14.1. According to Figure 14.1, the *formation* stage of the partnerships begins at the initiation of CSAP funding (Butterfoss et al., 1993). The agency that is granted the funding, the lead agency, convenes an ad hoc committee of local community leaders. The ad hoc committee nominates influential citizens to serve on committees representing business, education, religion, criminal justice, and other sectors of the community. Committee members are trained in ATOD prevention and health promotion goals, issues, and tasks. The *implementation* stage occurs as each of the committees conducts a needs assessment to determine the extent and nature of its constituents' concerns and the availability of resources to address alcohol, tobacco, and other drug abuse. The needs assessment consists of secondary data as well as written questionnaires, town meetings, focus groups, and interviews that are developed and conducted by the committees with input from the staff and evaluation team. Implementation continues with committees using the results of the needs assessment to develop a community-wide intervention plan. The *maintenance* stage consists of the monitoring and upkeep of the committees and their planned activities. The *outcome* stage consists of the impacts that result from the deployment of community-wide strategies.

Even a casual view of Figure 14.1 suggests the necessity of a substantial amount of expertise, time, coordination, and long-term

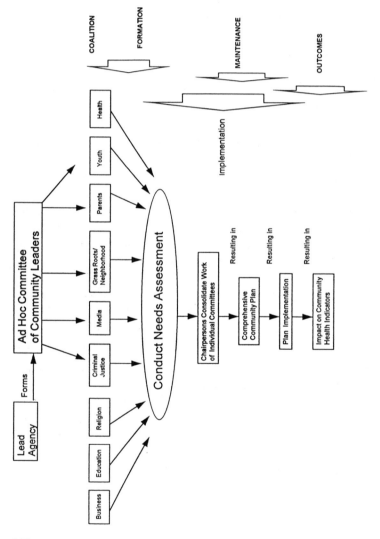

Figure 14.1. Overview of the Development of a Community Coalition

commitment in the implementation of such a complex model. Certainly, the lay citizenry who are charged with the task of coalition development need expert input, consultation, and guidance to develop the necessary capabilities for model implementation. Such circumstances inevitably had an impact on the evaluation approach, which increasingly focused on developing capacity to implement the model. The next section describes the general development of the PQI as an empowerment evaluation tool.

The Plan Quality Index (PQI): Stages of Development

To assure that the lay citizenry that sat on each partnership committee had guideposts for initiating complex tasks like needs assessment and plan development, the evaluation team developed several tools. First, to help guide an orderly needs assessment process, the evaluators developed a checklist of criteria, which appears in Appendix 14.A. The evaluators examined evidence that criteria for quality needs assessment were met by each committee and rated them in the table. Items that were given a missing or inadequate rating (–) became the focus for in-service training. Second, to ensure that needs assessment data were used properly to plan and develop logical and feasible strategies, the evaluation team developed a workbook for plan development for committee members and the Plan Quality Index (PQI) to assess plan quality.

The PQI is an index that was developed to rate community prevention plans on the basis of whether they meet given criteria that define quality plans. The PQI was refined over three developmental stages. In each stage, the PQI evolved to become a tool for empowering the coalitions to increase their capacity for program planning and implementation. In Stage 1, the PQI was developed to be used as a quantitative instrument for evaluation research purposes (Butterfoss, Goodman, & Wandersman, in press). In Stage 2, based upon negative reactions to the ratings, the evaluators added brief narratives to the scores to provide feedback to program managers about plan quality. In Stage 3, the evaluators eliminated the quantitative ratings, provided more narrative feedback, and added consultation and training on plan quality and implementation for program managers.

As the use of the instrument was refined from Stage 1 to Stage 3, the role of the evaluators also changed. First, as somewhat traditional consultants, the evaluators trained raters to use the PQI to review the plans and report numerical ratings of the plans to the coalitions. Then the evaluators served as technical assistants to analyze the PQI scores and provide feedback to program managers. Based on the reaction to that feedback, the evaluators transitioned to a more problem-solving, collaborative relationship with coalition staff. Finally, the evaluators served as coaches (Fetterman, 1994a) who, based on training and past experience, clarified communications and provided useful information, direction, and self-assessment tools to keep the planning effort on track. The following sections describe each stage in the development of the PQI to its emergence in Stage 3 as a bona fide tool for empowerment evaluation.

STAGE 1: THE PLAN QUALITY INDEX (PQI) AS A RESEARCH TOOL

In Stage 1, the evaluators worked with the first partnership and developed the PQI as a research tool. A review of the literature revealed a paucity of instruments for assessing the quality of prevention plans. Therefore, the primary goal in evaluating the quality of the coalition's plan concerned the development of an instrument and methodology for assessing plan quality. The intent was not to share the quality ratings with the coalition but to pilot test the instrument.

The PQI was developed on the basis of planning criteria from several sources (Chavis, Florin, Rich, & Wandersman, 1987; Florin, Mitchell, & Stevenson, 1993; Franchak & Norton, 1984; Kroutil & Eng, 1989; Nelson, 1986; Steckler, Dawson, & Herndon, 1980). With input from evaluation and community planning experts,[1] 18 of the original 25 criteria were used to measure the quality of the plans in the first partnership.

The PQI included 18 items: (a) 3 items that measured whether objectives and activities were clear, realistic, and reflected the goals and priorities identified in the need and resource assessments of the community; (b) 6 items that measured the scope of the plan that covered time lines, staff, targeted populations, and coordination with existing agencies and programs; (c) 3 items that measured the identification of resources in the community that would support the activities now and after the grant funding period; and (d) 6 items that measured the overall plan quality. (See Appendix 14.B for the PQI instrument.)

Three raters used the PQI to assess the quality of the plans submitted by committees in the coalition (four committees were not charged to develop plans). Prior to rating the plans, the raters were given a 2-hour workshop on the rating process that included rating an actual prevention plan from another partnership. Before using the PQI, the raters discussed all the items and reached a common understanding of their meaning.

Each of the three raters independently rated the 16 committee plans by using the PQI. They assigned a score of 0 to 5 for each item on the PQI instrument that corresponded to the categories of from 0 to 100% adequacy of the plan component (0 = 0% adequate; 1 = 1%-20%; 2 = 21%-40%; 3 = 41%-60%; 4 = 61%-80%; and 5 = 81%-100%). Each plan could receive a score ranging from 0 to 90 points. Through factor analysis, the instrument was considered to be unidimensional and the 18 items were added to provide mean scores for each plan. The interrater reliability for the 16 plans ranged from .58 to .86, and the final interrater reliability statistic was determined acceptable at .73.

As part of the evaluation of the committees' work, we also had administered a survey to the members and chair of each committee near the time the committees completed their plans.[2] By this time, committee membership had stabilized and characteristics of committee functioning could be related to the quality of the plan. The survey measures factors related to coalition effectiveness such as member satisfaction, member input in the planning process, and level of training. The primary goal of this survey was to determine whether member satisfaction with the plan created by the coalition, member input in the planning process, and member level of training were related to the quality of the plans created by the coalition.

The evaluators received surveys from 190 members (85%) of all 20 coalition committees. Members rated their satisfaction with the committee plan and with their input into the plan. Then, ratings of satisfaction and mean plan quality were grouped into high, medium, and low categories. Fisher exact tests with one-tailed tests of significance in the hypothesized direction were used to compensate for small cell values. The Fisher exact tests measured the association between levels of plan quality (high, medium, low), satisfaction with the plan (high, medium, low), and satisfaction with input into the plan (high, medium, low).

The survey also included items that asked coalition members to assess training needs (whether they had enough training in ATOD prevention to understand committee issues, whether they had the skills to perform a needs assessment, and whether they needed more training in ATOD). The responses were reported as percentages by committee.

Results. The mean values for satisfaction with the committee plans ranged from a low of 3.0 (neither satisfied nor dissatisfied) to a high of 4.2 (satisfied to highly satisfied). The mean values for satisfaction with level of input into the committee plans ranged from a low of 3.0 (neither satisfied nor dissatisfied) to a high of 4.5 (satisfied to highly satisfied). Neither satisfaction with, nor input into, the plan was, however, significantly associated with the rating of the quality of the plans.

The variance among the ratings of plan quality was relatively low. The mean scores of the committee plans ranged from 38.3 to 58.7 points. Thus, the ratings for the actual plans varied only within a 20-point range and the scores were low-moderate to moderate. No plans fell in the poor or high range of quality.

For ATOD prevention training needs, 49.3% of committees responded that they had adequate or highly adequate training to understand the issues being addressed by the coalition. Only 33.6% of committees indicated that they were skilled enough to do needs assessments. Finally, 88% of committees felt that they needed more training in ATOD prevention. These responses were consistent with information that we learned from coalition staff and members during our collaboration.

Discussion. One explanation for the lack of correlation between plan quality and committee satisfaction was that committees had little or no formal training in planning or evaluation and were not well equipped to produce a better plan. Thus, although committees were satisfied with planning efforts, the resulting plans were modest in quality.

Another set of explanations concerns the lack of variability among plans, which reduced the probability of finding significant associations between plan quality and committee satisfaction. First, although each plan was developed by a separate committee, committees had a high degree of communication *within* each county. The same staff supported each committee within the county, and committee chairpersons communicated regularly to compare progress and share ideas. The final plans were actually drawn up for each committee by the staff in a similar tabular format, which may account for the narrow range of plan scores and the lack of individual committee differences. Second, staff across counties collaborated in plan development as well. The ultimate goal of the coalition was to create compatible committee plans, combine them into three county plans, and coalesce them into a regional plan of prevention strategies. Finally, the PQI may need further refinement. The limitations of the instrument may have been responsible for ratings that were similar across plans.

In Stage 1, the coalition did not find the PQI ratings to be very useful. The mean rankings of plans were moderate to low and these modest scores dampened the enthusiasm of coalition members, who considered their work on plan development to be quite satisfactory. Moreover, the ratings provided the evaluators and coalition members with little guidance for improving plan quality and subsequent plan implementation. Because the Stage 1 rating process did not produce the type of useful data and coalition empowerment that the evaluators desired, we modified the procedure for Stage 2. The Stage 2 experience follows.

STAGE 2: TRANSITION OF THE PLAN QUALITY INDEX TO A CONSULTATION TOOL

The evaluators modified the use of the PQI in Stage 2 so that it was no longer a research tool but a consultation tool. The PQI was used again in a second CSAP partnership to rate its nine committee plans. Before rating the partnership's plans, however, the evaluators provided staff with a copy of the plan rating instrument. We believed that the items in the instrument would help guide the partnership's planning process and inform staff about what we would be looking for when rating their plans. Then, we offered the partnership a workbook[3] about how to write a prevention program plan using needs assessment data. The workbook included definitions and examples of goals, objectives, and activities that were relevant to ATOD abuse prevention. In addition, we offered to present a workshop that explained the workbook in greater detail. The partnership, however, did not think it needed the workshop. Implicit in our relationship with the coalition at this time was a sense that we were available for consultation but that we respected their decision about whether they needed a workshop.

After the nine committee plans were developed, raters were trained and the rating process proceeded as in Stage 1. In addition to the 18 criteria rated in the first partnership, 7 items were included to provide more comprehensive feedback (4 items under "community resources" and 3 items under "overall impression of plan"). This time, after the plans were rated, we provided the partnership's staff with tables that provided each rater's score on each item of the PQI. In addition, the evaluators included a brief written narrative describing general trends across ratings and specific suggestions for plan improvement. The following passage was taken from an actual committee plan and serves to illustrate the type of narrative feedback that accompanied the ratings:

GOAL—To assist in the coordination of community-based alternatives for youth and adults.

OBJECTIVES AND TASKS—Work with existing and developing community-based projects to:

1. Investigate various projects to see how we can work together.
2. Bring information to the committee to decide on directing efforts.
3. Become involved with one or more programs in the promotion of community-based alternatives for youth and adults. For example: The Community Outreach Program which is seeking to offer multiple programs in an effort to educate and build leadership in our community.

Evaluator feedback about this part of the plan was as follows:

Some of the language used was vague. The goal states that you want to work with "existing and developing community-based projects." What projects? Also, how will you work together? It is good that you mentioned the Demascus Community Outreach Program as an example; how would you work with them?[4]

We praised staff for citing examples, and pointed out that it would be beneficial to be more specific. Despite the presence of the narrative, the feedback emphasized the actual item scores for each plan.

After we presented this information to the partnership's staff, an intense discussion ensued about the plan-rating process. The staff was upset with the plan ratings and with us. They stated that the actual item scores did not sufficiently reflect their efforts. The staff related that the narrative was more helpful than the PQI scores and requested emphasis on the narrative in the future. In addition, the partnership staff wanted future feedback reports to be more balanced by including both positive and negative feedback. Initially, the plan-rating process was not sensitive enough to meet the needs of the partnership staff and, instead of empowering the committees, temporarily strained our working relationship with them. Through dialogue, however, we were able to address their concerns and repair our relationship. Our experience reinforced the understanding that empowerment evaluation is a challenging task.

The report of the PQI scores and the accompanying narrative had a significant impact on the partnership's activities. They decided to delay implementation and to revise and resubmit their committee plans to the evaluators for additional feedback. To meet the needs of the partnership better, the partnership staff and the evaluators collaboratively created a new rating system. The evaluators agreed to provide more balanced feedback and view the next drafts of their plans in a developmental context; that is, the plans would evolve into more detailed documents as implementation neared. Thus, the evaluation of the plan would provide feedback not only on what was done poorly but on what should be addressed in the future. Our original feedback to the partnership about the lack of time lines in their plans was as follows:

> The plans do not have a time line that projects a start and a finish time. Some of the plans state that they will have a certain activity completed by a certain date (e.g., "To involve parents and children in a combined sports activity, to promote healthy parenting by May 1994."), but the plans do not state both a starting date and a completion date.

Even with the new emphasis on narrative and suggestions for better implementation, the sites' reaction to the plan quality feedback was negative. Sites experienced the narrative as overly prescriptive and critical, and preferred more guidance in problem solving about plan implementation. We began to recognize that the partnerships had expectations of the evaluation beyond the mere raising of questions and suggestions for plan implementation. They desired real collaboration on building solutions and we were compelled to provide additional guidance and assistance. In response to the previous problem of a missing time line, for example, we added statements such as the following:

> Although many of these dates are not known at the present time, implementation is still several months away. When planning, it is helpful to have an idea of when you will begin and end your programs. We expect that as implementation of your program draws near, you will finalize these dates.

Adding this guidance informed the staff about what still needed to be done and why, rather than just telling them what they did wrong. By building on the positive and providing anticipatory guidance, members were empowered to produce a quality plan. To this end, we incorporated several additional tools into our approach to the Plan Quality Index in Stage 3, as described in the next section.

STAGE 3: THE CURRENT METHOD OF
CONSULTATION AND FEEDBACK: THE PQI AS
A TOOL FOR IMPROVING IMPLEMENTATION PLANS

In Stage 3, we refined the PQI feedback by expanding the narrative to include four major sections: those elements of the plan that were *well developed,* aspects of the plan determined to be *challenges for future development,* a series of questions to be considered *preparation for implementation,* and a *summary of the main points* of the committee plan evaluation. This narrative-based format was used to provide feedback on 27 committee plans developed by a third CSAP partnership. It was designed with recognition that committee plans are "developmental," are subject to ongoing refinement, and should be flexible in meeting the changes and challenges of the community partnership. The *operational definitions* of the four sections of the narrative-based feedback are as follows:

(1) Well-Developed Aspects. This section focused on the positive elements of the plan and those aspects that were well developed, according to principles of health promotion planning, implementation, and quality assurance (Green & Kreuter, 1991; Lowe, Windsor, & Valois, 1989; Windsor, Baranowski, Clark, & Cutter, 1994). In particular, the following were examined: links among needs assessment data, committee goal(s), objectives, and the planned activities; scope and sequence of activities; coordination, collaboration, and committee resources; and monitoring of progress by the committee. The major focus here was to begin the evaluation report on a positive note, recognize the committees' previous needs assessment and planning efforts, and encourage further refinement of plans.

(2) Challenges for Future Development. This section provided constructive suggestions that encouraged each committee to consider refining some of the major aspects of their plans as soon as possible. The narrative balanced the criticism of shortcomings within the major plan elements with praise (when possible) about the potential for ATOD prevention. Constructive feedback focused on refining objectives to be more realistic about the time line and target population, linking activity efforts with other community agencies, clarifying resources needed, forecasting for future planning and institutionalization of programs/activities, and encouraging refinement or establishment of a monitoring and evaluation system.

(3) Preparation for Implementation. This section focused on refinement of committee plans from a detail and time line perspective. This section was Socratic (learning by questions) in nature and provided specific questions for consideration as the committee developed its plan in greater detail. The partnership was instructed that they may not be able to answer these questions 8 months prior to implementation but that they would need to answer more of them as the implementation date drew closer. Although some queries were very specific to the committee, the activity, and the target population, more typical queries included the following:

Who will be responsible for . . .?

When will the activity be completed?

When will the planning for this activity begin and end?

Have you considered the barriers to this activity?

Have you considered media coverage for this effort?

Have you made arrangements for . . .?

What criteria will be used to . . .?

Where will this activity take place?

What evaluation/self-monitoring methods will be used?

How will you know if you met your objectives?

How will . . . be completed?

Have you considered a detailed budget for this activity?

(4) Summary. This section was designed to summarize the major strengths and weaknesses of each committee plan. Comments were focused first on the well-developed or strong aspects of the plan, then tempered with the imperatives for an improved plan. In closing this section, committees were often asked, "Will this activity have an impact on the ATOD problem in your community and how will you know?"

As in Stage 1, the 27 plans were rated independently by two members of the evaluation team, and numerical ratings and justifications for the ratings were provided. In the Stage 3 approach, a third evaluator compiled the information and wrote a four-part narrative for each plan, as described below. In this stage, the numerical ratings were used only internally by the third evaluator to guide the writing of his narrative. Thus, the emphasis was on constructive, qualitative feedback and not on numerical scores. The narrative format was well received by staff and coalition members. The objective, nonthreatening nature of the feedback allowed members to focus on challenges and improvement, without obsessing about scores and their negative connotations.

Expanding the narrative so that the feedback was more detailed provided a richer description of the challenges the coalition faced in implementing the plan. But it did not address how the evaluations could help to assure that the PQI feedback was implemented. To take this next step, the evaluators developed several tools to be used in conjunction with the PQI feedback including a *Pre-Implementation*

Plan Form, Definitions, and *Summary Checklist* to be completed by coalition staff and members (see Appendix 14.C). These tools were meant to help coalitions problem solve about the inevitable obstacles to implementation. They became central to an ongoing dialogue the evaluators maintained with the partnerships to increase effective plan implementation. The self-assessment tools were empowering because they (a) clarified what would be evaluated, (b) encouraged the coalition staff to document and track their progress in the planning process, and (c) reinforced the linkage among planning, implementation, and evaluation.

Future Directions for Research, Consultation, and Training

The Plan Quality Index was designed to help evaluate, understand, and guide the planning functions of coalitions. There are two essential issues to consider in developing the PQI as an *evaluation research tool.* First, we must continue our efforts to determine whether the quality of the plans is related to the quality and effectiveness of the interventions. Satisfaction, commitment, and input in the planning process alone do not *assure* either plan quality or coalition success (Butterfoss et al., in press; Rogers et al., 1993). The reported research was conducted during the formation and early maintenance stages of a coalition; plan quality should also be reconsidered later in the coalition life. Changing staff and membership patterns, resources and linkages with other organizations, and formal structures within the coalition may affect the outcomes and impacts of a coalition. The costs and benefits of an extensive needs assessment and planning process should also be considered. If needs assessments and planning are time- and resource-consuming and do not result in better outcomes, coalitions may decide to put their efforts elsewhere.

Second, testing this tool with other community partnerships and coalitions directed at disease prevention and health promotion is warranted. Are the same planning factors useful and generalizable to all such coalitions? Do these factors change when the focus of the health promotion effort shifts to another health problem or to another type of collaborative arrangement?

In developing the PQI as a *consultation and feedback tool,* we need to use and refine the instrument and our methods of collaborative planning within and across our communities. We should be open to developing new instruments or methods (such as the preimplementation checklists) as the need arises. Because volunteers may not be adequately prepared to plan community health promotion projects, we need to provide training on specific planning and program development skills to meet the demands of prevention efforts (Butterfoss et al., 1993; Butterfoss et al., in press; Goodman et al., 1993). The plans that are jointly developed by evaluators, staff, and members should be feasible, effective, and reflect the needs and resources of the community.

Conclusion

Effective community-based planning is challenging. Modest implementation of disconnected activities may influence the people who participate but will not necessarily have a large-scale community impact. The tools and methodology for planning prevention programs described in this chapter may be one way to improve the quality of community plans and boost the impact of prevention programs. The methods described here should help strengthen the relationships among program planning, implementation, and evaluation. As was evidenced during our three stages of instrument development, the use of the PQI instrument was enhanced from a research tool (Stage 1) that we used while remaining more "outside" of the project to a tool that promoted collaborative problem solving (Stage 3) "within" the coalition framework. Clearly, our provision of a numerical plan quality rating (Stage 1) to coalition staff was not enough to ensure an empowering planning and implementation process. A thoughtful, structured narration (Stage 3) and dialogue were needed that described the positive aspects of the plan, the challenges to be met, and points to consider before plan implementation. We also shared Fetterman's (1994a) vision of empowerment evaluation as "highly sensitive to the life cycle of the event and/or organization" and "geared toward the appropriate developmental level of implementation" (p. 7). Thus, we tempered our expectations of what coalitions could accomplish in needs assessment and planning at each stage of their development.

Even though our evaluation team believed in the concept of community empowerment, we started out with the primary aim of developing research tools and evaluation methodology. Gradually, the empowerment evaluation potential in our community work became evident, and a method for realizing that potential emerged. We learned firsthand what Zimmerman (1990) advocates, that "paradigm shifts and creative research strategies may be required to fully understand the construct" of empowerment (p. 170).

By emphasizing the positive, building on strengths, and providing the tools and process that encouraged coalition staff and members to self-evaluate, we enabled the empowerment process to unfold. When coalition staff and members actually participate in the plan evaluation process, levels of empowerment should be higher than when evaluators alone take this responsibility (Zimmerman & Rappaport, 1988). Evaluators and program planners need to continue to involve coalition staff and members in this kind of collaborative process to ensure useful evaluation and high-quality prevention plans.

Because many of these projects are still in their intervention phases, the relationship between plan quality, plan implementation, and effectiveness of the coalition remains to be seen. If community partnerships become empowered to plan well, however, then perhaps they can improve the health of our communities. Evaluators who are committed to the ideals of empowerment can play a vital role in helping coalitions and other community-based groups to achieve this goal.

APPENDIX 14.A

Needs Assessment Index

Name of Coalition: _____   Name of Rater: _____   Date: _____

ASSESSMENT SCHEME: Choose one for each question (1-23)

> \+   Specific strategy represented
> +*   No specific strategy represented but time line to develop strategy is present
> –   Inadequate strategy represented
> NA   Not applicable

	COMMITTEES					
NEEDS ASSESSMENT QUESTIONS	1	2	3	4	5	6

GENERAL

1. Have committee members been trained to conduct needs assessment?   ☐ ☐ ☐ ☐ ☐ ☐
2. Are methods for needs assessment adequate?   ☐ ☐ ☐ ☐ ☐ ☐
3. Is time line for conducting needs assessment adequate?   ☐ ☐ ☐ ☐ ☐ ☐
4. Is target population adequately sampled by committee?   ☐ ☐ ☐ ☐ ☐ ☐
5. Was sampling technique planned by competently trained individuals?   ☐ ☐ ☐ ☐ ☐ ☐
6. Are instruments available, valid, and reliable?   ☐ ☐ ☐ ☐ ☐ ☐
7. Were instruments adequately developed with committee input?   ☐ ☐ ☐ ☐ ☐ ☐

DATA COLLECTION

8. Are types of data to be collected by committee indicated?   ☐ ☐ ☐ ☐ ☐ ☐

COMMITTEES

	1	2	3	4	5	6
9. Will committee use other available health status indicators?	☐	☐	☐	☐	☐	☐
10. Are methods for using other available health status indicators indicated?	☐	☐	☐	☐	☐	☐
11. Are methods for data collection indicated (mail, interviews)?	☐	☐	☐	☐	☐	☐
12. Are data collectors indicated?	☐	☐	☐	☐	☐	☐
13. Is method for dealing with nonresponders indicated?	☐	☐	☐	☐	☐	☐
14. Are persons responsible for monitoring data collection indicated?	☐	☐	☐	☐	☐	☐
15. Are methods for storing and organizing questionnaires/interviews indicated?	☐	☐	☐	☐	☐	☐

DATA ANALYSIS

	1	2	3	4	5	6
16. Is method for data analysis adequate?	☐	☐	☐	☐	☐	☐
17. Are methods to code, input, and interpret data adequate?	☐	☐	☐	☐	☐	☐
18. Are competent individuals responsible for this activity?	☐	☐	☐	☐	☐	☐
19. Will data be presented to committees in user-friendly format?	☐	☐	☐	☐	☐	☐
20. Are methods for using results indicated?	☐	☐	☐	☐	☐	☐
21. Are data related to health status and social indicators?	☐	☐	☐	☐	☐	☐
22. Are data related to initial community needs assessment?	☐	☐	☐	☐	☐	☐
23. Is method provided for planning and prioritizing interventions based on data?	☐	☐	☐	☐	☐	☐

APPENDIX 14.B

Plan Quality Index (PQI)

Name of Coalition: _____   Name of Rater: _____

Name of Work Group or Committee: _____

Date of Rating: _____   Score: _____

RATING SCHEME: Check one choice for each component (1-18)

0	None of this plan component is adequate
1	Approximately less than 20% of this plan component is adequate
2	Approximately 20%-40% of this plan component is adequate
3	Approximately 41%-60% of this plan component is adequate
4	Approximately 61%-80% of this plan component is adequate
5	Approximately 81%-100% of this plan component is adequate

COMPONENTS OF COMMITTEE PLAN

Committee Goal(s), Objectives, and Activities	Rating (% adequate)						Score
	0	1-20	21-40	41-60	61-80	81-100	0-5
1. Goal(s) adequately reflect desired outcomes to problems/needs identified in needs assessment or community members.	☐	☐	☐	☐	☐	☐	☐
2. At least one relevant objective is stated for each goal.	☐	☐	☐	☐	☐	☐	☐
3. Specific, feasible activities are provided for each goal.	☐	☐	☐	☐	☐	☐	☐

SCOPE OF COMMITTEE PLAN

4. A time line projects start and completion of each activity.	☐	☐	☐	☐	☐	☐	☐
5. The agency/group/individual who will coordinate each activity is identified.	☐	☐	☐	☐	☐	☐	☐

	Rating (% adequate)						Score
	0	1-20	21-40	41-60	61-80	81-100	0-5

6. Sources of coordination/ collaboration among community agencies and groups are identified. ☐ ☐ ☐ ☐ ☐ ☐ ☐

7. Specific target populations are identified for each activity. ☐ ☐ ☐ ☐ ☐ ☐ ☐

8. New preventive activities are coordinated with existing community programs/activities. ☐ ☐ ☐ ☐ ☐ ☐ ☐

9. A strategy to develop community support and participation in planned activities is provided. ☐ ☐ ☐ ☐ ☐ ☐ ☐

COMMUNITY RESOURCES

10. A budget is provided that outlines sources of funding and expenses for activities. ☐ ☐ ☐ ☐ ☐ ☐ ☐

11. Staff is specified and available to coordinate and train volunteers. ☐ ☐ ☐ ☐ ☐ ☐ ☐

12. Facilities are specified and will be available for convening activities. ☐ ☐ ☐ ☐ ☐ ☐ ☐

13. Equipment and supplies for activities are specified and will be provided. ☐ ☐ ☐ ☐ ☐ ☐ ☐

14. Media coverage is planned to promote activities. ☐ ☐ ☐ ☐ ☐ ☐ ☐

15. Strategy is planned for seeking funding beyond grant period. ☐ ☐ ☐ ☐ ☐ ☐ ☐

16. Strategy is provided to monitor/ revise the plan. ☐ ☐ ☐ ☐ ☐ ☐ ☐

OVERALL IMPRESSION OF PLAN

17. The plan is written clearly and concisely. ☐ ☐ ☐ ☐ ☐ ☐ ☐

18. Plan represents state-of-the-art technology in education, prevention, and intervention. ☐ ☐ ☐ ☐ ☐ ☐ ☐

	Rating (% adequate)						Score
0	1-20	21-40	41-60	61-80	81-100		0-5

19. The plan is logically developed (i.e., priorities identified in needs assessment lead to goals, which lead to objectives, which lead to activities, which lead to resource requirements).

☐ ☐ ☐ ☐ ☐ ☐ ☐

20. The plan considers constraints in the community (e.g., sociocultural or political) that could limit implementation of AODA activities, and offers means to overcome them.

☐ ☐ ☐ ☐ ☐ ☐ ☐

21. The plan is feasible (i.e., activities can be set up by a small group working with a limited budget).

☐ ☐ ☐ ☐ ☐ ☐ ☐

22. Activities appear to be sufficient in duration to produce effects in the target population.

☐ ☐ ☐ ☐ ☐ ☐ ☐

23. Activities appear to be sufficient in intensity to produce effects in the target population.

☐ ☐ ☐ ☐ ☐ ☐ ☐

24. The plan is innovative (i.e., a creative approach to local circumstances).

☐ ☐ ☐ ☐ ☐ ☐ ☐

25. The activities are designed to become part of regular community practice (i.e., organizations in the community will take responsibility for maintaining at least 50% of the activities).

☐ ☐ ☐ ☐ ☐ ☐ ☐

APPENDIX 14.C
Pre-Implementation Plan and Summary Checklist

Date Completed: _____   Plan Name: _____

PRE-IMPLEMENTATION PLAN

Goal #	Outcome #	Target Group(s)	# Involved for Each Group	# of Sessions	Session length (hrs.)
_____	_____	_____	_____	_____	_____
_____	_____	_____	_____	_____	_____
_____	_____	_____	_____	_____	_____
_____	_____	_____	_____	_____	_____

Activity # ____	Activity # ____ Description	Resources Needed	Source of Resources	Person(s) Responsible for Implementation	Duties of Person(s) Responsible
_____	_____	_____	_____	_____	_____
_____	_____	_____	_____	_____	_____
_____	_____	_____	_____	_____	_____
_____	_____	_____	_____	_____	_____
_____	_____	_____	_____	_____	_____

Location of Activity # ____	Time Line/Dates	Budget for Activity # ____	Implementation Barriers Considered	Solutions to the Barriers	Collaborate With Whom & How
_____	_____	_____	_____	_____	_____
_____	_____	_____	_____	_____	_____
_____	_____	_____	_____	_____	_____
_____	_____	_____	_____	_____	_____
_____	_____	_____	_____	_____	_____

PRE-IMPLEMENTATION SUMMARY CHECKLIST

Date Completed: _____ Plan Name: _____

	Known Yet?	If No, write date when information will be known
1. Goal	Y N	_____
2. Outcome	Y N	_____
3. Target Group	Y N	_____
4. Number Involved	Y N	_____
5. Number of Sessions	Y N	_____
6. Sessions Length	Y N	_____
7. Activity	Y N	_____
8. Activity Development	Y N	_____
9. Resources Needed	Y N	_____
10. Source of Resources	Y N	_____
11. Person(s) Responsible for Implementation	Y N	_____
12. Duties of Person(s) Responsible	Y N	_____
13. Location of Activity	Y N	_____
14. Time Line/Dates	Y N	_____
15. Budget of Activity	Y N	_____
16. Implementation of Barriers Considered	Y N	_____
17. Solutions to Barriers	Y N	_____
18. Collaborate With Whom & How	Y N	_____

Notes

1. The following evaluation experts provided invaluable assistance in reviewing and critiquing the PQI instrument: S. Fawcett, University of Kansas; P. Florin and J. Stevenson, University of Rhode Island; M. Gwaltney, COSMOS Corporation, Washington, D.C.; K. Kumpfer, University of Utah, Salt Lake City; T. Rogers, University of California, Berkeley; and A. Steckler, University of North Carolina, Chapel Hill.

2. The instrument, survey methods, participant demographic characteristics, and overall results are available on request (Butterfoss et al., in press).

3. The planning and evaluation workbook is available on request (Butterfoss et al., 1993).

4. The name has been changed to ensure confidentiality.

References

Butterfoss, F., Goodman, R., & Wandersman, A. (1993). Community coalitions for prevention and health promotion. *Health Education Research, 8*(3), 315-330.

Butterfoss, F., Goodman, R., & Wandersman, A. (in press). Community coalitions for prevention and health promotion: Factors predicting satisfaction, participation and planning. *Health Education Quarterly.*

Centers for Disease Control and Prevention (CDC). (1993). *Planning and evaluating HIV/AIDS prevention programs in state and local health departments: A companion to Program Announcement #300.* Atlanta, GA: U.S. Department of Health and Human Services.

Chavis, D., Florin, P., Rich, R., & Wandersman, A. (1987). *The role of block associations in crime control and community development: The Block Booster Project.* Unpublished report to the Ford Foundation.

Davis, D. (1991). A systems approach to the prevention of alcohol and other drug problems. *Family Resource Coalition, 10,* 3.

Feighery, E., & Rogers, T. (1989). *Building and maintaining effective coalitions* (Guide No. 12 in the series, How-to Guides on Community Health Promotion). Palo Alto, CA: Stanford Health Promotion Resource Center.

Fetterman, D. M. (1994a). Empowerment evaluation. *Evaluation Practice, 15*(1), 1-15.

Fetterman, D. M. (1994b). Steps of empowerment evaluation: From California to Cape Town. *Evaluation and Program Planning, 17*(3), 305-314.

Fetterman, D. M. (1995). In response to Dr. Daniel Stufflebeam's: "Empowerment evaluation, objectivist evaluation, and evaluation standards: Where the future of evaluation should not go and where it needs to go." *Evaluation Practice, 16*(2), 321-338.

Florin, P., Mitchell, R., & Stevenson, J. (1993). Identifying training and technical assistance needs in community coalitions: A developmental approach. *Health Education Research, 8,* 417-432.

Franchak, S., & Norton, E. (1984). *Business, industry and labor involvement: Guidelines for planning and evaluating vocational educational programs.* Columbus: Ohio State University, National Center for Research in Vocational Education.

Goodman, R., Steckler, A., Hoover, S., & Schwartz, R. (1993). A critique of contemporary community health promotion approaches: Based on a qualitative review of six programs in Maine. *American Journal of Health Promotion, 7*(3), 208-221.

Goodman, R., & Wandersman, A. (1994). FORECAST: A formative approach to evaluating community coalitions and community-based initiatives. *Journal of Community Psychology* (Monograph Series—CSAP Special Issue), pp. 6-25.

Green, L., & Kreuter, M. (1991). *Health promotion planning: An educational and environmental approach* (2nd ed.). Mountain View, CA: Mayfield.

Green, L., & Kreuter, M. (1992). CDC's planned approach to community health as an application of PRECEED and an inspiration for PROCEED. *Health Education, 23*, 140-144.

Hawkins, D., & Catalano, R. (1992). *Communities that care: Action for drug abuse prevention*. San Francisco: Jossey-Bass.

Klitzner, M. (1991). *National evaluation plan for fighting back*. Unpublished report to the Robert Wood Johnson Foundation.

Kroutil, L., & Eng, E. (1989). Conceptualizing and assessing potential for community participation: A planning method. *Health Education Research, 4*, 305-319.

Linney, J., & Wandersman, A. (1991). *Prevention Plus III: Assessing alcohol and other drug prevention programs at the school and community level: A four-step guide to useful program assessment*. Rockville, MD: U.S. Department of Health and Human Services, Office for Substance Abuse Prevention.

Lowe, J., Windsor, R., & Valois, R. (1989). Quality assurance methods for managing employee health promotion programs: A case study in smoking cessation. *Health Values, 13*(2), 12-23.

Morton, M. (1992). *Assessing alcohol and other drug prevention programs: A trainer's manual*. Pensacola, FL: Prevention Consultants and Administrators.

Nelson, S. (1986). *How healthy is your school? Guidelines for evaluating school health promotion*. New York: National Center for Health Education (NCHE) Press.

Rogers, T., Howard-Pitney, B., Feighery, E., Altman, D., Endres, J., & Roeseler, A. (1993). Characteristics and participant perceptions of tobacco control coalitions in California. *Health Education Research, 8*, 345-358.

Sorensen, G., Glasgow, R., & Corbett, K. (1991). Promoting smoking control through worksites in the Community Intervention Trial for Smoking Cessation (COMMIT). *International Quarterly of Community Health Education, 11*, 239-257.

Steckler, A., Dawson, L., & Herndon, S. (1980). Analysis of health education sections of health systems plans. *Health Education Quarterly, 7*, 186-202.

Steckler, A., Orville, K., Eng, E., & Dawson, L. (1992). Summary of a formative evaluation of PATCH. *Journal of Health Education, 23*, 174-178.

U.S. Department of Health and Human Services. (1988). *Reducing the health consequences of smoking: 25 years of progress* (CDC Publication No. 89-8411). Atlanta, GA: Centers for Disease Control.

Wandersman, A., & Goodman, R. (1991). Community partnerships for alcohol and other drug abuse prevention. *Family Resource Coalition, 10*, 8-9.

Wandersman, A., Goodman, R., Butterfoss, F., & Imm, P. (1992). *CSAP quarterly evaluation report for fighting back*. Columbia: University of South Carolina, Department of Psychology, School of Public Health, Department of Health Promotion and Education.

Windsor, R., Baranowski, T., Clark, N., & Cutter, G. (1994). *Evaluation of health promotion, health education and disease prevention programs* (2nd ed.). Mountain View, CA: Mayfield.

Zimmerman, M. (1990). Taking aim on empowerment research: On the distinction between individual and psychological conceptions. *American Journal of Community Psychology, 18*(1), 169-177.

Zimmerman, M., & Rappaport, J. (1988). Citizen participation, perceived control, and psychological empowerment. *American Journal of Community Psychology, 16*(5), 725-750.

15

Building Community Capacity With Evaluation Activities That Empower

STEVEN E. MAYER

At Rainbow Research, we didn't start off doing "empowerment evaluation." Instead, we came to it after discovering that evaluation done in the traditional style did not have the impact we hoped our evaluations would have.

Our purpose as a nonprofit organization is "to assist socially concerned communities and organizations in responding more effectively to social problems." Already one can see an activist bias: We hope our findings are to be acted upon in ways that improve community viability and vitality.

The projects we choose to evaluate are those that have a purpose of "building community capacity." To us, *community capacity* is "the sum total of commitment, resources, and skills that a community can mobilize and deploy to address community problems and strengthen community assets" (Mayer, 1994). We consider building community capacity fundamental to the concept of empowerment.

We prefer to involve ourselves with programs that intend to build capacity because we feel they represent the best use of scarce public, private, and philanthropic resources, and the best hope for communities.

We're eager to learn how they work so that we can communicate the lessons learned to other communities. Our purpose is to get valuable information about what works into the hands of people who can use it.

Not all projects have capacity building as a purpose. Indeed, far too few of the projects conceived and supported in the nonprofit and public sectors have much capacity building going on, and as a result communities too often do not show an increment of strength after receiving and spending scarce resources. The paradigm of development is only just beginning to shift from the "needs or deficits model" to the "strengths or assets model" in which capacity building has a more explicit purpose (Kretzmann & McKnight, 1993).

Just as projects vary in their potential for building community capacity, so too does evaluation. We believe that evaluation can assist capacity building, especially when it gives the intended beneficiaries of a project the opportunity to get involved in its evaluation. Evaluation that allows the project's intended beneficiaries to get involved in the evaluation process in ways that give them more commitment, resources, and skills could be said to fit the description of "empowerment evaluation" (see Fetterman, 1994a, 1994b).

We think that empowerment evaluation is consistent with empowerment theory (Zimmerman, in press). At a minimum, this means, in our view, that we must listen to, respect, and act on what the project's intended beneficiaries have to say about how they are benefiting from efforts made allegedly in their behalf.

Not all evaluation includes the simple act of listening to the intended beneficiaries in the community. As a result, it is very easy and possibly the norm for evaluation results to be ignored. As stated at the outset of this chapter, Rainbow Research did not start off intending to do empowerment evaluation. Instead, we discovered that, for evaluation to be useful and used, certain voices have to be included.

Traditional evaluation tends not to pay much attention to the real voices of real people, preferring to seek the alleged precision that counting or scoring events that form a dependent variable allegedly provides, and testing a quantitatively formulated hypothesis using the principles of inferential statistics and the scientific method. Unfortunately, the scientific method was designed for use in situations under the scientist's control, and most humans and communities don't meet that condition.

As a result, except in extremely unusual circumstances, evaluations done in the allegedly scientific manner tend to be (a) costly, typically as a result of attempts at scientific control; (b) unsatisfying to audiences, typically because only a very few dependent variable are used; (c) problematic, typically because actions taken to achieve the necessary scientific control can easily be offensive or injurious to humans or communities; (d) useless, typically because the scientist's victory— rejection of a null hypothesis at a statistically significant level—provides communities with no information on how to proceed with their task of strengthening themselves. (For a further critique of the limitations of the application of traditional scientific approaches, such as experimental design, to evaluation, see Conrad, 1994; Fetterman, 1982.)

The people who could most benefit from reading or hearing an evaluation report are *not* typically other evaluators or social scientists, so it is not necessary to design evaluations that conform to their expectations. The people who could benefit most are those who are already involved in the program being evaluated, those similar to them in other parts of the community near and far, and those who work with and support them directly.

It is these groups who are in a position to heed the findings and adopt the ideas and recommendations suggested in an evaluation. It is these groups, then, whose curiosities and sensibilities ought be included in the evaluation process if the findings are to be heeded and progress is to be made. It is these groups who are to be empowered in the evaluation process, and thereby in the community-building process itself.

What can evaluators do to involve intended beneficiaries in evaluation, so that they gain in commitment, resources, and skills—the basic ingredients of community capacity? The basic perspective here is that communities are strengthened when their capacities are developed, that is, when their commitment is increased, when their resources are increased, and when their skills are increased. If evaluation is to help in this endeavor, the evaluation process too should provide opportunities for community strengthening.

This position suggests that communities are strengthened when they can build upon their strengths and assets rather than be made to focus on their needs or deficits. Evaluation, too, can be constructed from this perspective. We suggest three key features that can ally evaluation with the tasks of community capacity building.

1. HELP CREATE A CONSTRUCTIVE ENVIRONMENT FOR THE EVALUATION

A constructive environment is one conducive to action that helps the community use the evaluation process to develop its commitment, resources, and skills.

Codiscoverers. Minimize the distance between evaluator (as expert) and program participants (as ignorant). We like to engage program participants in ways that let us say that we are "codiscoverers" with them in efforts to learn about the merits of their program.

Risk Containment. We recommend a policy in which negative findings from an evaluation should not lead directly to punishment by program funders or directors. Negative consequences can be made to happen if a remedial plan drawn up to fix revealed shortcomings is not followed.

Partnership With Funders and Other Supporters. The intention of an evaluation should be to strengthen community responses, not punish. The evaluation should not be about fault finding but should identify opportunities for improvement and constructive roles for stakeholders, including funders. (It may be legitimate to undertake evaluation with the intent to decide if a program's support should be cut off, but such an evaluation could not be called empowering or capacity building.)

2. ACTIVELY INCLUDE THE VOICES OF INTENDED BENEFICIARIES

Capacity-building projects are to lead to improvements in the systems for serving communities. It stands to reason, then, that the voice of community members should be included in the evaluation process.

Include Their Sense of Legitimate Inquiry. Intended beneficiaries of a program should have a meaningful voice in deciding the purposes of the evaluation (how findings are to be used and by whom), the styles of inquiry, sources of information, and interpretation of findings. If the evaluation process does not pass muster with those who are supposed to benefit, findings can be dismissed as illegitimate.

Include Their Experience, Wisdom, and Standards of Excellence. Intended beneficiaries of a program can be considered the ultimate source for assessing the merits of a program as well as the standards on which programs are judged. Without their considerable input, findings could be dismissed as merely hearsay.

Include Those Not Normally Included. For the most part, programs are still operated by a somewhat-professionalized class of "helpers" who tend not to look very much like those they say they want to help. This "disconnect" typically extends to the evaluators as well. It is easy, therefore, for an evaluation to overlook or discount the more marginalized of the community, yet it is they who are typically designated to be intended beneficiaries of community programs—people of color, the older and younger, and the less abled.

3. HELP COMMUNITIES USE EVALUATION
FINDINGS TO STRENGTHEN COMMUNITY RESPONSES

My organizer friends remind me that you cannot give people a voice—they already have a voice, thank you very much—but if you want to be helpful, you can give them an ear, or help others hear their voice. This means evaluators and others wanting to help communities build capacity can help make sure that community voices are heard, not just in designing and conducting the evaluation but in helping communities and other audiences (such as policymakers) move forward with the findings.

Help Spread the Lessons Learned. Evaluation findings that stay on the shelf unread and unheeded are worthless, no matter how legitimate the process for discovery. All media should be considered for disseminating worthwhile findings—print, broadcast, workshops, storytelling, and electronic bulletin boards.

Help Create Links Among People Who Can Use the Information. For various reasons, community-building work has been compartmentalized into systems, regions, agencies, and professional groups. Evaluations, if done well, may have implications across these boundaries. Evaluators and their colleagues should consider their audience broadly, and help findings penetrate boundaries that normally separate worthwhile efforts.

Help Communities and Their Organizations Build on Gains. Recommendations should be written that allow community organizations to mobilize and strengthen the commitment they bring to their work, increase the financial and other resources usable for strengthening their work, and further develop the skills needed to make their work effective.

Examples

WOMEN'S EMPOWERMENT LOGBOOKS

A program that works with mothers trying to get off welfare asked us to help them evaluate their program in a way that would educate themselves and public officials about life on (and getting off) welfare. As part of this, we developed a way for mothers to notice and journal their daily efforts to overcome barriers to greater self-sufficiency. These journals will be used to help educate county welfare officials on difficulties faced by women on welfare as well as small, manageable opportunities for overcoming those difficulties, as actually experienced. They'll be used by program staff to help other mothers notice and take advantage of opportunities to make gains. And they'll be used by mothers themselves as they continue to educate themselves and each other in support groups about what works for them.

The Acts of Empowerment Evaluation Logbook is presented in Appendix 15.A. Citing the empowerment features presented above, we installed these mothers as the voice of authority on successful efforts at getting off welfare and helped them notice and evaluate the effectiveness of their own behavior. We also invited in policymakers, staff, and each other to be their audience.

AFFORDABLE HOUSING EVALUATION TOOLBOX

A nonprofit housing developer asked us to develop tools that their developments' residents and boards could use to assess their quality of life (and housing). These tools come as ready-to-administer surveys, instructions for conducting focus groups, and sample report outlines for reporting outcomes. Administrators can use them to monitor costs and problem areas. The field can use them to guide housing and human development efforts.

A large section of one of the toolboxes, including instructions and several of the ready-to-administer tools, is presented in Appendix 15.B. In this case, we helped residents and owners form a partnership in strengthening the quality of life in "their" affordable housing development. Inquiries are conducted in areas known to be of concern to both residents and owners, and inquiries involve the participation of both parties. The tool is meant to minimize conflict and maximize mutual understanding, and can be used by the entire field.

LEADERSHIP PROGRAM FOR COMMUNITY FOUNDATIONS

The purpose of the leadership program evaluation was to discover the ways in which a select group of community foundations grow and develop when given a fairly sizable infusion of financial, technical, and nontechnical support. The point in discovering this was to formulate findings and principles that the rest of the community foundation field could learn from. After 5 years of codiscovering with the various stakeholders of 18 different community foundations and frequent communications on preliminary learnings, we wrote a book for distribution to the field. And because the work of community foundations touches on the work of so many other institutions, distribution efforts focus on getting the book into these other networks.

The introductory letter we sent to participants struck the tone of codiscovery and set up expectations for site visits and the evaluation workbook itself. In this case, we created a safe environment for the mutual discovery of common and unique lessons on the dynamics of growth, insisting there is no "one way" to grow, and then created a permanent form for sharing the lessons with the community foundation and adjacent fields.

CRIME PREVENTION ASSESSMENT

A highly evolved neighborhood organization undertook a substantial community crime prevention program in partnership with the city's police department. The evaluation was charged with evaluating how such a partnership works (for possible replication in other parts of the city) and with learning what gains were being made in crime prevention in the perspective of community residents. We spoke with people with all kinds of stakes in the community: homeowners, renters, absentee landlords, street residents, business owners, workers, police, and other city officials. Special efforts were made to listen to those not frequently heard from or found (including minority business owners and tenants) but with a voice that can speak to the quality and effectiveness of this partnership, allegedly undertaken in their name, as they experienced it.

In this case, we worked to present the project as well as the evaluation as jointly owned by all possible stakeholders, and to include people in the evaluation that are normally hard to include. The report presented conditions we believed important for successful replication elsewhere, allowing other neighborhood groups and other cities' police departments to consider this model. The availability of the report was announced in law enforcement circles and neighborhood development circles.

DRUG ABUSE PREVENTION ASSESSMENT

A rural community of 10,000 had mounted a comprehensive, school-based drug abuse prevention program. We orchestrated an evening-long "prevention forum" that showcased the findings of a committee of concerned citizens who sought a variety of evidence on the merits and demerits of the program. Presentations included results of interviews with school personnel; surveys of parents of schoolchildren; a review of pertinent literature including what other communities had discovered; and testimony by the police, state drug officials, and local recovering celebrities. The forum played to a packed high school gymnasium on a weekday night, well attended by parents in both their public and their private roles.

In this case, we worked to create local ownership of the evaluation, built and strengthened links among different kinds of community participants (schools, parents, and state officials), helped them develop their critical abilities and perspective on drug abuse as a local issue, and created a basis for them to build further on the gains made from this evaluation.

Evaluation done with the intent of contributing to community capacity is not easy to do. It presents its own challenges, not only of design but also of implementation. Each of the three features of empowerment evaluation discussed above—creating a constructive environment, actively including key voices, and helping communities use findings—requires regular vigilance and attention to issues of integrity and fairness, as does the larger issue of striking the right balance between compassion for activists' intentions and dispassionate inquiry and analysis.

Yet the rewards are greater, we find. Because this approach intends to listen to beneficiaries and those who work on their behalf, we are accorded trust and rewarded with access to the heart and soul of community-building work. And if it's obvious that we've listened well, our analyses are more likely to be heeded, and communities are thereby rewarded.

APPENDIX 15.A

WE NEED: Women's Empowerment Group
Acts of Empowerment Evaluation Logbook

ACTS OF EMPOWERMENT

These are actions a group member might take—as a result of her participation in the empowerment group—that indicate self-sufficiency: a willingness or capacity to take control of her life, improve her life.

HOW TO USE THIS LOGBOOK

1. Individual participant progress. Keep an Empowerment Logbook for each participant. Update periodically (e.g., every 2 weeks, or after every group meeting, or after every contact).

2. Total program impact. Total up the check marks in each category for all participants to describe impact on participants overall.

Participant: _____

Dates Involved in Empowerment Group: From _____ to _____

A. Academic/Career Education
_____ Conduct a self-assessment to prioritize educational needs
_____ Use resource to help do this (e.g., discussed in group, met with outside consultant, took aptitude or interest or competence tests): _____
_____

_____ Investigate educational resources (e.g., called schools, visited schools, talked to school representatives, talked to knowledgeable friends or other consultants/resources): _____
_____

Enroll in educational program:
_____ Pre-GED _____ Postsecondary: _____
_____ Other: _____
Financial aid and scholarships:
_____ Investigate availability, eligibility
_____ Earn eligibility for merit grants and scholarships
_____ Apply for
Participate in educational program:
_____ Attend class
_____ Do homework
_____ Take exams
_____ Pass courses
_____ Obtain certificate or diploma

B. Mobility
Learner's permit
_____ Take test to obtain learner's permit
_____ Pass test, obtain permit
Driver's license
_____ Enroll
_____ Participate
_____ Complete driver training
_____ Practice driving on her own
_____ Take driver's license exam
_____ Written
_____ Driving
_____ Pass exam, obtain license

Car management

Learn about

_____ Car maintenance. Describe: _____

_____ Smart ways to shop for car, insurance, maintenance.
Describe: _____

_____ Do own car maintenance tasks. Describe (before and after):
_____

_____ Use improved car consumer expertise. Describe examples
of smarter shopping/bargaining for car services: _____
_____

_____ Increased use of public transportation
How often: Before (when): _____
After (when): _____

C. Increased Engagement in the World

Use of

_____ Child care
Type (e.g., friend, relative, baby-sitter, day care at home,
co-op exchange, day care at center): _____
How often: Before (when): _____
After (when): _____

_____ Improvement/enrichment classes (e.g., exercise, cooking,
household management, parenting): _____

_____ Increased volunteer activities/responsibilities: _____
_____

_____ Register to vote for first time
_____ Voted. When: _____

_____ Decreased use of food shelf:
Before: _____ times per quarter in _____ (time frame)
After: _____ times per quarter in _____ (time frame)

_____ Skills bartering: _____
_____

D. Parenting

_____ Do things with child(ren): _____
_____

_____ Attend child's activities (e.g., school plays or concerts):
_____
_____

_____ Enroll child in enrichment/development activities (e.g.,
dance lessons, recreation programs, Scouts, Sunday school)

_____ Take steps that enable child to participate (e.g., provide transportation, bag lunch): _____

_____

E. Economic Empowerment
_____ Apply for Food Stamps
_____ Apply for other financial or health benefits: _____

_____

_____ Apply for a job (what, what pay, where, when): _____

_____

_____ Get a job, or a better job (what, what pay, where, when):

_____

_____ Start or expand a business (describe business and steps taken): _____

Banking, credit, and personal financial management:
_____ Open an account. Checking: _____ Savings: _____
_____ Apply for a loan. Amount and purpose: _____

_____

_____ Pay back loan on time
_____ Responsible use of credit cards (describe "before" and "after" use): _____

_____

_____ Pay bills on time
_____ Prepare family budget
_____ Spend within budget guidelines

F. Home and Household
_____ Increased stability of housing. Describe: _____

_____

_____ Improved housing. Describe: _____

_____

_____ Improved maintenance of household
_____ Cleaner: _____

_____

_____ Less breakage and wear and tear: _____

_____

_____ Increased self-sufficiency in maintenance (do own repairs): _____
_____ Increased energy efficiency of home: _____

_____

_____ Improved diet and nutrition: _____

_____

      _____ Participate in community or private garden: _____

_____

      _____ Increased canning of homegrown produce

G. Individual Acts of Empowerment in the Group Setting
      _____ Take on additional responsible role(s) beyond "group member" (e.g., child care liaison, corresponding secretary, recording secretary, treasurer): _____

_____

      _____ Initiate or lead group activities or projects: _____

_____ _____

      _____ Bring, maintain notebook regarding what you're learning in group
      _____ Participate in group activities:
           _____ Role-playing (describe roles): _____

_____

           _____ Other: _____
Acts on behalf of the group:
      _____ Help raise funds (describe actions, e.g., draft proposal, edit proposal, practice presentation, make oral presentation to potential funders, organize grassroots fund-raising activity): _____

_____

      _____ Obtain in-kind assistance for group (e.g., network with agencies, churches, and other external resources for jars, books, produce, training, etc.): _____

_____

      _____ Recruit additional members: _____
      _____ Present the program to other potential sites (e.g., other food shelves): _____

H. Legal Empowerment (Use of Legal System Resources)
      _____ Use Tenants Union: _____

_____

      _____ Use Legal Aid: _____

_____

      _____ Attend divorce clinic: _____
      _____ Learn/exercise rights as debtor under bankruptcy and collection statutes: _____

Use court system:
_____ Small claims court: _____
_____ Divorce: _____
_____ Custody: _____
_____ Restraining order: _____
_____ Make/update will: _____
_____ Other: _____

I. Self-Image and Interpersonal Skill Development
_____ Participate in counseling: _____
_____
_____ Participate in support groups: _____
_____ Participate in other activities to build self-esteem and interpersonal skills: _____
_____ Evidence of improved self-image or improved interpersonal skills: _____

ADDITIONAL NOTES, OBSERVATIONS (documenting acts of empowerment the group participant did at least partially as a result of something she got from the empowerment group process):

APPENDIX 15.B

Evaluation Toolbox[1]

RESIDENT SERVICES PROGRAMS

CONTENTS

INSTRUCTIONS
A. Introduction
B. Evaluation Areas and Indicators
C. Time Line for Implementing Tools
D. Survey Guidelines
E. Focus Group Guidelines
F. Personal and Telephone Interview Guidelines
G. Guidelines for Selecting Respondents

REPORT SECTIONS
A. Self-Sufficiency and Quality of Life
    (1) Staff records on independence
    (2) Resident's personal interviews on independence and quality of life
B. Awareness of, Involvement in, and Satisfaction With Resident Services
    (3) Residents' survey on involvement in and satisfaction with resident services
    (4) Residents' focus group on involvement in and satisfaction with resident services
    (5) Staff records on number of residents attending resident services programs and events
    (6) Staff records on number of residents served individually by resident services staff
C. Property Management Costs
    (7) Staff records on costs of property management
D. Use of Other Human Service Providers
    (8) Staff records on use of other service providers
    (9) Service providers' focus group on effective use of services providers and perceptions of residents
E. Perception of Community Members
    (10) Community members' telephone interview on perceptions of the housing site

INSTRUCTIONS

INTRODUCTION

This Evaluation Toolbox has been designed to help Westminster Corporation assess, document, and describe the effectiveness of the resident services programs they provide at affordable housing sites. Each tool has been developed to collect information on the potential outcomes of the resident services programs that were identified by Westminster staff and other interested audiences. The evaluation areas and indicators that correspond to the potential outcomes identified by Westminster staff and other audiences for the evaluation are listed after this introduction.

At this time, the Toolbox is a working draft to be implemented by Westminster Corporation in a field test. The field test will allow Westminster to collect preliminary evaluation data and identify areas where the Toolbox can be strengthened and revised. At the end of the field test, the Toolbox will be finalized and used by Westminster to learn about the outcomes of their programs.

The Toolbox has been divided into three sections:

1. The guidelines present overall directions for implementing the tools—how to arrange for a focus group, how to select a sample of respondents, and how to administer a survey.
2. The report sections are examples of the type of evaluation statements that can be made based on the information collected by each tool.
3. The attachments contain the individual tools—surveys, focus groups, telephone interviews, and other data collection instruments—along with directions for implementing each tool, samples to include, and suggested time lines.

To begin working with the Toolbox, select which tools to use, identify an individual to coordinate the data collection and analysis for each tool, and set a time line for using each tool. During the field test, we recommend that each tool be used at least once—even if it is only with a small sample of respondents. We have attached a possible time line for using each of the tools during the field test.

The Toolbox has been created to provide Westminster with a variety of tools from which to make selections. In using the Toolbox, Westminster can decide which tools to implement and whether to make any modifications in those tools. To reduce the magnitude of the evaluation, Westminster can

1. implement only some of the tools within an evaluation section;

2. eliminate some of the questions from a particular tool; or

3. combine several tools to collect more information from each respondent.

For example, within the self-sufficiency evaluation area of the Resident Services Toolbox, staff records could be used to collect evaluation data without the personal interviews. If the personal interviews were implemented, some of the questions on independence could be left out. Or the focus group on residents' satisfaction with resident services could be combined with the focus group on residents' satisfaction with the board in the Owner Services Toolbox.

EVALUATION AREAS AND INDICATORS

EVALUATION AREAS	INDICATORS
Greater Resident Self-Sufficiency and Improved Quality of Life	Income
	Employment
	Use of public assistance programs
	Use of emergency/nonemergency funding programs
	Self-report of changes in independence/self-sufficiency/ quality of life
Involvement in and Satisfaction With Resident Services	Attendance at social gatherings and resident service programs
	Satisfaction with resident services
	Number served by resident services staff as an individual client
Reduced Costs of Property Management	Vacancy rates
	Operating cost for property management
	Replacement and maintenance costs
	Operating cost for resident services
Effective Use of Other Human Service Providers	Service providers used
	Types of services provided
	Number of residents served
	Report of service providers on effective use by Westminster
	Changes in service providers' perceptions of affordable housing residents
Community Perceptions	Community members' perceptions of the housing site

FIELD TEST TIME LINE

EVALUATION TOOL	IMPLEMENTATION TIME LINE
(1) Staff Records on Independence	Baseline data collected on a sample of 30 units in February and July
(2) Residents' Personal Interviews on Independence and Quality of Life	Five personal interviews in June
(3) Residents' Survey on Involvement in and Satisfaction With Resident Services	Sample of 10 respondents with surveys in March
(4) Residents' Focus Group on Involvement in and Satisfaction With Resident Services	One focus group in May
(5) Staff Records on Number of Residents Attending Resident Services Programs and Events	Attendance sheets completed continuously with summary in July
(6) Staff Records on Number of Residents Served Individually by Resident Services Staff	Staff records completed monthly with summary in July
(7) Staff Records on Costs of Property Management	Begin collecting baseline data in February with summary in July
(8) Staff Records on Use of Other Service Providers	Begin compiling data in February with summary in July
(9) Service Providers' Focus Group on Effective Use of Service Providers and Perceptions of Residents	One focus group in April
(10) Community Members' Telephone Interviews on Perceptions of the Housing Site	Five interviews in May (same interviews used for owner services)

SURVEY GUIDELINES

To use the surveys in this Toolbox, we suggest taking the following steps, in order:

☐ Select the sites and the individuals at each site from whom you would like to collect information (see the guidelines on selecting respondents).

☐ Review the survey and make any revisions or additions that are necessary. For example, you will need to add a list of issues considered by

the boards at each housing site to the survey on the relationship
between the board and the residents in the Owner Services Toolbox.

☐ Send each individual a postcard notifying them of the upcoming
survey (optional).

☐ Send each individual the survey along with a return envelope and a
cover letter indicating whom to contact with questions. If the survey
is to be returned by mail, be sure to stamp the return envelopes.

☐ Send each individual a postcard reminding him or her to complete
the survey (optional).

☐ Send each individual who has not returned a survey a follow-up
cover letter with another copy of the survey and a return envelope.
If you have not coded the surveys to track who has returned a survey,
you will need to send this mailing to everyone (optional).

Because the individuals who do not return the survey may differ from those
who do return the survey, it is generally best to make a special effort to
encourage everyone to respond. Some suggestions for increasing the number who respond include the following:

1. Offer those who return the survey some type of incentive.
2. If you mail the survey, use stamps rather than a postage meter.
3. Use personalized letters.
4. Stop at their homes to pick up the surveys.
5. Complete the optional steps listed above.

FOCUS GROUP GUIDELINES[2]

These guidelines have been prepared to help you prepare for a focus
group, facilitate the discussion, and analyze the data collected through the
focus group.

WHAT IS A FOCUS GROUP?

"A focus group is a carefully planned discussion, with five to ten participants, designed to obtain perceptions about a specific topic in a permissive
and nonthreatening environment" (Krueger, 1994).

WHOM SHOULD YOU INCLUDE?

The people you invite to the focus group should be, in some way, similar
to each other. For example, the focus groups in this Toolbox bring together
groups of residents or groups of service providers. Typically, you would
not invite both residents and service providers to the same focus group.

You should, however, invite people with a range of experiences and perspectives so that you can discover the range of possible responses to your questions. For example, in a focus group of residents, you may want to invite residents from different sites with different types of resident services programs.

In some situations, you may not want to invite people to the focus group who know each other. People may be reluctant to answer certain types of questions or provide honest responses if there are others in the group that they know.

WHAT TYPE OF INFORMATION WILL YOU COLLECT?

Focus groups are generally used to discover the perceptions, feelings, or thinking of respondents about a particular topic. The purpose of the focus group is to find out about the range of possible responses and identify common patterns in the responses. The focus group discussion should not be used to reach consensus, provide specific recommendations, or make decisions. Focus groups produce qualitative rather than quantitative information.

HOW DO YOU MODERATE A FOCUS GROUP?

The moderator of the focus group provides a permissive, open environment and asks questions to guide the discussion. It is important that the moderator listen carefully to responses without giving his or her opinions, answers, or values.

For each of the focus groups in this Toolbox, a brief introduction and list of focus group questions have been provided. In all focus groups, the moderator should briefly review the purpose of the focus group, ensure participants that their comments will remain confidential, and remind participants that different perspectives are important and consensus is not necessary.

WHAT PREPARATIONS ARE
NEEDED BEFORE THE FOCUS GROUP?

Extend a telephone invitation to participants approximately 10 to 14 days before the focus group meeting. Describe the purpose of the focus group and the types of questions that will be asked. This will help participants to prepare for the questions.

Send a personal letter to each participant, after the initial telephone call, confirming the date, time, and location of the focus group. The letter should also indicate the purpose of the focus group and key questions that will be asked. The day before or the day of the focus group, give each participant a telephone call reminding him or her of the time and location of the focus group.

The focus group should be held in a neutral setting where respondents can freely state their answers without being overheard by others who are not participating in the focus group. Participants should sit around a table or in a circle to encourage everyone to participate.

In addition to taking notes, the focus group should be tape-recorded. In the introduction, the moderator should indicate that the focus group is being recorded, but that all comments will remain confidential.

If possible, you may want to send each respondent a brief letter after the focus group, thanking him or her for participation.

PERSONAL AND TELEPHONE INTERVIEW GUIDELINES

These guidelines have been provided to help you prepare for and conduct the personal interviews and telephone interviews included in the Toolbox.

WHOM SHOULD YOU INCLUDE?

There are two ways to select individuals to participate in a telephone or personal interview:

1. Select a random or systematic sample as described in the Guidelines for Selecting Respondents, below.
2. Select individuals whose comments will be representative of a group of individuals you would like to hear from. For example, you might interview the president of the local Chamber of Commerce to learn the business community's perspective.

WHAT TYPE OF INFORMATION WILL YOU COLLECT?

The personal and telephone interviews in this Toolbox will generate primarily qualitative information, although some quantitative information will also be collected. In many cases, qualitative information can be strengthened by using follow-up questions (otherwise called "probing questions" or "clarifying questions"). Follow-up questions can be used to encourage the respondent to elaborate on or clarify an answer. They can also be used to help focus the interview. Some examples follow:

1. "Tell me more about [some aspect of the respondent's answer]."
2. "Could you describe [some aspect of the respondent's answer] more fully?"
3. Simply repeat the respondent's answer as a question. For example, "So you felt that [repeat respondent's answer]."
4. Allowing for a moment of silence before asking the next question often leads the respondent to further elaborate on his or her previous answer.

WHAT PREPARATIONS ARE NEEDED?

Begin by reviewing the questions to be used in the telephone or personal interview. Make any modifications or additions to the questions that are necessary. You might also want to highlight any key follow-up questions to ask respondents.

For personal interviews, contact each respondent in advance to describe the purpose of the interview and arrange a date, time, and location. If the interview involves sensitive topics, it is best to find a location where the respondent will be able to talk freely without being interrupted or overheard by others. You may also want to send each respondent a letter confirming your arrangements and describing the type of questions that will be asked.

For telephone interviews, contact each respondent and ask whether or not this would be a good time for the interview. If not, set up another time to call that would be more convenient for them.

OTHER SUGGESTIONS

Other suggestions that you may want to consider in conducting telephone and personal interviews:

1. Establish a comfortable, friendly rapport with respondents so that they can enjoy the interview.
2. Be completely neutral in your comments and questions.
3. In making phone calls, consider what the best time would be to reach respondents.
4. If possible, you may want to send respondents a brief thank you letter after the interview.

GUIDELINES FOR SELECTING RESPONDENTS

These guidelines are designed to help you identify the housing sites and the individuals at each housing site to select as respondents when implementing a particular tool.

☐ Select which housing sites to collect information from. Depending on the purpose of the tool, you may decide to include

1. sites with both resident services and owner services programs;
2. sites with only resident services or only owner services programs;
3. sites with specific types of resident services programs or specific owner services training; and
4. sites with residents that have particular characteristics.

Most of the tools within this Toolbox have been designed for family sites with resident services programs and/or owner services programs.

☐ Identify any important subgroups within the sites selected. Comparisons can be made between these subgroups later in the analysis. By identifying these subgroups ahead of time, you will ensure that they are included in the sample. Important subgroups may include

1. sites with and without certain types of resident services and/or owner services programs;
2. sites with programs that are just beginning and sites with established programs; and
3. sites that have residents with certain characteristics that may affect the analysis and sites that have residents with different characteristics.

☐ Decide how many individuals to select from each site. We recommend the following:

1. For sites with a large number of units, the goal should be to collect information on 30 residents.
2. For sites with only a few units, the goal should be to collect information from almost all of the units.

If the evaluation tool asks you to collect information more than once from the same respondents—possibly to examine changes during the evaluation period—select a larger sample to allow for attrition. For example, if you want a final sample of 30 respondents and you expect that one third of the units will have a turnover in residents, select a sample of 40 rather than 30.

☐ Identify which individuals to select from each site. In general, you will want to select individuals in two different ways:

1. For some tools, you will want to select individuals with certain characteristics. For example, you may want to select individuals who have received "case management" assistance through the resident services programs. Once you have identified the important characteristic, you can either select everyone with that characteristic or select a sample of individuals with that characteristic (see below).
2. For other tools, you will want to select a random or systematic sample.
   a. A random sample can be selected by attaching an ID number to every individual in a particular housing site and drawing random numbers.

b. For a systematic sample, divide the total number of residents in the housing site by the size of the sample. If a housing site contains 20 residents and you would like a sample of 5, divide 20 by 5 to obtain 4. Then use that number to select every fourth (or whatever number you arrive at) resident.

REPORT SECTIONS

REPORT OUTLINE   TOOL 1

EVALUATION AREA: Greater self-sufficiency of residents
INDICATORS: Income, employment, use of public
                                    assistance, use of emergency
                                    funding programs
DERIVED FROM TOOL: Staff records on independence

INTRODUCTION

Although the resident services programs provided at each site differ, their common goal is to increase the independence and quality of life of residents. This section of the evaluation report examines the extent to which residents' independence and quality of life have, indeed, changed. The indicators of independence and self-sufficiency used include income, employment, use of public assistance, and use of emergency funding programs.

For each of the indicators, information is provided from across all the units at Westminster housing sites. Because these measures may be influenced by turnover in the residents rather than changes in independence or quality of life, a sample of residents who have remained in the housing site throughout the evaluation period has also been used to generate a second set of measures.

Income

1. The average household income of residents (increased/decreased/stayed the same) from $—— to $—— during the evaluation period. This represents an (increase/decrease/remained constant) of —%.

2. For the sample of residents who have remained in the housing site throughout the evaluation period, their average household income (increase/decrease/stayed the same) by —%.

3. The percentage of Westminster residents with household income at or below the poverty level changed from —% to —% during the evaluation period or — percentage points.

4. For the sample of residents who have remained in the housing site throughout the evaluation period, the percentage with household in-

come at or below the poverty level has (increased/decreased/stayed the same) by — percentage points.

Employment

5. The percentage of households where one or more person is employed (increased/decreased/stayed the same) from —% to —% during the evaluation period or — percentage points.

6. For the sample of households who have remained in their housing site, the percentage of households with one or more person employed (grew/fell/remained constant) from —% to —% during the evaluation period.

7. The percentage of income that Westminster households receive from employment (increased/decreased/stayed the same) from —% to —% in during the evaluation period.

8. For the sample of households who have remained in their housing site, the percentage of income received from employment (increased/decreased/remained constant) by — percentage points.

Use of Public Assistance

9. The percentage of Westminster households who receive public assistance (increased/decreased/remained constant) from —% to —% during the evaluation period.

10. For the sample of residents who have remained in their housing site, the percentage of households who receive public assistance (increased/decreased/remained constant) from —% to —% during the evaluation period.

11. The percentage of household income that Westminster households receive from public assistance (increased/decreased/remained constant) from —% to —% during the evaluation period.

12. For the sample of residents who have remained in their housing site, the percentage of household income received from public assistance (increased/decreased/remained constant) by — percentage points.

REPORT OUTLINE   TOOL 2

EVALUATION AREA:	Greater independence/self-sufficiency Improved quality of life
INDICATORS:	Self-report of changes in independence and quality of life
DERIVED FROM TOOL:	Residents' personal interviews on independence and quality of life

INTRODUCTION

Westminster Corporation provides and coordinates resident service programs at —% of its housing sites. Although the programs and services available at each site differ, their common goal is to increase the independence and the quality of life of residents. We have identified 12 key areas where residents may experience a change in their independence and quality of life. These areas include

1. Income
2. Employment
3. Education/training
4. Self-esteem/emotional health
5. Physical health
6. Budgeting/financial management
7. Family/parenting issues
8. Interpersonal relations
9. Volunteer activities/responsibilities
10. Transportation/mobility
11. Daily household activities
12. Safety

To assess the extent to which residents at the housing sites managed by Westminster have experienced a change in their independence or quality of life, we completed personal interviews with a sample of residents from each site who have participated in resident services programs.

In the interviews, we asked residents to (a) report the areas where they have experienced a change in their self-sufficiency or quality of life, (b) give examples or evidence of the changes they have made, and (c) indicate any ways that living in the housing site and/or participating in its programs contributed to the change. A total of — residents were interviewed.

The next section of this report presents findings from the interview.

FINDINGS

Improvements in Independence or Quality of Life

1. Approximately —% of the respondents were able to cite at least — areas where they experienced an improvement in their independence or quality of life and provide supporting evidence. Over —% of the respondents indicated — or more areas of improvement and supplied supporting evidence.

2. Respondents were most likely to cite improvements in ——, ——, and ——. Examples of their increased independence or quality of life included ——, ——, and ——. The areas where respondents were least likely to cite improvements included ——, ——, and ——.

3. Over —% of the improvements in quality of life or independence cited by respondents were linked to their housing site.

4. Of the respondents who reported an improvement in one or more areas of their quality of life or independence, respondents with an improvement in —— were most likely to credit their housing site with contributing to that improvement. They stated that living in the housing site helped them by ——, ——, and ——.

5. Improvements in —— were linked to programs offered at the housing site by —% of the respondents who reported an increase in this area. They stated that the resident services programs on —— gave them ——.

6. Although many residents felt that their —— had improved, only —% indicated that their housing site had contributed in some way. Resident service programs in this area currently operate in only — of the housing sites with resident services programs.

7. Over —% of the residents from housing sites with resident services programs reported that their use of emergency funding programs had (increased/decreased/stayed the same) since moving into the housing site.

Reductions in Independence or Quality of Life

8. Close to —% of the respondents were able to cite at least — areas where they experienced a reduction in their independence or quality of life and provided supporting evidence. Approximately —% of the respondents indicated — or more areas where their quality of life or independence had decreased and supplied supporting evidence.

9. Respondents were most likely to cite worsening conditions in ——, ——, and ——. Respondents were least likely to cite worsening conditions in ——, ——, and ——. The examples they provided of decreased independence or self-sufficiency included ——, ——, and ——.

10. Approximately —% of the reductions in quality of life or independence cited by respondents were linked to their housing site.

11. Of all the respondents who reported a reduction in one or more areas of their quality of life or independence, respondents with reductions in ——— were most likely to indicate that their housing cite was a contributing factor. They felt that living in the housing site resulted in ———, ———, and ———.

12. Although —% of the housing sites offer programs in ———, close to —% of the respondents cited a reduction in their independence or quality of life in this area.

13. Over —% of the respondents cited a reduction in ———, an area where Westminster currently does not offer resident services program.

REPORT OUTLINE   TOOL 3

EVALUATION AREA: Involvement in and satisfaction
                             with resident services
INDICATORS: Awareness, participation in, and
                             satisfaction with resident services
DERIVED FROM TOOL: Residents' survey on involvement
                             in and satisfaction with resident
                             services

INTRODUCTION

For residents to benefit from the resident services programs, they must be aware of and actively participate in the programs that are available at their housing site. Beyond awareness and participation, the satisfaction of residents with the resident services programs is critical to the overall success of the programs.

We conducted a survey of the residents at ——, ——, and —— housing sites to learn from answers to the following questions:

1. How aware are residents of the services and programs available at their housing site?
2. What percentage of the residents participate in the services and programs?
3. How satisfied are residents with the services and programs available at their site?
4. How could the resident service programs be strengthened?
5. What additional services or programs, not currently available, would residents find useful?

Approximately — residents returned the survey. This represents —% of the total number of residents living at those housing sites. The next section of this evaluation report presents findings from the survey.

FINDINGS

Awareness and Participation

1. Overall, a (high/low) percentage of respondents were aware of the —, —, and — programs available at their housing site, while a (low/high) percentage of respondents were aware of the —, —, and — programs.
2. Respondents had the highest rates of participation for the —, —, and — programs. The —, —, and — programs had relatively low rates of participation.

Satisfaction

3. Among those who had participated, most respondents reported being somewhat or very satisfied with the —, —, and — programs. The —, —, and — programs had the highest rates of dissatisfaction among respondents.

Outcomes

4. The program outcomes for respondents varied with different programs. As a result of the — program, several respondents reported that they were able to ————. The — program led some respondents to . . .

Strengths and Limitations

5. Respondents appreciated being able to attend the programs without arranging for child care and transportation (example). They found the programs to be an important way to meet other people living in their housing site and the surrounding community (example).

6. Respondents felt that the programs needed more ————. Some respondents indicated that the programs kept them from ————.

Recommendations

7. Respondents suggested that more programs be offered in the evening hours and that child care be arranged for the programs most highly attended (example).

8. Additional programs that respondents would find useful include . . .

EVALUATION AREA:	Involvement in and satisfaction with resident services
INDICATORS:	Participation in and satisfaction with resident services
DERIVED FROM TOOL:	Residents' focus group on involvement in and satisfaction with resident services

This focus group and the survey in this evaluation area are interchangeable (see directions for using the tool).

INTRODUCTION

For residents to benefit from the resident services programs, they must be aware of and actively participate in the programs that are available at their housing site. Beyond awareness and participation, the satisfaction of residents with the resident services programs is critical to the overall success of the programs.

We conducted a series of focus groups with residents of the housing sites to discover the following:

1. In what ways have residents benefited from the resident services programs?
2. What are the strengths and weaknesses of the resident service programs?
3. How could the resident service programs be more effective?
4. What additional resident service programs, not currently provided, would residents find useful?

This section of the evaluation report presents findings from our focus groups with residents of the housing sites.

FINDINGS

Use of the Resident Services Programs

1. (Each/most/a few) of the respondents reported that they had used one or more of the resident service programs provided to their housing site. The most frequently cited programs included ——, ——, ——.
2. (A few/none) of our respondents indicated that they had participated in the ——, ——, —— programs.

   a. When asked whether they were aware of these programs, (most/ several/a few) of our respondents indicated that they did not know about the —, —, — programs.

   b. (Most/several/a few) of the respondents were aware of the —, —, — programs. They stated, however, they were unable to attend these programs because —.

Outcomes of the Resident Services Programs

3. (Most/several/a few) of the respondents who had participated in the — programs reported that the program helped them to —, —, —. (Most/several/a few) other respondents felt that the programs had —, —, and —.

4. (Most/several/a few) respondents stated that, although they had participated in the — programs, the programs had limited results for them.

Strengths of the Resident Services Programs

5. (Most/several/a few) of the respondents enjoyed the resident services programs because — and —. They appreciated the programs for their —.

6. (Most/several/a few) respondents felt that they were more likely to use the —, —, — programs because —.

7. (Most/several/a few) respondents praised the individuals brought in for the —, —, — programs as being —, —, —.

Weaknesses of the Resident Services Programs

8. (Many/some/a few) respondents felt that the programs could be more useful. They stated that the programs could be improved by —, —, or —.

9. (Many/some/a few) respondents felt that the — programs were not at all useful. They stated that the programs —. They did not, however, have any suggestions for increasing their usefulness.

RECOMMENDATIONS

10. (Most/several/a few) respondents felt that the resident services programs could be improved by offering . . .

11. (Most/some/a few) respondents also suggested that . . .

12. (Many/some/a few) of the respondents felt that additional resident service programs could be developed to address —, —, and —.

REPORT OUTLINE   TOOL 5

EVALUATION AREA: Participation of residents in
                                    resident services programs
INDICATORS: Number of individual residents
                                    served by resident services
                                    programs and attendance at
                                    resident services programs
DERIVED FROM TOOL: Staff records on number of
                                    residents attending resident
                                    services programs and events

INTRODUCTION

The active participation of residents is necessary for resident services programs to be effective. The number of residents attending the events and programs available at their site serves as a sign of the interest and involvement of residents.

FINDINGS

1. On average, the social events held at the housing sites during the past year were attended by —% of the residents of that housing site. This percentage (increased/decreased) from —% to —% during the evaluation period.

2. The resident services programs on ——————— were highly attended by residents. On average, —% of the housing site residents attended the programs on ———————.

3. Resident services programs on ———, ———, and ——- were also highly attended. On average, —% of the residents of the housing site attended the programs on ———————, —% attended the programs on ———————, and —% attended the programs on ———————.

4. The resident services programs on ———, ———, and ——— were attended by a relatively low percentage of the housing site residents. The programs on —— were attended by —% of the residents while the programs on —— and —— were attended by —% and —%, respectively. The relatively low attendance at these programs could be explained by ———, ———, or ———————.

5. In addition to serving residents of the housing sites, several of the resident services programs also served other residents of the local community. The resident services programs on — were attended by an average of — local community residents. This number has (increased/decreased) by —% during the evaluation period.

REPORT OUTLINE   TOOL 6

EVALUATION AREA: Involvement in and satisfaction
                           with resident services
INDICATORS: Number served through individual
                           contacts by resident services staff
                           members
DERIVED FROM TOOL: Staff records on number of
                           residents served individually by
                           resident services staff

INTRODUCTION

The resident services staff at Westminster provide assistance to individual residents at the housing sites where resident service programs are available. The staff is available on-site regularly and is available "on call" as needed by residents. In individual meetings and phone conversations with a resident, staff members assess the resident's needs, provide information, make referrals to other service providers, and monitor the resident's ongoing status.

FINDINGS

1. During the past year, the resident services staff provided assistance to — residents on a continuous basis or approximately — residents per site.
2. In addition, the resident services staff worked with an additional — residents on a "one-time contact" or approximately — residents per site.
3. A majority of these residents received assistance with ——, ——, or ——. Over —% of these residents received assistance with ——. Close to —% of these residents were provided help with —.

REPORT OUTLINE   TOOL 7

EVALUATION AREA:	Reduced costs of property management
INDICATORS:	Vacancy rates, operating costs for property management, replacement and maintenance costs, operating costs for resident services
DERIVED FROM TOOL:	Staff records on costs of property management

INTRODUCTION

By increasing the quality of life and the independence of residents, the resident services programs may also reduce the vacancy rates, the cost of replacement and maintenance, and the cost of property management at the housing sites with resident services programs. This cost reduction must be considered along with the additional cost of providing resident services programs.

This section of the evaluation report presents information on the vacancy rates, the cost of property management, the cost of replacement and maintenance, and the cost of resident services programs at Westminster housing sites.

FINDINGS

Unit Vacancies

1. The percentage of potential rent revenue that was lost due to vacancies (increased/decreased) from —% to —% per year across all the Westminster housing sites during the evaluation period.

2. At housing sites where the resident services programs were initiated or expanded during the evaluation period, the percentage of potential rent revenue lost due to vacancies (increased/decreased) from —% to —% per year during the evaluation period.

Cost of Property Management

3. The average cost for property management at the Westminster housing sites was approximately $— per unit per year. The cost per unit ranged from a high of $—— to a low of $—— per year.

4. At housing sites where the resident services programs were initiated or expanded during the evaluation period, the average cost for property

management (increased/decreased) from —% to —% per year during the evaluation period. This compares with a —% increase for housing sites managed by other property management providers.

Replacement and Maintenance Costs (to be revised)

5. The average cost of replacement and maintenance due to vandalism, destruction, and negligence by residents at the Westminster housing sites was $— per year at the beginning of the evaluation period and $— per year at the end of the evaluation period.

6. At ——— housing site, where the a resident services program focusing on ——— was implemented during the past year, the cost of replacement and maintenance due to resident vandalism, destruction, and negligence fell from $——— to $———.

Cost of Resident Services

7. On average, the resident services programs cost approximately $— per unit per year. This cost includes Westminster staff time and direct expenses. It does not include the dollar value of services from outside providers.

   a. Because each site has a unique set of programs for residents, the expenses of each site vary. The cost for an extensive resident services program is approximately $— per unit at a housing site with programs in ———, ———, and ———.

   b. The cost for a smaller resident services program is around $— per unit at a housing site with programs in only ———, ———, and ———.

REPORT OUTLINE   TOOL 8

EVALUATION AREA: Use of other service providers
INDICATORS: List of service providers, the types
                          of services provided, and the
                          number of residents served
DERIVED FROM TOOL: Staff records on use of other service
                          providers

INTRODUCTION

Westminster's residents services programs strive to develop effective linkages with other service providers in the community and to use fully their services and expertise. During the past year, Westminster has worked with an extensive number of providers to bring a wide range of services into the housing sites on a one-time or continuous basis. These providers and the services they have brought to the housing sites include the following:

Service Provider	Type of Services Provided	Estimated Number of Residents Served
_____	_____	_____
_____	_____	_____
_____	_____	_____
_____	_____	_____
_____	_____	_____
_____	_____	_____
_____	_____	_____
_____	_____	_____
_____	_____	_____
_____	_____	_____
_____	_____	_____
_____	_____	_____
_____	_____	_____

REPORT OUTLINE   TOOL 9

EVALUATION AREA: Effective use of other human
                            service providers
INDICATORS: Effective use of other human
                            service providers, changing
                            perceptions among service
                            providers, and an opportunity
                            assessment for increased
                            effectiveness
DERIVED FROM TOOL: Service providers' focus group on
                            effective use of services providers
                            and perceptions of residents

INTRODUCTION

As part of the resident services program, Westminster develops linkages with other service providers in the communities where their housing sites are located and arranges for those service providers to work with residents of the Westminster housing sites. As a result, residents gain access to services that may otherwise be unavailable to them. In addition, Westminster gains access to the expertise and resources of other service providers.

We conducted a series of focus groups with service providers working with the residents of Westminster housing sites to discover the following:

1. To what extent are outside service providers being used effectively by Westminster?
2. What opportunities exist for strengthening the types of services that are provided and how services are provided to residents at Westminster housing sites?
3. In what ways, if any, has working with residents of the Westminster housing sites changed the perspective that outside service providers have of affordable housing residents?
4. Have any changes in the perspectives held by outside service providers resulted in any changes in the types of services provided or the way services are provided to residents of the housing sites?

This section of the evaluation report presents findings from those focus groups.

FINDINGS

Outcomes

1. Respondents reported that as a result of their work on-site at the Westminster housing sites, residents have been able to ——, ——, and ——.

2. Other respondents stated that their work with residents of the Westminster housing sites has resulted in ——, ——, and ——.

Strengths

3. (Most/several/a few) respondents felt that their services were particularly effective at the Westminster housing sites because ——.

4. Other respondents indicated that their work with residents of the housing sites has been effective because ——.

5. (Most/several/a few) respondents stated that their working relationship with Westminster has benefited the housing residents by ——. They appreciated Westminster's ——.

Limitations

6. (Many/some/a few) respondents indicated that the effectiveness of their services at the housing site and with its residents was limited by ——.

7. (Many/some/a few) respondents stated that their working relationship with Westminster limited the effectiveness of their work by ——.

Changes in the Perspectives Held by Outside Service Providers

8. For (many/several/a few) respondents, working at the Westminster housing site or with Westminster residents has changed their ——. Prior to working with Westminster residents, they felt that ——. But now, after working with Westminster residents, they feel that ——.

9. (Many/some/a few) respondents stated that they are providing more —— services to Westminster residents and less —— services as a result of changes in their perspective on the residents and the needs of residents.

10. (Some/A few) respondents also indicated that they had changed —— in their service provision as a result of changes in their perspective of Westminster residents.

Opportunities for Strengthening Effectiveness

11. (Several/a few/one) respondent(s) suggested that, in addition to the current services provided to residents, services in ———, ———, and ——— would be beneficial.

12. (Several/a few/one) respondent(s) felt that they could increase the effectiveness of their services by ———————.

13. (Several/a few/one) respondent(s) suggested that ——— services be provided through ——————— rather than the current method of service provision.

REPORT OUTLINE   TOOL 10

EVALUATION AREA: Perception of community members
INDICATORS: Community members have a
                           positive perception of the
                           housing site
DERIVED FROM TOOL: Community members' telephone
                           interviews on perceptions of the
                           housing site

INTRODUCTION

Westminster strives to build strong connections between each housing site and its local community. This linkage allows residents of the housing site to benefit from and contribute to their surrounding community and allows the community to benefit from and contribute to the housing site. As a key aspect of this linkage, Westminster asked community members about their awareness and perception of the housing site located in their community.

Telephone interviews were conducted with approximately (number) community members in (list of communities), where the (list of housing sites) are located. Of these housing sites, Westminster provides both resident services and owner services to (list of housing sites), while only property management services are provided to (list of housing sites). For the telephone interviews, we selected community members who are (groups of community members). The next section of our evaluation report presents the findings from our interviews with community members.

FINDINGS

Community Member Awareness

1. Overall, —% of our respondents indicated that they were aware of the housing site located in their community.

   a. In communities where Westminster supports both owner services and resident services, —% of the respondents were aware of the housing site.

   b. In contrast, —% of the respondents were aware of the housing site in communities where Westminster provides only property management services.

   c. In communities where Westminster has added or substantially increased owner services and resident services during the past year, the percentage of respondents who reported being aware of West-minster (stayed the same/jumped/fell) from —% to —%.

2. Respondents were most likely to be aware of the housing sites through ————, ————, and ————. Table — provides the percentage of respondents that reported that they were aware of the housing site in their community through various information sources.

   a. Respondents from the communities where Westminster supports both owner services and resident services were more likely to be aware of the housing site in their community through ——, ——, and —— than respondents from communities where Westminster provides only property management.

   b. In communities where Westminster has added or substantially increased owner services and resident services during the past year, the percentage of respondents who reported that they were aware of the housing site through —— (stayed the same/jumped/fell) from —% to —%.

3. In communities where Westminster provides resident services and owner services, —% of the respondents were aware of one or more of the services or programs available to residents. Respondents were most familiar with the ———— programs.

Overall Perceptions, Strengths, and Limitations

4. Approximately —% of the respondents reported that their overall per-ception of the housing site in their community was somewhat or very positive while —% reported that their overall perception was somewhat or very negative.

   a. Of the respondents from communities where resident services and owner services are provided, —% reported that their overall per-ception was somewhat or very positive.

   b. In contrast, —% of respondents from communities where resident services and owner services are not provided reported having somewhat or very positive perceptions.

   c. While —% of the respondents from communities where resident services and owner services are provided reported that they had neither a positive or negative perception, —% of the respondents from communities where only property management is provided had neither a positive nor a negative perception.

5. In the communities where resident services and owner services have been added or substantially increased during the past year, most respondents reported that their perception of the housing site in their community had (stayed the same/become more positive/become more negative). They attributed this change primarily to ————, ————, and ————.

6. Overall, —% of the respondents reported that they were happy that the housing site was located in their community while —% stated that they would prefer that the housing site was located elsewhere.

   a. Close to —% of the respondents from communities where resident services and owner services are provided stated that they were happy that the housing site was located in their community.

   b. Approximately —% of the respondents from communities where only property management is provided indicated that they were happy that the housing site was located in their community.

7. When asked to indicate the strengths of the housing site in their community, respondents most frequently cited the ——, ——, and —— of the housing sites.

   a. Respondents from communities where Westminster provides owner and resident services were highly supportive of ——, ——, and ——.

   b. Respondents from communities where Westminster only provides property management tended to cite —— and generally did not include ——.

8. When asked to indicate the limitations of the housing site in their community, respondents most frequently cited ——, ——, and ——.

   a. Respondents from communities where Westminster only provides property management tended to report more —— and —— than respondents from communities where owner services and resident services are provided.

   b. Respondents from communities where Westminster provides property management and owner services also indicated that ——.

Suggestions

9. Respondents reported a wide variety of changes they would make to the housing sites in their community. The most frequent responses included ——, ——, and ——. Other responses are listed below by site.

Notes

1. The Evaluation Toolbox in Appendix 15.B was developed for Westminster Corporation Resident Services Programs by Rainbow Research, Inc., Minneapolis, MN 55408 (February 1990).

2. The information presented in the Focus Group Guidelines section of Appendix 15.B is from *Focus Groups: A Practical Guide for Applied Research* (Krueger, 1994).

References

Conrad, K. J. (1994). *Critically evaluating the role of experiments* (New Directions for Program Evaluation, No. 63). San Francisco: Jossey-Bass.

Fetterman, D. M. (1982). Ibsen's baths: Reactivity and insensitivity (A misapplication of the treatment-control design in a national evaluation). *Educational Evaluation and Policy Analysis, 4*(3), 261-279.

Fetterman, D. M. (1994a). Empowerment evaluation. *Evaluation Practice, 15*(1), 1-15.

Fetterman, D. M. (1994b). Steps of empowerment evaluation: From California to Cape Town. *Evaluation and Program Planning, 17*(3), 305-313.

Kretzmann, J. P., & McKnight, J. L. (1993). *Building communities from the inside out: A path toward finding and mobilizing a community's assets.* Evanston, IL: Northwestern University, Center for Urban Affairs and Policy Research.

Krueger, R. (1994). *Focus groups: A practical guide for applied research.* Thousand Oaks, CA: Sage.

Mayer, S. E. (1994). *Building community capacity: The potential of community foundations.* Minneapolis: Rainbow Research, Inc.

Zimmerman, M. A. (in press). Empowerment theory: Psychological, organizational, and community levels of analysis. In J. Rappaport & E. Seldman (Eds.), *Handbook of community psychology.* New York: Plenum.

CONCLUSION

Conclusion

Reflections on Emergent Themes and Next Steps

DAVID M. FETTERMAN

Empowerment evaluation is more than a recommendation for the future, it is a statement about the present—a crystallization of many similar approaches. Empowerment evaluation has already taken hold in government, foundations, nonprofits, and academe. It has been adopted in programs throughout the world because it is responsive to fundamental needs and helps to improve program practice.

Empowerment evaluation's fluidity can be frightening for external evaluators who have not lived with program participants, but can feel natural to those who work in and with social programs on a daily basis. A comparison with fuzzy logic in the computer industry might be useful. At first shunned because it was not an objective—yes or no—binary patterned response to questions, fuzzy logic has value precisely because it does not model computer logic. Instead, it models the learning process. It allows the researcher to evaluate a situation by degrees rather than as an absolute. It continually samples the environment to determine its state at any given point in time. Thus, a body of water may be a puddle, a swamp, a pond, a lake, or a cloud, depending on the moment it is perceived. Fuzzy logic takes the rules directly from the data, much like grounded theory. A fuzzy-logic approach probes the environment, continually recording data, and responding accordingly at any given time, acknowledging the fact that "reality" continually changes.

Empowerment evaluation, too, is designed to operate in a real world of perpetual change and thus requires a high tolerance of chaos and ambiguity. Traditional evaluators may consider empowerment evaluation as much an oxymoron as fuzzy logic seems at first glance (see Patton's, 1994, discussion). Certainly this new evaluation approach challenges traditional evaluators to explore new ways of understanding and responding to the world. This challenge is also empowerment evaluation's greatest strength. It is responsive to rapid and unexpected shifts in program design and operation because it requires continual collection, description, reflection, and feedback of information about a group or organization in all its complexity.

Emergent Themes and the Next Steps

Numerous themes emerging from this collection merit additional attention. Empowerment values and theory crosscut almost every chapter. Values embodied by these works include improvement, an emphasis on building strengths (rather than focusing on problems and finding fault), collaboration, participation, and self-determination. In these various programs, participants control and are actively engaged in the design and execution of program evaluation and improvement. Evaluators are program and community resources, serving as coaches and facilitators who work with participants. Evaluators and the program participant are on a more equal footing than in traditional evaluation.

Empowerment theory provides principles and a framework to guide practice. Theory focuses on both processes (the means by which participants or citizens attempt to gain greater control over their environments or their lives) and outcomes (the consequences of these attempts). Empowerment theory also instructs the evaluator to view empowerment on multiple levels. In a discussion of outcomes, for example, Zimmerman (in press) explains:

Empowered outcomes also differ across levels of analysis. When we are concerned with individuals, outcomes might include situation-specific perceived control, skills, and proactive behaviors. When we are studying organizations, outcomes might include organizational networks, effective resource acquisition, and policy leverage. When we are concerned with community level empowerment, outcomes might include evidence of pluralism, the existence of organizational coalitions, and accessible community resources.

Zimmerman (in press; Zimmerman, Israel, Schulz, & Checkoway, 1992; Zimmerman & Rappaport, 1988), Rappaport (1987), Mithaug (1991, 1993), and Fetterman (1989, 1993, 1994a, 1994b, 1995) have helped to build a theoretical and philosophical foundation for empowerment evaluation that is apparent throughout this collection.

The importance of capacity building is another theme that runs through each chapter. The empowerment evaluator is committed to building capacity on many levels to institutionalize evaluation as part of the normal, everyday practice of organizational planning and management. Evaluation knowledge and tools are shared and constructed with program participants. The process does not run exclusively from the evaluator to the participant. Technical assistance and workshops require a dialogue and an exchange that reflects the experiential wisdom and cultural context of the program and community. Training is part of the evaluation process—part of the design and implementation phases of evaluation practice. It is iterative and ongoing, as additional training is needed at different developmental stages of the program's life cycle (and as participants' expertise evolves).

Common developmental steps and stages were evident in each of the chapters. Most were variations on the pattern of taking stock, establishing goals, developing strategies to accomplish goals, and documenting progress toward goals. In this collection, training, facilitation, advocacy, illumination, and liberation are all facets of empowerment evaluation. The investigation of worth or merit and plans for program improvement were typically viewed as the means by which self-determination is fostered, illumination generated, and liberation actualized.

Many self-critical questions were raised in these discussions concerning objectivity, quality, and the expertise of participants to conduct their own evaluations. In addition, tensions were highlighted, such as that between improvement and accountability. Such concerns and questions are characteristic of a healthy learning organization or, in this case, association of empowerment evaluators. They also point the way for future work and improvement.

A useful theoretical framework has been developed. It must now be refined and elaborated, based on practice in the field. Safeguards, including open group participation, dialogue, critique, and norming sessions, have been established to respond to questions of objectivity and participant bias. Reconciling and/or reporting disparate views needs additional attention. In addition, serious attention has been paid to quality, and specific steps are built into the process to establish and encourage quality work. Nevertheless, most of the environments in which empowerment evaluation is adopted are characterized by an almost frenetic pace, extraordinary challenges, and environmental constraints. This "keep your head above water" atmosphere often mitigates against maintaining standards of high quality. The risk always exists that in a chaotic and uncertain program environment, an internalized evaluation system could become a product of that culture rather than a reforming tool or agent to help remedy a maladaptive environment. On balance, internalizing evaluation has been more productive than not, but this area, too, requires additional scrutiny. The issue of evaluation expertise is linked to methodological expertise through training and technical assistance, but it is also linked to political credibility. How credible is the effort for supervisors and sponsors who have not been socialized in the values and philosophy of this approach? This collection offers some methods, such as including supervisors in training sessions to establish a common ground of understanding, but others are needed. The issues of what levels of participant expertise are needed for specific types of evaluation activities and what quality in the effort is appropriate or required for specific tasks merit additional investigation.

Finally, the tension between improvement and accountability must be addressed and ways found to resolve it more thoroughly. Empowerment evaluation has successfully served these two masters simultaneously (out of necessity)—CSAP work throughout the country is an excellent example. There are also examples of programs dissolving or merging on their own accord as a result of their own reviews of themselves. Many programs throughout the country that have adopted empowerment evaluation are taking responsibility for their own actions and holding themselves accountable, in a credible fashion, to supervisors and sponsors. The inherent tension remains, however, between focusing on improvement and enhancing wellness or accountability. As Cronbach and associates (1980) have pointed out: "Accountability emphasizes looking back in order to assign praise or blame; evaluation is better used to understand events and processes for the sake of guiding future activities" (p. 4). These are all issues that empowerment evaluators recognize, confront, and manage, as evidenced in this collection. They are not settled, however, and will require more work to improve practice.

Evaluation Is Basic

To make my own position about evaluation explicit, I believe that evaluation is basic—like reading, writing, and arithmetic. I believe that evaluation should be a fundamental skill, an integral part of any educated citizen's repertoire. I also believe that anyone can learn the basic skills of evaluation, as demonstrated in empowerment evaluation. This does not deny the education and expertise of professional evaluators, it simply reaffirms the right of every citizen to use evaluation to foster improvement and self-determination within a context of social justice. We need every tool we can find to respond to the pressing social and environmental problems we face. Evaluation has an instrumental role to play in helping us respond to our problems, adapt, and build the future.

References

Cronbach, L. J., Ambron, S. R., Dornbusch, S. M., Hess, R. D., Hornik, R. C., Phillips, D. C., Walker, D. F., & Weiner, S. S. (1980). *Toward reform of program evaluation: Aims, methods, and institutional arrangements.* San Francisco: Jossey-Bass.

Fetterman, D. M. (1989). *Ethnography: Step by step.* Newbury Park, CA: Sage.

Fetterman, D. M. (1993). *Speaking the language of power: Communication, collaboration, and advocacy (Translating ethnography into action).* London: Falmer.

Fetterman, D. M. (1994a). Empowerment evaluation [American Evaluation Association presidential address]. *Evaluation Practice, 15*(1), 1-15.

Fetterman, D. M. (1994b). Steps of empowerment evaluation: From California to Cape Town. *Evaluation and Program Planning, 17*(3), 305-313.

Fetterman, D. M. (1995). In response to Dr. Daniel Stufflebeam's: "Empowerment evaluation, objectivist evaluation, and evaluation standards: Where the future of evaluation should not go and where it needs to go." *Evaluation Practice, 16*(2), 177-197.

Mithaug, D. E. (1991). *Self-determined kids: Raising satisfied and successful children.* New York: Macmillan (Lexington imprint).

Mithaug, D. E. (1993). *Self-regulation theory: How optimal adjustment maximizes gain.* New York: Praeger.

Patton, M. (1994). Developmental evaluation. *Evaluation Practice, 15*(3), 311-319.

Rappaport, J. (1987). Terms of empowerment/exemplars of prevention: Toward a theory for community psychology. *American Journal of Community Psychology, 15,* 121-148.

Zimmerman, M. A. (in press). Empowerment theory: Psychological, organizational, and community levels of analysis. In J. Rappaport & E. Seldman (Eds.), *Handbook of community psychology.* New York: Plenum.

Zimmerman, M. A., Israel, B. A., Schulz, A., & Checkoway, B. (1992). Further explorations in empowerment theory: An empirical analysis of psychological empowerment. *American Journal of Community Psychology, 20*(6), 707-727.

Zimmerman, M. A., & Rappaport, J. (1988). Citizen participation, perceived control, and psychological empowerment. *American Journal of Community Psychology, 16*(5), 725-750.

Author Index

Subject Index

About the Editors

DAVID M. FETTERMAN is a Professor, Director of Research and Evaluation in the School for Transformative Learning at the California Institute of Integral Studies, and Director of the M.A. Policy Analysis Program at Stanford University. He was formerly a Principal Research Scientist at the American Institutes for Research and a Senior Associate and Project Director at RMC Research Corporation. He received his Ph.D. from Stanford University in educational and medical anthropology. He has conducted fieldwork in both Israel (including living on a kibbutz) and the United States (primarily in inner cities across the country). He works in the fields of educational evaluation, ethnography, and policy analysis, and focuses on programs for dropouts and gifted and talented education.

Dr. Fetterman is a past president of the American Evaluation Association and the American Anthropological Association's Council on Anthropology and Education. He has also served as the program chair for each of these organizations and has been elected a fellow of the American Anthropological Association and the Society for Applied

Anthropology. In addition, he has received numerous awards for his contribution to evaluation.

He has consulted for a variety of federal agencies, foundations, corporations, and academic institutions, including the U.S. Department of Education, National Institute of Mental Health, Centers for Disease Control, W. K. Kellogg Foundation, Rockefeller Foundation, Walter S. Johnson Foundation, Syntex, the Independent Development Trust in South Africa, and universities throughout the United States and Europe.

Dr. Fetterman is the General Editor for Garland Publication's Studies in Education and Culture Series. He has contributed to a variety of encyclopedias, including the *International Encyclopedia of Education* and the *Encyclopedia of Human Intelligence.* He is also the author of *Speaking the Language of Power: Communication, Collaboration, and Advocacy; Ethnography: Step by Step; Qualitative Approaches to Evaluation in Education: The Silent Scientific Revolution; Excellence and Equality: A Qualitatively Different Perspective on Gifted and Talented Education; Educational Evaluation; Ethnography in Theory, Practice, and Politics;* and *Ethnography in Educational Evaluation.*

SHAKEH JACKIE KAFTARIAN (Ph.D.) is a health psychologist. She is currently the Deputy Director of the Office of Scientific Analysis at the Center for Substance Abuse Prevention (CSAP) and an adjunct Research Professor at the Uniformed Services University of the Health Sciences. During her short tenure at the federal government, she has been instrumental in the initiation of multiple-site, comprehensive, and rigorous evaluation research projects for CSAP-sponsored community-based prevention grant programs, which involve significant conceptual, methodological, and practical challenges.

She has served on a number of national and international advisory committees for mental health promotion and substance abuse prevention. She has been a pioneer member of the USA/USSR Telemedicine

Spacebridge committee, which helped NASA in the initiation of the first collaborative scientific and health promotion endeavor between the two superpowers at the end of the cold war era.

ABRAHAM WANDERSMAN (Ph.D.) is Professor of Psychology at the University of South Carolina—Columbia. He was interim Codirector of the Institute for Families in Society at the University of South Carolina. He performs research and evaluation on citizen participation in community organizations and coalitions and on interagency collaboration. He is a coauthor of *Prevention Plus III* and of many books and articles. He serves or has served on a number of advisory committees for prevention including the U.S. Conference of Mayors' Advisory Committee on HIV Community Prevention Planning, Technical Assistance Committee of the National Evaluation of CSAP Community Partnerships, Technical Support Group for the CSAP evaluation of Training and Technical Assistance, the Prevention Working Group of the Center for Mental Health Services. Dr. Wandersman is affiliated with the Environmental Hazards Assessment Program of the Medical University of South Carolina.

About the Contributors

ALETHA AKERS is a graduate of Bryn Mawr College, where she earned a B.A. in both biology and chemistry. Recently, she spent a year in Africa studying the role of traditional medicine in African health care systems on a Thomas J. Watson Fellowship. She has also conducted international public policy research with the Harry S. Truman Foundation and research on the Superfund Act with the Agency for Toxic Substances and Disease Registry. She will enter the Johns Hopkins School of Medicine to pursue an M.D. and master's in public health with a focus on international health, program planning, and management. Her research interests include the applications of traditional medicine in rural health care delivery, health care financing, and public policy development in Africa, Latin America, and the Caribbean.

ARLENE BOWERS ANDREWS (Ph.D., L.I.S.W.) is Associate Director of the University of South Carolina (USC) Institute for Families in Society and Associate Professor at the USC College of Social Work. A social worker and community psychologist, she is a magna cum laude graduate of Duke University and completed her graduate degrees at the University of South Carolina. She is the author of *Victimization and Survi-*

vor Services and several articles and book chapters regarding family violence prevention and community systems development. She recently wrote the handbook *Helping Families Survive and Thrive: Ways Citizens Can Help to Strengthen Families in Their Communities.* She was the founding executive director of Sistercare: Services to Abused Women and of the Council on Child Abuse and Neglect, Inc. She has served as a consultant and volunteer with numerous developing organizations, including the Nurturing Center, the Alliance for South Carolina's Children, the Rape Crisis Network, and Human Services Associates (HSA), a therapeutic family foster care agency.

KAREN BASS (P.A.-C.) is a full-time Clinical Instructor at the University of Southern California's Physician's Assistant Program. She is a licensed physician's assistant. She is the Executive Director of the Community Coalition for Substance Abuse Prevention and Treatment serving the South Los Angeles community. She has over 20 years of experience in community activism, has received numerous awards for her work in this area, and is a member of a number of professional organizations and community groups.

JANNETTE Y. BERKLEY (B.S.E.E.) is pursuing her Ph.D. in Developmental and Child Psychology in the Human Development and Family Life Department at the University of Kansas in Lawrence. She is a Research Associate of the Work Group on Health Promotion and Community Development of the Schiefelbusch Institute for Life Span Studies. Her work involves evaluating and providing technical assistance and community research to community health initiatives. Her research interests include community psychology, adolescent health issues, and promotion of academic and personal success of African American adolescents.

DIDRA L. BROWN (M.A.) is pursuing her Ph.D. in clinical psychology at the California School of Professional Psychology in Los Angeles. She has a master's in clinical psychology from California State University, Dominguez Hills, and specializes in Afrocentric applications in mental health service delivery and research. Her current research focuses on the relationship between environmental issues and alcohol consumption patterns among African Americans. Her current projects include studies of the relationship between alcohol availability and consumption patterns in urban African American communities, malt liquor beer consumption patterns among African Americans, and the concentration of tobacco and alcohol advertisements in African American communities.

FRANCES DUNN BUTTERFOSS (Ph.D.) is Assistant Professor of Pediatrics at the Center for Pediatric Research, Eastern Virginia Medical School, and Children's Hospital of The King's Daughters. She is currently the Principal Investigator on an evaluation grant for the community-based Virginia Caring Program and coinvestigator of a CDC immunization project (CINCH) under which she coordinates and evaluates coalition development. Her publications and presentations address both research and practice issues surrounding coalition effectiveness. She is a nationally recognized consultant on the development and maintenance of coalitions for immunizations; nutrition; alcohol, tobacco, and other drug abuse prevention; and breast and cervical cancer prevention.

MATTHEW J. CHINMAN is currently completing his clinical internship at the Psychology Section of the Department of Psychiatry at Yale University. Previously, he was pursuing his doctorate at the University of South Carolina (USC) in the Ph.D. Clinical/Community Psychology program. While at USC, he worked for 4 years as an evaluator of a Center for Substance Abuse (CSAP)-funded prevention project in Spartanburg, South Carolina, and helped design the evaluation methodology used there. His interests include evaluation methodology, benefits and costs of voluntary participation, and adolescent empowerment.

MARGRET A. DUGAN is the owner of Redhawk Research, a firm specializing in participatory and empowerment evaluation practices as well as prevention programming for at-risk youth. She has held senior administration positions concerned with child advocacy and family well-being for two decades. Trained as a social psychologist, she has completed research investigation on resilience and protective factors in children of adversity. Her current research deals with children in crisis and self-assessment evaluation. She publishes, speaks, and consults nationally. She was honored by former first lady Barbara Bush as an outstanding child advocate in 1991.

STEPHEN B. FAWCETT is Professor of Human Development and Director of the Work Group on Health Promotion and Community Development of the Schiefelbusch Institute for Life Span Studies at the University of Kansas. In his work, he uses behavioral and community research methods to help understand and improve health and social concerns of importance to communities. A former VISTA volunteer, he has worked as a community organizer in public housing and low-income neighborhoods. He holds an endowed professorship, the Kansas Health Foundation Professorship for Community Leadership. He is coauthor of nearly 100 articles and book chapters in the areas of health promotion, community development, empowerment, self-help, independent living, and public policy.

JACQUELINE L. FISHER (M.P.H., M.S.) is Coordinator for Site Development for the School/Community Sexual Risk Reduction Replication Initiative and a Research Assistant of the Work Group on Health Promotion and Community Development of the Schiefelbusch Institute for Life Span Studies at the University of Kansas in Lawrence. Her work involves providing technical assistance and community research to community health initiatives. Her research interests include issues in the areas of public health, adolescent health, health promotion and education, and maternal and child health.

PAUL FLORIN received his Ph.D. in clinical/community psychology from George Peabody College of Vanderbilt University in 1981. He is currently Professor of Psychology at the University of Rhode Island. For the past 15 years, he has been involved as both researcher and practitioner in the areas of citizen participation and community development. He has received grants from the National Science Foundation and Ford Foundation to systematically investigate citizen participation and community development. He has provided consultation, training, and technical assistance to communities, agencies, and governmental units wishing to plan and implement community approaches to health promotion programming. He has written on citizen participation, organizational effectiveness in community development, and training and technical assistance needs of community coalitions.

VINCENT T. FRANCISCO is a Research Associate with the Work Group on Health Promotion and Community Development, Institute for Life Span Studies, University of Kansas. He is primarily interested in research in community development, especially in research that enhances community integration and support and works toward empowerment of marginalized groups. He is interested in the provision of technical support for the development of coalitions as well as evaluation of community-based intervention programs focusing on reduction of risk for substance abuse, assaultive violence, cardiovascular disease, and teen parenthood. He is coauthor of several research articles in prevention of substance abuse and cardiovascular disease. He is also a coauthor of several manuals for community development in the areas of strategic planning and evaluation of prevention initiatives. He has been a consultant for a variety of organizations, including private foundations, community coalitions and advocacy organizations, and governmental agencies.

ELLEN GOLDSTEIN is the Community Liaison for the University of California, San Francisco, Center for AIDS Prevention Studies (CAPS) and AIDS Clinical Research Center (ACRC) Department of General Internal Medicine. In this capacity, she is working to make research more accessible to AIDS service providers and program planners and make community resources and expertise more accessible to researchers. At CAPS, she manages a collaboration between scientists, funders (Northern California Grantmakers), and 11 CBOs in an effort to evaluate prevention efforts and promote science/service interaction. She has recently completed data collection on a national survey of HIV prevention program managers regarding the sources and types of information accessed for program planning. Additionally, she consults with the University of California's University-wide AIDS Research Program on the development of a statewide Prevention Initiative. As a volunteer, she has facilitated groups and trained people to provide support for people with AIDS. As a consultant, she has conducted an evaluation of the James Irvine Foundation's HIV/AIDS Grantmaking program and developed AIDS Awareness seminars for the Federal Reserve Bank and the YMCA. Her primary goal is to get people talking to each other.

CYNTHIA A. GÓMEZ (Ph.D.) is a Research Specialist at the Center for AIDS Prevention Studies (CAPS) in the Department of Epidemiology and Biostatistics at the University of California, San Francisco. She received her master's in psychology from Harvard University and her Ph.D. in clinical psychology from Boston University. Prior to coming to CAPS, she spent 12 years working in community health settings, including 5 years as director of a child and family mental health center in Boston. In addition, she consulted at schools and community agencies regarding HIV/AIDS prevention models, and facilitated long-term support groups for physicians working with HIV-/AIDS-infected persons. Currently, she works as coinvestigator on projects geared primarily toward HIV/AIDS prevention in the Latino population, including projects focused on women and inmates as well as school-based HIV prevention curricula targeting sixth graders.

ROBERT M. GOODMAN is Associate Professor in the Department of Health Promotion and Education, School of Public Health, University of South Carolina. His areas of study include organization and community development, and the evaluation of these enterprises. He is currently evaluating projects for the National Cancer Institute, Centers for Disease Control and Prevention, the Center for Substance Abuse Prevention, as well as for several state health departments. In 1992, he received the Early Career Award from the Health Education and Health Promotion section of the American Public Health Association, and in 1994, he received the Health Promotion and Education Advocacy Award jointly sponsored by the Centers for Disease Control and the Association for State and Territorial Directors of Public Health Education.

CHERYL N. GRILLS (Ph.D.), a graduate of Yale University and UCLA, is Associate Professor of Psychology and Alcohol and Drug

Studies at Loyola Marymount University. She is also a Research Associate of the UCLA Drug Abuse Research Center. Her research interests and current projects include research on an African-centered model of treatment engagement with African American substance abusers, research on traditional medicine in West Africa, and community partnership evaluation research. She is a member of the Association of Black Psychologists, is a founding member of the Institute for the Study of African Centered Intervention, is a licensed clinical psychologist in California, and consults on a number of prevention and treatment issues particularly regarding matters of cultural competence, multiculturalism, and Africentric interventions.

KARI JO HARRIS is a Research Assistant for the Work Group on Health Promotion and Community Development of the Schiefelbusch Institute for Life Span Studies at the University of Kansas. She has a master's degree in organization development and is currently pursuing a master's degree in public health and a Ph.D. from the University of Kansas. She is the lead site evaluator on an initiative to reduce adolescent pregnancy and on a replication of a school health project to reduce elementary school-aged children's risk for chronic diseases.

NANCY F. JACOBS (Ph.D.) is the Executive Director of the Criminal Justice Research Center at the John Jay College of Criminal Justice. Her most recent experiences have focused on the development of technical assistance program evaluations that are designed in concert with criminal justice agencies, community-based organizations, and private not-for-profits. Recent activities have included formative, summative, and empowerment evaluations for several community partnerships funded by the federal government's Center for Substance Abuse Prevention (CSAP). Her work has also included a survey of private foundation grantmaking trends in criminal justice and an evaluation of a gun control project. In addition to completing a pilot study of a school-based drug prevention program (SPECDA) for the NYCPD and the New York City Board of Education, she has completed a study for a Special Task Force, appointed by the New York City Commissioner of Correction, to examine the use of force in the city's jails.

JOYCE KELLER received her first degree from the University of Dallas, an M.M. in piano from the St. Louis Institute of Music, an M.P.A. in accounting, and a Ph.D. in educational psychology from the University of Texas at Austin. She taught piano at Incarnate Word College in San Antonio and accounting at the University of Texas at Austin, where she received several teaching awards. She has evaluated numerous programs, including educational interventions in several Dallas-Fort Worth school districts, the education-job training offerings of the Texas JOBS program, and the effectiveness of the Texas probation system. Her efforts at empowerment evaluation at the Texas adult probation system and, most recently, at the Texas Department of Human Services, are described in this volume. In conjunction with the Texas Office of the State Auditor, she is currently writing a manual to further assist Texas state agencies in ongoing evaluation, including the assessment of the costs and benefits of interventions.

HENRY M. LEVIN is the David Jacks Professor of Higher Education and Economics at Stanford University. He is also Director of the Center for Educational Research at Stanford (CERAS) and was the founding Director of the Institute for Research on Educational Finance and Governance (IFG). He received the Ph.D. in economics from Rutgers University in 1966. Prior to his arrival at Stanford in 1968, he was a Research Economist at the Brookings Institution. He has been a Fellow at the Center for Advanced Studies in the Behavioral Sciences, a Fulbright Professor at the University of Barcelona, and a Distinguished Visiting Professor at the University of Beijing. He is a specialist in the economics of education and human resources. His work has focused specifically on cost-effectiveness, educational finance, educational and workplace productivity, and investment strategies for educationally at-risk students.

RHONDA K. LEWIS (M.A.) is a doctoral student in the Human Development and Family Life Department and a Research Associate of the Work Group on Health Promotion and Community Development of the Schiefelbusch Institute for Life Span Studies at the University of Kansas. She uses behavioral and community research methodologies to help contribute to understanding how communities work, and promotes health strategies to improve the health among people living in Kansas. Her research interests include self-help groups, disadvantaged populations, adolescent health issues, and the elderly.

JEAN ANN LINNEY is Professor of Psychology and Associate Dean at the University of South Carolina. She received her doctoral degree in 1978 from the University of Illinois at Urbana-Champaign. For nearly 20 years, she has been active in research on prevention and school-based intervention, the development of community-based programs, and work with community agencies. Her current research interests are in the primary prevention of substance abuse among adolescents. She is a past President of the Society for Community Research and Action, and a Fellow of the American Psychological Association.

CHRISTINE M. LOPEZ is a doctoral student in the Human Development and Family Life Department. She is also a Research Associate for the Work Group on Health Promotion and Community Development of the Schiefelbusch Institute for Life Span Studies at the University of Kansas. She provides technical assistance to community initiatives that address issues of health promotion, such as reducing adolescent substance abuse. She is interested in community-based research in the areas of substance abuse, youth violence, and health promotion in rural settings.

STEVEN E. MAYER earned his bachelor's degree from George Washington University, his master's from Ohio State, and his Ph.D. in industrial and organizational psychology from the University of Minnesota in 1973, specializing in the question, "How does one know when an organization is working well?" His expertise includes the

design of program evaluation projects and activities that result in stronger organizations and programs. He is skilled in quantitative and qualitative research methodology and in participatory strategies of evaluation. He is the Founder and Executive Director of Rainbow Research, a Minneapolis-based nonprofit organization with a mission "to assist socially concerned organizations in responding more effectively to social problems." He has served on the faculties of the University of Minnesota and University of Georgia, and is currently adjunct faculty at the University of St. Thomas in St. Paul.

RICARDO A. MILLETT, Director of Evaluation for the W. K. Kellogg Foundation, monitors the development and implementation of evaluation strategies for foundation programming. His efforts focus on improving projects through greater communication, team building, and using evaluation as an integral part of programming. He also reviews proposals and makes site visits to existing and potential projects. Prior to joining the foundation, he served as senior vice president of planning and resource management for the United Way of Massachusetts Bay in Boston. He has been a leader in major collaboration initiatives that have brought community and corporate actors and their respective institutions together to support program activities in housing, anti-drug and -violence programs, and child care. He has also published a book and several articles on the subject of citizen participation and community capacity building.

ROGER E. MITCHELL (Ph.D.) received his Ph.D. in clinical psychology from the University of Maryland and did postdoctoral work at the Social Ecology Laboratory at Stanford University Medical Center. He is currently Assistant Professor (Research) of Community Health at

Brown University, and Coordinator of Prevention Research at the Center for Alcohol and Addiction Studies. Trained as a clinical-community psychologist, his research interests and publications have included the areas of program evaluation, development of informal helping networks, and application of stress, social support, and coping paradigms to examining psychological well-being. As a member of the Community Research and Services Team, he currently serves with colleagues as the evaluation team for three CSAP community partnership grants and one CSAP high-risk youth grant.

DENNIS E. MITHAUG (Ph.D.) is Professor and Chair of the Department of Special Education at Teachers College, Columbia University. He has conducted and directed research and demonstration projects in special education since 1968, first at the University of Washington's Experimental Education Unit, Child Development and Mental Retardation Center; later, as Professor of Education and Director of the Center for Educational Research at the University of Colorado—Colorado Springs; and now as director of a federally funded project to develop self-determination skills in youth with disabilities. He received a B.A. in psychology from Dartmouth College and an M.A. and Ph.D. in sociology and an M.Ed. in education from the University of Washington. He is a former Dean of Education and a past President of the Council of Exceptional Children's Division for Research. He has published numerous books and articles, including two books, *Self-Determined Kids: Raising Satisfied and Successful Children* and *Self-Regulation Theory: How Optimal Adjustment Maximizes Gain*.

ADRIENNE PAINE-ANDREWS is Program Director for the Project Freedom Replication Initiative and Program Co-director for the School/Community Sexual Risk Reduction Replication Initiative. She is also courtesy Assistant Professor in the Department of Human Development and Associate Director of the Work Group on Health Promotion & Community Development with the Schiefelbusch Institute for Life Span Studies at the University of Kansas. She is primarily interested in research that promotes community development, enhances community integration and social support, and works to empower marginal groups. She is coauthor of several articles in the areas of self-help, community development, and health promotion.

KIMBER P. RICHTER is a Research Assistant for the Work Group on Health Promotion and Community Development of the Schiefelbusch Institute for Life Span Studies at the University of Kansas. She provides technical assistance to projects that seek to address such issues as reducing substance abuse among adolescents, increasing the use of advance directives, and reducing adolescents' risk for cardiovascular disease.

JERRY A. SCHULTZ is Assistant Research Professor in the Work Group for Health Promotion and Community Development, which is a research, teaching, and technical assistance program of the Schiefelbusch Institute for Life Span Studies and the Department of Human Development, and he is an Adjunct Assistant Professor in the Department of Anthropology, at the University of Kansas. He received his Ph.D. in anthropology from the University of Kansas and has taught anthropology at the University of Nebraska. He is currently directing a rural health promotion initiative in Kansas and engaged in the development of an electronic technical assistance system for health promotion and community development. He has also worked extensively and published in the area of Native American education.

JOHN F. STEVENSON received his Ph.D. in psychology from the University of Michigan in 1974. He is currently Professor of Psychology at the University of Rhode Island, Associate Professor (Research) in the Department of Psychiatry and Human Behavior, Brown University Medical School, and a member of the Training Faculty at the Brown University Center for Alcohol and Addiction Studies. He has been teaching and conducting evaluation research since 1977, when he received a National Science Foundation Faculty Fellowship to pursue this interest. His current research interests include evaluation of treatment and prevention programs for alcohol and other drug abuse.

As a member of the Community Research and Services Team, he currently serves with colleagues as the evaluation team for three CSAP community partnership grants and one CSAP high-risk youth grant.

ROBERT F. VALOIS (M.S., Ph.D., M.P.H.) is Associate Professor of Health Promotion & Education and Family & Preventive Medicine in the Schools of Public Health & Medicine at the University of South Carolina (USC), and also Associate Director of the Prevention Center at the USC School of Public Health. He holds a B.S. degree in health science from the State University of New York at Brockport. His M.S. degree in school health and his Ph.D. in community health/educational psychology are from the University of Illinois at Urbana-Champaign. He completed a postdoctoral fellowship, a U.S. Public Health Service Traineeship, and the M.P.H. degree in Health Behavior at the University of Alabama Medical Center, School of Public Health, at Birmingham. He has served as a consultant to the National Institutes of Health, the U.S. Centers for Disease Control and Prevention, and the U.S. Public Health Service. His research is focused on adolescent health, sexuality education, tobacco, alcohol, and other drug abuse prevention, health promotion program planning, and evaluation.

ELLA L. WILLIAMS is currently a doctoral student in the Department of Human Development and has been a Research Associate for the Work Group on Health Promotion and Community Development of the Schiefelbusch Institute for Life Span Studies at the University of Kansas. She has experience in multicultural issues as they relate to community development and health promotion. She also has expertise in the area of substance abuse and drug-related violence.

ROBERT K. YIN is President of COS-MOS Corporation, a firm that provides applied research, technological support, and management assistance aimed at improving public policy, private enterprise, and collaborative ventures. Currently, at COSMOS, he is the Project Director for the Phase II National Evaluation of the Community Partnership Demonstration Grant Program. He is the author of numerous books and articles. His book on the case study method, *Case Study Research: Design and Methods,* has had three editions (1984, 1989, 1994). He is a former member of the RAND Corporation (1970-1978) and a member of the Cosmos Club. He received his B.A. (magna cum laude) from Harvard College in 1962 (in history) and his Ph.D. in 1970 from the Department of Brain and Cognitive Sciences, Massachusetts Institute of Technology.